Lecture Notes in Computer Science 9594

Commenced Publication in 1973
Founding and Former Series Editors:
Gerhard Goos, Juris Hartmanis, and Jan van Leeuwen

More information about this series at http://www.springer.com/series/7407

Malcolm I. Heywood · James McDermott
Mauro Castelli · Ernesto Costa
Kevin Sim (Eds.)

Genetic Programming

19th European Conference, EuroGP 2016
Porto, Portugal, March 30 – April 1, 2016
Proceedings

 Springer

Editors
Malcolm I. Heywood
Dalhousie University
Halifax, NS
Canada

Ernesto Costa
University of Coimbra
Coimbra
Portugal

James McDermott
University College Dublin
Dublin
Ireland

Kevin Sim
Edinburgh Napier University
Edinburgh
UK

Mauro Castelli
Universidade Nova de Lisboa
Lisboa
Portugal

ISSN 0302-9743   ISSN 1611-3349 (electronic)
Lecture Notes in Computer Science
ISBN 978-3-319-30667-4   ISBN 978-3-319-30668-1 (eBook)
DOI 10.1007/978-3-319-30668-1

Library of Congress Control Number: 2016932316

LNCS Sublibrary: SL1 – Theoretical Computer Science and General Issues

Printed on acid-free paper

This Springer imprint is published by SpringerNature
The registered company is Springer International Publishing AG Switzerland

Preface

The 19th European Conference on Genetic Programming (EuroGP) took place between March 30 and April 1, 2016. Porto, Portugal, was the setting, and the *Seminário de Vilar Rua Arcediago Van Zeller* was the venue.

The unique character of GP has been recognized from its very beginning. To date, GP is essentially the only approach that has demonstrated the ability to automatically generate, repair, and improve computer code in a wide variety of problem areas. It is also one of the leading methodologies that can be used to "automate" science, helping researchers to induce hidden complex models from observed phenomena. Furthermore, GP has been applied to many problems of practical significance, and has produced human-competitive solutions. Collectively, over 11,000 articles now appear in the online GP bibliography[1].

EuroGP is a mature event, the only conference exclusively devoted to the evolutionary generation of computer programs. It attracts scholars from all over the world. EuroGP has had an essential impact on the success of the field, by serving as an important forum for expressing new ideas, meeting fellow researchers, and starting collaborations. Indeed, EuroGP represents the single largest venue at which GP results are published. Many success stories have been witnessed by the 18 editions of EuroGP.

EuroGP 2016 received 36 submissions from around the world. The papers underwent a rigorous double-blind peer-review process, each being reviewed by at least three members of the international Program Committee from 23 countries. The overall quality of submissions was very high, and therefore not all good papers could be accepted. The selection process resulted in this volume, with 11 papers accepted for full-length oral presentation (30.6 % acceptance rate) and eight for short talks (52.8 % global acceptance rate for both categories combined).

The wide range of topics in this volume reflects the current state of research in the field. Thus, we see topics as diverse as semantic methods, recursive programs, grammatical methods, coevolution, Cartesian GP, feature selection, metaheuristics, evolvability, and fitness predictors; and applications including image processing, one-class classification, SQL injection attacks, numerical modelling, streaming data classification, creation and optimization of circuits, multi-class classification, scheduling in manufacturing and wireless networks. The results presented here represent the state of the art in this exciting field.

Together with three other co-located evolutionary computation conferences (EvoCOP 2016, EvoMusArt 2016, and EvoApplications 2016), EuroGP 2016 was part of the Evo* 2016 event. This meeting could not have taken place without the help of many people.

---

[1] Maintained at http://liinwww.ira.uka.de/bibliography/Ai/genetic.programming.html by William B. Langdon

First to be thanked is the great community of researchers and practitioners who contributed to the conference by both submitting their work and reviewing others' as part of the Program Committee. Their hard work, in evolutionary terms, provided both variation and selection, without which progress in the field would not be possible!

The papers were submitted, reviewed and selected using the MyReview conference management software. We are sincerely grateful to Marc Schoenauer of Inria, France, for his great assistance in providing, hosting, and managing the software.

We would like to thank the local organizing team: Penousal Machado and Ernesto Costa, University of Coimbra, Coimbra, Portugal.

We thank Kevin Sim from the Institute for Informatics and Digital Information, Edinburgh Napier University, for creating and maintaining the official Evo* 2016 website, and Pablo García-Sánchez (Universidad de Granada, Spain) for being responsible for Evo* 2016 publicity.

We would also like to express our sincerest gratitude to our invited speakers, who gave the inspiring keynote talks: Richard Forsyth, a GP pioneer, and Dr. Kenneth Sorensen, a leader in Operations Research and Metaheuristics and Research Professor at the Faculty of Applied Economics, the University of Antwerp, Belgium.

We especially want to express our genuine gratitude to Jennifer Willies of the Institute for Informatics and Digital Innovation at Edinburgh Napier University, UK. Her dedicated and continued involvement in Evo* since 1998 has been and remains essential for building the image, status, and unique atmosphere of this series of events.

April 2016   Malcolm I. Heywood
                                                                         James McDermott
                                                                          Mauro Castelli
                                                                          Ernesto Costa
                                                                             Kevin Sim

Organization

Administrative details were handled by Jennifer Willies, Institute for Informatics and Digital Innovation, Edinburgh Napier University, Scotland, UK.

Organizing Committee

Program Co-chairs

Malcolm I. Heywood   Dalhousie University, Canada
James McDermott   University College Dublin, Ireland

Publication Chair

Mauro Castelli   Universidade Nova de Lisboa, Portugal

Local Chair

Ernesto Costa   University of Coimbra, Portugal

Webmaster

Kevin Sim   Edinburgh Napier University, UK

Program Committee

Alexandros Agapitos	University College Dublin, Ireland
Lee Altenberg	University of Hawaii at Manoa, USA
R. Muhammad Atif Azad	University of Limerick, Ireland
Ignacio Arnaldo	MIT, USA
Douglas Augusto	LNCC/UFJF, Brazil
Wolfgang Banzhaf	Memorial University of Newfoundland, Canada
Mohamed Bahy Bader	University of Portsmouth, UK
Helio Barbosa	LNCC/UFJF, Brazil
Heder Bernardino	LNCC/UFJF, Brazil
Anthony Brabazon	University College Dublin, Ireland
Nicolas Bredeche	Université Pierre et Marie Curie, France
Stefano Cagnoni	University of Parma, Italy
Ernesto Costa	University of Coimbra, Portugal
Luis Da Costa	Université Paris-Sud XI, France
Antonio Della Cioppa	University of Salerno, Italy
Grant Dick	University of Otago, New Zealand
Federico Divina	Pablo de Olavide University, Spain
Marc Ebner	Ernst-Moritz-Arndt Universität Greifswald, Germany
Aniko Ekart	Aston University, UK
Francisco Fernandez de Vega	Universidad de Extremadura, Spain

Gianluigi Folino   ICAR-CNR, Italy
James A. Foster   University of Idaho, USA
Christian Gagné   Université Laval, Québec, Canada
Steven Gustafson   GE Global Research, USA
Jin-Kao Hao   LERIA, University of Angers, France
Inman Harvey   University of Sussex, UK
Erik Hemberg   MIT, USA
Malcolm I. Heywood   Dalhousie University, Canada
Ting Hu   Dartmouth College, USA
David Jackson   University of Liverpool, UK
Colin Johnson   University of Kent, UK
Ahmed Kattan   Um Al Qura University, Saudi Arabia
Graham Kendall   University of Nottingham, UK
Michael Korns   Korns Associates, USA
Jan Koutnik   IDSIA, Switzerland
Krzysztof Krawiec   Poznan University of Technology, Poland
Jiri Kubalik   Czech Technical University in Prague,
                                  Czech Republic
William B. Langdon   University College London, UK
Kwong Sak Leung   The Chinese University of Hong Kong, China
John Levine   University of Strathclyde, UK
Evelyne Lutton   Inria, France
Penousal Machado   University of Coimbra, Portugal
James McDermott   University College Dublin, Ireland
Andrew McIntyre   Dalhousie University, Canada
Jorn Mehnen   Cranfield University, UK
Julian Miller   University of York, UK
Alberto Moraglio   University of Exeter, UK
Xuan Hoai Nguyen   Hanoi University, Vietnam
Quang Uy Nguyen   Military Technical Academy, Vietnam
Miguel Nicolau   University College Dublin, Ireland
Julio Cesar Nievola   Pontificia Universidade Catolica do Parana, Brazil
Michael O'Neill   University College Dublin, Ireland
Una-May O'Reilly   MIT, USA
Fernando Otero   University of Kent, UK
Ender Ozcan   University of Nottingham, UK
Andrew J. Parkes   University of Nottingham, UK
Gisele Pappa   Federal University of Minas Gerais, Brazil
Tomasz Pawlak   Poznan University of Technology, Poland
Clara Pizzuti   Institute for High Performance Computing
                                  and Networking, Italy
Thomas Ray   University of Oklahoma, USA
Peter Rockett   University of Sheffield, UK
Denis Robilliard   Université Lille Nord de France, France
Conor Ryan   University of Limerick, Ireland
Marc Schoenauer   Inria, France

Contents

Full Presentations

One-Class Classification for Anomaly Detection with Kernel Density Estimation and Genetic Programming

Van Loi Cao[✉], Miguel Nicolau, and James McDermott

Natural Computing Research and Application Group,
University College Dublin, Dublin, Ireland
loi.cao@ucdconnect.ie, {miguel.nicolau,james.mcdermott2}@ucd.ie
http://ncra.ucd.ie

Abstract. A novel approach is proposed for fast anomaly detection by one-class classification. Standard kernel density estimation is first used to obtain an estimate of the input probability density function, based on the one-class input data. This can be used for anomaly detection: query points are classed as anomalies if their density is below some threshold. The disadvantage is that kernel density estimation is lazy, that is the bulk of the computation is performed at query time. For large datasets it can be slow. Therefore it is proposed to approximate the density function using genetic programming symbolic regression, before imposing the threshold. The runtime of the resulting genetic programming trees does not depend on the size of the training data. The method is tested on datasets including in the domain of network security. Results show that the genetic programming approximation is generally very good, and hence classification accuracy approaches or equals that when using kernel density estimation to carry out one-class classification directly. Results are also generally superior to another standard approach, one-class support vector machines.

Keywords: Anomaly detection · One-class classification · Kernel density estimation

1 Introduction

Anomaly detection is the problem of detecting samples or patterns in data that are different from expected behavior [3]. The nonconforming patterns are referred to a variety of names in different application domains, but the terms *anomalies* and *outliers* are common. Anomaly detection is applied in many fields such as intrusion detection, credit card fraud detection, insurance, health care, fault detection in safety critical systems. Anomaly detection plays an important role in a variety of application domains because anomalies in data often translate to critical, actionable information or potentially dangerous situations [1,3].

In network security, anomaly detection means the discrimination of illegal and malicious activities from normal connections or expected behavior of systems

© Springer International Publishing Switzerland 2016
M. Heywood et al. (Eds.): EuroGP 2016, LNCS 9594, pp. 3–18, 2016.
DOI: 10.1007/978-3-319-30668-1_1

[13,14]. The role of automated anomaly detection has become increasingly important in network security due to the widespread use of computer networks in recent years [13]. However, there are some issues that make constructing anomaly detection models challenging. One of the major issues is that anomalies are continuously evolving over time. The model built from current data may not be able to capture attacks or unauthorized accesses in the future [6]. Collecting the anomalous data is extremely difficult due to the privacy and security concerns of computer networks and the shortage of intrusive network traffic in host logs and events. Moreover, labeling such data is also a challenging and time-consuming task for experts in the domain and has potential problems [24]. Thus, in many situations only the normal class is available.

Because of these issues, one-class learning or novelty detection is a common method for anomaly detection. In *one-class classification* (OCC), only one class (the target) is available for constructing a classifier. The classifier is then used to distinguish whether a test instance belongs to the target class or the non-target class [29]. In this work, we use the terms target and non-target to refer to normal and anomaly respectively. More details about one-class learning and recent one-class methods are discussed in Sect. 2.

In this paper, we propose a one-class learning method by combining Genetic Programming (GP) with Kernel Density Estimation (KDE) (described in detail in Sect. 3.2). KDE has the ability to directly estimate density from data. It can thus be used as a one-class classifier by imposing a density threshold: points in low-density areas are classed as anomalies. However the computational cost of KDE at query time is potentially high, scaling with the number of training points. Fortunately, GP has the ability to approximate density. The result is a model where the computational cost at query time depends on the number of nodes in the GP tree. Therefore, in our system KDE is first used to estimate the density from target examples, and GP is then employed to approximate this density. The resulting one-class classifier not only yields high accuracy in detecting anomalies, but has reduced computational cost relative to KDE. Another potential advantage, not pursued in this paper, is that the resulting models are interpretable GP trees. The method is described in detail in Sect. 4.

The rest of this paper is organized as follows. In the next section, we briefly review some work related to one-class classification. In Sect. 3, we give a short introduction to GP for classification, and KDE. This is followed by a section proposing OCC using KDE and GP. Experiments, and Results and Discussion are presented in Sects. 5 and 6 respectively. The paper concludes with highlights and future directions.

2   Related Work

The concept of one-class classification was originated by Moya et al. [18]. One-class classification has rapidly emerged in a variety of fields from document classification [17], concept learning [9], novelty detection [2] to anomaly detection [8,20]. Based on the availability of training data, one-class classification can be categorized into three groups [11]: (1) learning from only target examples; (2) learning from target examples and unlabeled data; (3) learning from target examples and a small number of non-target examples. In terms of anomaly detection, we concentrate on (1), that is the one-class learning techniques that construct a model when non-target data is absent.

Tax and Duin [28, 29] proposed a method called Support Vector Data Description (SVDD) to solve the problem of OCC. In the method, a hypersphere with minimum radius around the target examples in feature space is found, which encompasses almost all target instances. In order to achieve a sufficient decrease in the volume of the hypersphere, it possibly rejects some fraction when training this model. This illustrates a theme present in all one-class classification research, the trade-off between false positive and false negative rates. Tax [27] introduced different kernel functions to SVDD that make the method more flexible, and the Gaussian kernel was found to be the most suitable for many datasets. However, this technique requires a large number of target examples, and becomes inefficient in high dimension [11].

Instead of finding a hypersphere, Schölkopf [21, 22] presented an alternative approach called one-class SVM. The one-class classifier is achieved by searching for a hyperplane with a maximum margin between the region containing target data and the origin in feature space. The target data is again mapped to feature space via a kernel. The efficacy of the method is evaluated on a handwritten digit dataset, and the results show that the classifier performs well.

Recently, several evolutionary algorithm approaches have been proposed for one-class classification problem [4, 5, 31]. Heywood and Curry have a long line of research, e.g [4, 5] on multi-objective one-class genetic programming. In order to establish a more precise boundary, a large number of artificial data is generated to construct a two-class classifier. They combine a multi-objective fitness function with a local membership function to improve the search for specific regions of the target distribution. Its performance was compared to one-class v-SVM, bottleneck neural network (BNN) and two-class SVM, and the results show that the one-class GP performs consistently overall.

Cuong To [31] proposed a different approach to one-class problem by calculating the sum of the Euclidean distance from a target point to all target examples with an assumption that the distribution of the summed distance is normal. This is similar to an inversion of a kernel method since a kernel is a similarity whereas a Euclidean distance is a dissimilarity. They then used GP to approximate this sum. They then used two thresholds at 2.5 % and 97.5 % on training dataset to classify new examples. New points whose sum of distances was below the 2.5 percentile or above the 97.5 percentile of the data were classed as anomalies. Effectively, the lower threshold meant that points with extremely low sums of distances (i.e. directly in the middle of the target class) were wrongly classed as anomalies. It is possible that malfunctions in their GP approximation made this necessary. Our method is similar but avoids this error by using a single, upper threshold. We also gain flexibility by using a kernel instead of Euclidean distance.

In this work, we present a new approach for one-class classification problem. This is to combine Kernel Density Estimator and Genetic Programming to take advantage of their different strengths. While KDE can estimate the density directly from data with unknown distribution, GP has the ability to search for a function mapping from data to its density. Thus, the classifier not only produces accuracy as good as KDE, but also reduces time consuming on testing stage.

3   Preliminaries

In this section we briefly introduce Genetic Programming and its application in classification (Sect. 3.1), and introduce Kernel Density Estimation and its application in one-class classification (Sect. 3.2).

3.1   Genetic Programming

GP was popularized by Koza in the 1990s [12]. It is an evolutionary paradigm that is inspired by biological evolution. It aims to find good solutions to a diverse spectrum of problems in the form of computer programs. GP has the ability to represent solutions to many problems since the representation is highly flexible.

In particular, GP methods can be adapted to classification problems [16]. For two-class classification problems, a typical approach is to evolve real-valued GP trees, and then translate the numeric (real) values returned by them into class labels using a threshold. Typically, the threshold is zero. Similar methods can be applied for one-class classification (OCC): artificial data acts as the non-target class [4,5]. However, in this paper we propose a different GP approach to OCC. This is to use GP to approximate the density of the training set (and artificial data) originally given by KDE.

3.2   Kernel Density Estimation

Kernel density estimation is one of the most attractive non-parametric methods in the statistical literature for estimating a probability density function from a sample of points [32]. It estimates the density function directly from data with no assumption about the underlying distribution. It produces asymptotic convergence to any density function [23]. These advantages make KDE a general approach for many problems that do not assume any specific distribution of the density function.

Let $x_1, x_2, ..., x_n$ be a set of d-dimensional samples in $\mathbb{R}^d$ drawn from an unknown distribution with density function $p(x)$. An estimate $\hat{p}(x)$ of the density at x can be calculated using

$$\hat{p}(x) = \frac{1}{n} \sum_{i=1}^{n} K_h (x - x_i) \tag{1}$$

where $K_h : \mathbb{R}^d \to \mathbb{R}$ is a kernel function with a parameter h called the *bandwidth*.

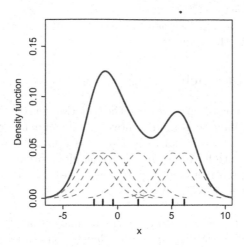

Fig. 1. The density distribution estimated by KDE. Figure reproduced from [34]

Two factors, kernel function and bandwidth play vital roles in KDE. A variety of kernel functions with different properties are typically chosen for KDE; eg. Gaussian, Uniform, Epanechnikov, Exponential, Linear and Cosine. The Gaussian kernel (Eq. 2) is probably the most common in applications and is the only one used in this paper. As illustrated in Fig. 1, in KDE each point contributes a small "bump" to the overall density, with its shape controlled by the kernel and bandwidth. The bandwidth parameter h controls the trade-off between bias of the estimator and its variance. This means that a large bandwidth leads to a very smooth (i.e. high-bias) density distribution while the density distribution is less smooth (i.e. high-variance) with a smaller bandwidth.

$$K_h\left(x\right) = \exp\left(-\frac{x^2}{2h^2}\right) \tag{2}$$

In terms of OCC, we construct a model from only the target class, and we typically do not assume any particular parametric distribution (e.g. Normal or uniform) on the target class. Thus, KDE is a suitable approach for estimating the density distribution of the target class. This can then be used to define a classifier by imposing a threshold in terms of density: new (query) points are classified as anomalies if they have a density lower than (say) 95 % of the training set. The choice of threshold in practice depends on the relative frequency of anomalies in the domain and the relative costs of false positives and false negatives.

A main drawback of KDE is its computational cost. KDE "remembers" all training data in order to compute the density of each new point. Thus, the larger the training data sample size, the greater the computational cost of querying new points. The computational cost of KDE can be markedly improved using space-partitioning trees, but the extent of the improvement depends on the bandwidth: for large values of bandwidth (which appear to give the best results in our experiments, described in Sect. 5), space-partitioning trees cannot eliminate as many points, hence the improvement is less. In any case, the complexity still depends on the size of the training data [7].

4   Proposed Approach

In this paper we use GP to approximate the function learned by KDE, in order to speed up the querying stage and remove the dependence on the size of the training data. A second potential benefit of this approach is that the density is expressed as a potentially interpretable symbolic expression, although we do not use this interpretation in this paper. Therefore, the GP classifier not only inherits the advantages of KDE, but also can reduce the computational cost of the querying stage.

However, if training data is drawn from a small area of the feature space, the classifier will lack the ability to predict density on points outside that area. One solution is to generate artificial data in the low-density regions of the space. Ideally, generating new data near the "border" between high and low-density regions will particularly improve the approximation in important areas. This will strengthen the classifier's discrimination on both target and non-target examples. More details of the proposed one-class genetic programming technique is described in the next subsection.

4.1   Description of Method

Let $T = \{x_1, x_2, ..., x_m\}$ be the target training set, where $x_i \in \mathbb{R}^n$ is a target instance and m is the number of samples in the target training set. We assume that T has been standardized, that is shifted and scaled to give zero mean and unit variance.

1. Kernel density estimation is employed to estimate the density distribution of the target training set. Using the kernel density estimator, we compute the density $d(x_i)$ for every target sample $x_i \in T$.
2. An artificial data set $A = \{x_{m+1}, x_{m+2}, ..., x_{m+q}\}$, is generated as described in Sect. 4.2. The KDE is used to compute the density $d(x_i)$ for each artificial example $x_i \in A$ against the target density distribution, where $i \in \{m+1, m+2, ..., m+q\}$.
3. Let $TS = T \cup A = \{x_1, x_2, ..., x_m, x_{m+1}, ..., x_{m+q}\}$, and $d(x_i)$ is the density of each point $x_i \in TS$, where $i \in \{1, 2, ..., m, m+1, ..., m+q\}$. We use GP symbolic regression with a root mean square error to search for a function f that satisfies the criterion in Eq. 3.

$$f(x_i) \approx d(x_i) \tag{3}$$

Therefore, the fitness function is given by

$$Fitness = \sqrt{\frac{\left(\sum_{i=1}^{m+p} (f(x_i) - d(x_i))^2\right)}{(m+p)}} \tag{4}$$

where m and p are the numbers of the target training set and the artificial data set respectively, $d(x_i)$ is the density of sample $x_i \in TS$.

Now we have a function f that, like KDE, can predict density on query data but with less computational cost. To construct a classifier, we impose a threshold on f, just as in the case of KDE, for example at a level that will classify the 5 % of the training data with the lowest density values as anomalies. The choice of threshold depends on the domain and dataset.

4.2 Generating Artificial Data

The aim is to generate artificial data around the target training set and in the low-density regions of the target training set. As mentioned in Subsect. 4.1, our input data is standardized hence it is centered at the origin. However, it is infeasible when we only use standard methods (e.g. Gaussian, Uniform) to generate data in high dimensional feature spaces [4,5,30]. The probability of an artificial point being in or around the boundary of the target distribution is very small. Therefore, we propose a method to generate artificial data with more points in and around the target distribution.

There are two main steps to generate the artificial data in our method. We generate data in a hypersphere centered at the origin, uniformly in terms of radius. We then sample only points that are around the target training set and in the low-density regions of the target training set. More details of these steps are described below:

1. Generate uniformly in terms of radius
   (a) Data X is generated from Gaussian distribution in n dimensions with zero mean and unit standard deviation, see Fig. 2(a)

$$X \sim \mathcal{N}(0,1) \tag{5}$$

   (b) Relocate all samples in X to the surface of the unit hypersphere, see Fig. 2(b). The direction of each point in X does not change, thus X' is uniformly distributed on the surface of the unit hypersphere:

$$X' = \frac{X}{\|X\|} \tag{6}$$

   where $\|X\|$ is the Euclidean distance from each sample in X to the origin.

   (c) Uniformly generate U in one dimension with a range $[0, R]$

$$U \sim \mathcal{U}(0,R) \tag{7}$$

   where R is the maximum radius of a hypersphere. R is chosen to give the desired distribution, including points in low-density regions. The value for each dataset is given in the next section.

(d) Rescale all objects in X' with a factor U:

$$X'' = U.X' \qquad (8)$$

Now X'' are uniform in terms of radius to the origin in the hypersphere with maximum radius R, see Fig. 2(c).

2. Sample data in and around the target training set
   (a) The density of each point in X'' is computed against the target training set by KDE.
   (b) Any point in X'' whose density is greater than a density value that is determined by a threshold $t_{overlap}$ is discarded. Threshold $t_{overlap}$ determines what percentage of the target training set is overlapped by the artificial data in terms of density. This is because in high-density regions we already have target examples.
   (c) Randomly sample a data set T from X'' with a proportion p, where p is proportional to the number of examples in target training set.

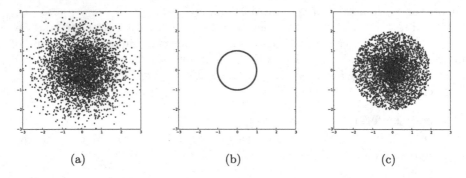

(a)   (b)   (c)

Fig. 2. Generate artificial data uniformly in terms of radius. (a) Gaussian distribution. (b) Uniform distribution on the surface of the unit hypersphere. (c) Distribution in an arbitrary hypersphere, uniform in terms of radius.

5   Experiments

5.1   Datasets

The goal of the experiments is to evaluate our method on one-class datasets. Thus, we choose datasets that have one class considered as target class and other classes treated as a non-target class [4,31]. Four datasets in UCI Machine Learning Repository [15] and NSL-KDD dataset [25] are employed for our experiments. In Wisconsin Breast Cancer Database (WBC), each instance contains 9 real-value attributes, one id and one class attribute that divides data into two classes, benign (458 instances) and malignant (241 instances). We remove

16 examples that contain missing data. There are 375 benign examples and 212 malignant examples in Wisconsin Diagnostic Breast Cancer (WDBC). Each instance is presented by 32 attributes (id, class attribute and 30 real-valued attributes). The third dataset is Cleveland heart disease (C-heart) that contains 164 examples for heart disease level 0, and 139 examples for heart disease level 1–4. There are 14 attributes (13 real-valued attributes and one class attribute). The last UCI dataset used in our experiments is the Australian Credit Approval dataset. There are 690 instances and each belongs to approval class or risk class (383 approval, 307 risk instances respectively). Each instance is described by 14 real-valued attributes and one class attribute.

For the four datasets, we randomly sample 50 percent of the target class for the target training set and the other 50 percent for the target testing set. However, all examples from non-target class is used for the non-target testing set. More details are presented in Table 1.

NSL-KDD dataset [26] is a filtered version of the KDD Cup 1999 dataset [10], which is in the domain of network security, after removing all redundant instances and making the task more difficult. In NSL-KDD, a connection is represented by 41 attributes (38 numeric continuous and discrete, and 3 categorical attributes). Each record is labeled as either normal or as a specific kind of attack belonging to one of the four main categories: Denial of Service (DoS), Remote to Local (R2L), User to Local (U2R) and Probe. NSL-KDD consists of two datasets: $KDDTrain^+$ and $KDDTest^+$ which are drawn from different distributions.

In this work, we plan to conduct our experiments on the R2L attack group. This is because the aim of the search is to reduce computational cost, thus it is efficient to reduce redundant features [33]. Moreover, the records from the R2L group are slightly similar to normal connections due to the fact that they are based on some content features of network traffic. This makes them more difficult to classify than DoS or Probe attack groups [13,14,24]. Based on a previous feature selection research [33], we choose only a subset of 10 features from 41 features in NSL-KDD for detecting the R2L attack category in our experiment. Several of the variables chosen are categorical or discrete. We simply treat them as real-valued (e.g. we map *service* values TCP, UDP and ICMP to values 1, 2 and 3). As shown in Sect. 6 this gives good results.

In our experiments, we randomly sample 2000 normal instances from $KDDTrain^+$ for the target training set, whereas 2000 normal examples and 2000 R2L examples are randomly selected from $KDDTest^+$ for the target testing set and the non-target testing set respectively. More details are presented in Table 1.

We use the proposed method in Sect. 4.2 to generate artificial data for our experiments. The threshold $t_{overlap}$ for the artificial data was set so that 10 % of the target training set was below it in terms of density. Figure 3 shows the density distribution of the target training set and artificial data with 10 % overlap. The size of the artificial training set was chosen to be approximately double the size of the target training set, except for datasets where this was small (< 100) or large (> 1000). Too large a number could lead to a good GP approximation in the low-density region and a poor GP approximation in the high-density region, while too small a number would lead to the opposite. The numbers of examples in target training set, artificial dataset, testing sets are shown in Table 1.

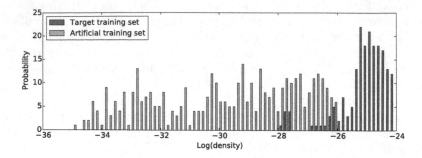

Fig. 3. The density distribution of target training set and artificial data

Table 1. One-class classification datasets

Dataset	Features	Training set		Testing set	
		Target	Artificial	Target	Non-target
C-heart	13	80	350	80	137
Australia credit card	14	191	400	192	307
WBC	9	222	500	222	239
WDBC	30	178	400	179	212
R2L (NSL-KDD)	10	2000	1000	2000	2000

5.2 Experimental Settings

One preliminary experiment and one main experiment on GP, KDE and one-class SVM are conducted in order to evaluate the proposed one-class GP on accuracy and runtime of the resulting models. The five datasets in Table 1 are used for the experiments. We use the terms OCGP, OCKDE, and OCSVM to refer to one-class GP, KDE, and SVM classifiers, respectively. The choice of threshold for classifier in practice varies from domain to domain, but in our experiments we evaluate AUC using many values of the threshold (and we do the same for all methods, OCGP, OCKDE, OCSVM). More details of the two experiments are described below:

In the preliminary experiment we investigated the effect of the bandwidth parameter. We chose the Gaussian kernel and used cross-validation to choose its bandwidth value for KDE. That is, the accuracy of the density estimation of the input data (not the accuracy on a one-class classification task) was used to choose an optimal value for bandwidth. Accuracy was measured as mean integrated squared error (MISE). The resulting bandwidth was in all cases low (about 0.8). When used in a one-class classification task, this bandwidth achieved poor performance. Similar results were found for OCSVM. Therefore, for remaining experiments we chose a larger bandwidth value (2.0) that gave worse results on density estimation, but better results on the one-class classification task.

The main experiment first plans to investigate KDE in terms of one-class classification. This is a basic framework to evaluate OCGP. We set up the same kernel function and bandwidth as described in the preliminary experiment. Kernel density estimator is run on five datasets in Table 1, and we calculate the AUC values for OCKDE on every dataset.

Secondly, the experiment aims to examine how efficiently OCGP performs on these datasets. Evolutionary parameters and the parameters for generating artificial data are presented in Table 2. We calculate the mean of AUC values over 50 runs to evaluate OCGP, and select the individual with median AUC value over 50 runs to draw ROC curves against OCKDE and OCSVM. The computational cost of OCGP on testing stage is calculated as the mean of the numbers of nodes in the best-of-run GP trees over 50 runs.

Finally, the experiment will examine one-class SVM [22] in order to compare its performance to OCGP. We set up the same kernel function and bandwidth as KDE, and $\nu = 0.5$. We use one-class SVM from sklearn [19] to run experiments over the five datasets. The AUC values and the number of support vector are computed. The results from the three experiments on the five datasets in Table 1 are presented in Tables 3 and 4, and Fig. 4.

Table 2. Parameter settings

Generate artificial data	
R	10
$t_{overlap}$	10 percent
KDE and OC SVM parameters	
Bandwidth	2.0
Kernel function	Gaussian
GP parameters	
Population size	400
Number of generation	500
Crossover probability	0.9
Mutation probability	0.1
Selection	Tournament
Tournament size	3
Function set	$\{+, -, \times, /, \exp, sqr, sqrt\}$

6   Results and Discussion

This section presents the experimental results of evaluating the proposed one-class GP classifier on the five datasets. The performance of the one-class GP classifier is evaluated along two measurements, Area Under ROC Curve (AUC) and computational cost on querying new instances. The results are shown in Tables 3 and 4, and in Fig. 4 the ROC curves of OCGP median over 50 runs on the five datasets are shown against those of OCSVM and OCKDE.

Table 3 illustrates the AUC values when applying the three one-class classifiers on the five datasets. It can be seen from the table that OCGP performs as well as OCKDE, and better than OCSVM in most of datasets in terms of accuracy.

The mean of AUC values from OCGP are close to OCKDE on C-heart, Australian Credit Card, WBC and R2L (see Table 3). In comparison to OCc, the AUC values from OCSVM are lower than those from OCGP on C-heart, Australian Credit Card and R2L. Conversely, OCGP does not produce a good accuracy on WDBC.

In general, these results suggest that the proposed OCGP tends to perform efficiently on the datasets that are low dimension (around 10 or 14 dimensions), but in high dimension (the WDBC dataset) the classifier may produce a poorer classification accuracy than OCKDE and OCSVM.

The ROC curves are demonstrated in Fig. 4. In this figure, we draw the ROC curve of $OCGP_{median}$ against the ROC curves of OCKDE and OCSVM. The term $OCGP_{median}$ refers to the GP individual that produces the median AUC

Table 3. The AUC results from three classifiers

Dataset	AUC		
	OCKDE	OCSVM	$OCGP_{mean}$
C-heart	0.7731	0.7557	0.7874
Australian credit card	0.8355	0.8201	0.8286
WBC	0.9908	0.9907	0.9904
WDBC	0.9529	0.9500	0.9204
R2L	0.9001	0.8592	0.8727

Table 4. The computational cost for different techniques at query time

Dataset	Feature	Training points	Support vectors	GP nodes
C-heart	13	80	53	237.86
Australia credit card	14	191	112	218.36
WBC	9	222	115	207.8
WDBC	30	178	121	182
R2L	10	2000	1001	197.2

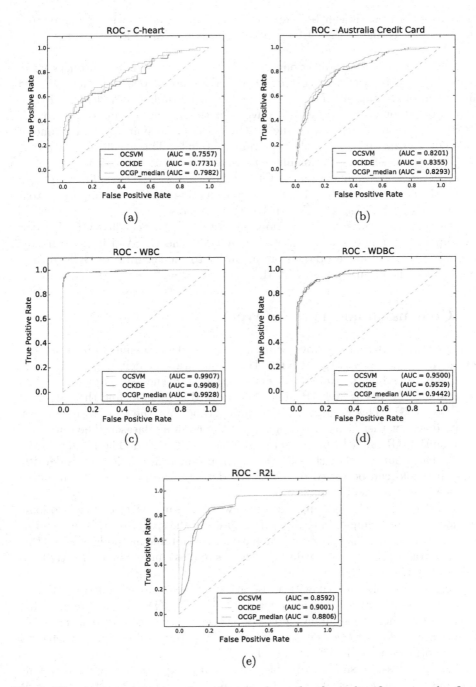

Fig. 4. The receiver operating characheristic from the three classifiers over the five datasets. (a) ROC on C-heart. (b) ROC on Australia credit card. (c) ROC on WBC (d) ROC on WDBC. (e) ROC on R2L

value over 50 runs. It can be seen from the figure that the ROC curve from $OCGP_{median}$ is higher than ROC curves from OCSVM in four datasets again with the single exception of WDBC.

Table 4 shows the computational cost for the three techniques at query time. Overall, the computational cost of OCGP does not depend on the size and dimension of training set. For OCGP, the average number of nodes in GP tree is around 200 on all five datasets. Conversely, the number of support vectors of OCSVM tends to increase with the number of training points. The runtime of OCKDE depends on the number of training points. This result suggests that OCGP is the least computationally expensive method at query time.

However, the training time of OCGP is much slower than OCKDE and OCSVM. It is approximately 10 min for constructing OCGP in comparison to a few seconds for building OCKDE or OCSVM models.

Overall, the results in this section suggest that the one-class GP classifier has superior scaling of query runtime to OCKDE and OCSVM in performance at query time, and its accuracy is often similar to the accuracy of OCKDE and higher than the accuracy of OC SVM.

7   Conclusion and Further Work

This paper has presented a novel approach to one-class classification. It aims to retain the biggest advantage of KDE, that is accurate modeling of the training data, while avoiding the main disadvantage, that is slowness at query time, when the training dataset is large. It aims to achieve this by modeling the output of KDE using GP. This requires extra training time, but the time to query new points does not then depend on the size of the training dataset. The output of the resulting GP model can be regarded as a density value. The model can be used to carry out one-class classification by imposing a threshold on this density. This threshold can be varied to reflect the desired false positive/false negative balance in any particular application.

In order to make the GP approximation work, it is necessary to generate some artificial training samples in low-density regions of the space, since otherwise GP has no knowledge of these regions. We have found suitable methods of doing this for our datasets. The density values for these samples can be found directly from the already-trained KDE.

Results have shown that with these artificially-generated samples, GP succeeds in approximating the density function given by KDE. While the resulting one-class classification accuracy cannot be expected to exceed that of using KDE, it often approaches or equals it. It is also often superior to the accuracy achieved using another standard method, one-class SVM.

The one-class classification problem has many important applications. One example is in network security. In this domain, query points are streamed in at a high rate, hence the performance gain achieved using the GP approximation is a valuable contribution. We have shown that our method can provide good accuracy in this domain.

Our method still has some limitations. In particular, we wish to adapt the method of generating artificial data to work well in higher-dimensional spaces and to deal with binary or categorical variables. We will also investigate alternative methods of setting the bandwidth automatically. We will apply and tailor our method to other network security datasets. Finally, we will report on the measured improvement in runtime on large datasets, and compare our method to alternatives such as the Autoencoder.

Acknowledgements. This work is funded by Vietnam International Education Development (VIED) and by agreement with the Irish Universities Association.

References

1. Aggarwal, C.C.: Outlier Analysis. Springer Science & Business Media, New York (2013)
2. Bishop, C.M.: Novelty detection and neural network validation. In: IEE Proceedings on Vision, Image and Signal Processing, vol. 141, pp. 217–222. IET (1994)
3. Chandola, V., Banerjee, A., Kumar, V.: Anomaly detection: A survey. ACM computing surveys (CSUR) **41**(3), 1–58 (2009)
4. Curry, R., Heywood, M.: One-class learning with multi-objective genetic programming. In: ISIC 2007 IEEE International Conference onSystems, Man and Cybernetics, pp. 1938–1945. IEEE (2007)
5. Curry, R., Heywood, M.I.: One-class genetic programming. In: Vanneschi, L., Gustafson, S., Moraglio, A., De Falco, I., Ebner, M. (eds.) EuroGP 2009. LNCS, vol. 5481, pp. 1–12. Springer, Heidelberg (2009)
6. Fiore, U., Palmieri, F., Castiglione, A., De Santis, A.: Network anomaly detection with the restricted Boltzmann machine. Neurocomputing **122**, 13–23 (2013)
7. Gray, A.G., Moore, A.W.: Nonparametric density estimation: toward computational tractability. In: SDM, pp. 203–211. SIAM (2003)
8. Hido, S., Tsuboi, Y., Kashima, H., Sugiyama, M., Kanamori, T.: Statistical outlier detection using direct density ratio estimation. Knowl. Inf. Syst. **26**(2), 309–336 (2011)
9. Japkowicz, N.: Concept-learning in the absence of counter-examples: an autoassociation-based approach to classification. Ph.D. thesis, Rutgers, The State University of New Jersey (1999)
10. KDD Cup Dataset (1999). http://kdd.ics.uci.edu/databases/kddcup99/kddcup99.html
11. Khan, S.S., Madden, M.G.: A survey of recent trends in one class classification. In: Coyle, L., Freyne, J. (eds.) AICS 2009. LNCS, vol. 6206, pp. 188–197. Springer, Heidelberg (2010)
12. Koza, J.R.: Genetic Programming: On the Programming of Computers by Means of Natural Selection, vol. 1. MIT press, Cambridge (1992)
13. Lee, W., Stolfo, S.J.: A framework for constructing features and models for intrusion detection systems. ACM Trans. Inf. Syst. Secur. (TiSSEC) **3**(4), 227–261 (2000)
14. Lee, W., Stolfo, S.J., Mok, K.W.: A data mining framework for building intrusion detection models. In: Proceedings of the 1999 IEEE Symposium on Security and Privacy, 1999, pp. 120–132. IEEE (1999)

15. Lichman, M.: UCI machine learning repository (2013). http://archive.ics.uci.edu/ml
16. Loveard, T., Ciesielski, V.: Representing classification problems in genetic programming. In: Proceedings of the 2001 Congress on Evolutionary Computation, 2001, vol. 2, pp. 1070–1077. IEEE (2001)
17. Manevitz, L.M., Yousef, M.: One-class SVMs for document classification. J. Mach. Learn. Res. **2**, 139–154 (2002)
18. Moya, M.M., Koch, M.W., Hostetler, L.D.: One-class classifier networks for target recognition applications. Technical report, Sandia National Labs., Albuquerque, NM (United States) (1993)
19. Pedregosa, F., Varoquaux, G., Gramfort, A., Michel, V., Thirion, B., Grisel, O., Blondel, M., Prettenhofer, P., Weiss, R., Dubourg, V., Vanderplas, J., Passos, A., Cournapeau, D., Brucher, M., Perrot, M., Duchesnay, E.: Scikit-learn: machine learning in Python. J. Mach. Learn. Res. **12**, 2825–2830 (2011)
20. Perdisci, R., Gu, G., Lee, W.: Using an ensemble of one-class SVM classifiers to harden payload-based anomaly detection systems. In: ICDM 2006, Sixth International Conference on Data Mining, pp. 488–498. IEEE (2006)
21. Schölkopf, B., Williamson, R., Smola, A., Shawe-Taylor, J.: SV estimation of a distributions support. Adv. Neural Inf. Process. Syst. **12** (1999)
22. Schölkopf, B., Williamson, R.C., Smola, A.J., Shawe-Taylor, J., Platt, J.C.: Support vector method for novelty detection. NIPS **12**, 582–588 (1999)
23. Scott, D.W.: Multivariate Density Estimation: Theory, Practice, and Visualization. John Wiley & Sons, New York (2015)
24. Shafi, K., Abbass, H.A.: Evaluation of an adaptive genetic-based signature extraction system for network intrusion detection. Pattern Anal. Appl. **16**(4), 549–566 (2013)
25. Tavallaee, M., Bagheri, E., Lu, W., Ghorbani, A.A.: A detailed analysis of the KDD cup 99 data set. In: Proceedings of the Second IEEE Symposium on Computational Intelligence for Security and Defence Applications 2009 (2009)
26. Tavallaee, M., Bagheri, E., Lu, W., Ghorbani, A.A.: NSL-KDD dataset (2012). http://www.iscx.ca/NSL-KDD
27. Tax, D.M.: One-class classification. Delft University of Technology (2001)
28. Tax, D.M., Duin, R.P.: Data domain description using support vectors. In: ESANN, vol. 99, pp. 251–256 (1999)
29. Tax, D.M., Duin, R.P.: Support vector domain description. Pattern Recogn. Lett. **20**(11), 1191–1199 (1999)
30. Tax, D.M., Duin, R.P.: Uniform object generation for optimizing one-class classifiers. J. Mach. Learn. Res. **2**, 155–173 (2002)
31. To, C., Elati, M.: A parallel genetic programming for single class classification. In: Proceedings of the 15th Annual Conference Companion on Genetic and Evolutionary Computation, pp. 1579–1586. ACM (2013)
32. Wand, M.P., Jones, M.C.: Kernel Smoothing. CRC Press, Boca Raton (1994)
33. Wang, W., Gombault, S., Guyet, T.: Towards fast detecting intrusions: using key attributes of network traffic. In: ICIMP 2008, The Third International Conference on Internet Monitoring and Protection, pp. 86–91. IEEE (2008)
34. Wikipedia: Kernel density estimation – Wikipedia, the free encyclopedia (2015). https://en.wikipedia.org/w/index.php?title=Kernel_density_estimation&oldid=690734894

Evolutionary Approximation of Edge Detection Circuits

Petr Dvoracek[✉] and Lukas Sekanina

Faculty of Information Technology, IT4Innovations Centre of Excellence,
Brno University of Technology, Božetěchova 2, 612 66 Brno, Czech Republic
{idvoracek,sekanina}@fit.vutbr.cz

Abstract. Approximate computing exploits the fact that many applications are inherently error resilient which means that some errors in their outputs can safely be exchanged for improving other parameters such as energy consumption or operation frequency. A new method based on evolutionary computing is proposed in this paper which enables to approximate edge detection circuits. Rather than evolving approximate edge detectors from scratch, key components of existing edge detector are replaced by their approximate versions obtained using Cartesian Genetic Programming (CGP). Various approximate edge detectors are then composed and their quality is evaluated using a database of images. The paper reports interesting edge detectors showing a good tradeoff between the quality of edge detection and implementation cost.

Keywords: Edge detection circuits · Cartesian genetic programming · Evolutionary computation

1 Introduction

Reduction of energy consumption is one of the key issues of current society. For example, widely popular battery-powered personal electronics requires energy-efficient computing to reduce the need for battery recharging and big data and supercomputing centers require energy-efficient computing to minimize their operation costs. In recent years, a new approach to reducing the energy consumption has been adopted—*approximate computing*. It exploits the fact that some applications are inherently *error resilient* which means that the errors in their outputs can safely be exchanged for energy consumption reduction. This is a typical feature of multimedia applications in which some errors are not recognizable because human perception capabilities are limited.

Edge detection is an important pre-processing step in advanced image processing applications such as feature detection and feature extraction. The goal of edge detection is to find sharp changes in image brightness. As edge detection is performed very often it makes sense to optimize its implementation. This paper deals with efficient circuit implementations of edge detection based on the Sobel edge detector.

© Springer International Publishing Switzerland 2016
M. Heywood et al. (Eds.): EuroGP 2016, LNCS 9594, pp. 19–34, 2016.
DOI: 10.1007/978-3-319-30668-1_2

Evolutionary computing has been employed to develop approximate implementations of existing circuits or to evolve approximate implementations from scratch. The objective of this paper is to propose and evaluate a method based on evolutionary computing which will enable to approximate edge detection circuits. Rather than evolving approximate edge detectors from scratch, we propose to approximate key components of existing edge detectors. In particular, Cartesian genetic programming (CGP) is used to generate approximations of adders which are basic components of Sobel edge detectors. Various approximate edge detectors are then composed of the approximate adders and their quality is evaluated using a database of images. The implementation cost is measured as the number of used gates. It has been shown in the literature that this measure provides a good estimate of power consumption [24].

The rest of the paper is organized as follows. Section 2 surveys relevant work. The proposed method is presented in Sect. 3. Experimental results are reported in Sect. 4. Conclusions are given in Sect. 5.

2   Relevant Work

2.1   Edge Detectors

The majority of edge detection methods is based on the computation of image gradients. These gradients are often estimated to reduce the computation requirements. The gradient magnitude is then compared with a predefined threshold and used as an indicator whether edges are present or not at an image point. Detailed description of various edge detection algorithms can be found in [19].

The Sobel operator is one of the most popular edge detectors. It uses two convolution kernels (each as a 3×3 pixel window) to estimate gradients in an image. Let A be the input image. The horizontal and vertical derivative approximations are computed as

$$X = \begin{pmatrix} -1 & 0 & +1 \\ -2 & 0 & +2 \\ -1 & 0 & +1 \end{pmatrix} * A, \qquad Y = \begin{pmatrix} -1 & -2 & -1 \\ 0 & 0 & 0 \\ +1 & +2 & +1 \end{pmatrix} * A,$$

where $*$ is the convolution operator. At each point of the image, the gradient magnitude is given by

$$G = \sqrt{X^2 + Y^2}. \tag{1}$$

In order to reduce the computational requirements, the gradient magnitude computed using the square root function is often replaced with a calculation of the absolute value, i.e.

$$G' = |X| + |Y|. \tag{2}$$

Edge detection algorithms have often been accelerated in hardware in order to meet real time constrains of a given application, see, for example, a fast stereo vision system in FPGA [21].

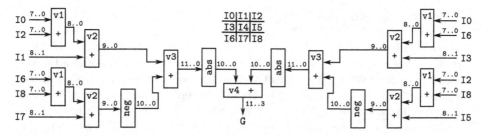

Fig. 1. A hardware implementation of Sobel edge detector. Symbols v denote the depth of addition. $I0, \ldots, I8$ are input pixels.

Figure 1 shows an example of a hardware implementation of Sobel edge detector which operates according to formula 2. In total, this edge detector consists of twenty NOT gates, twenty XOR gates, four 8-bit adders, four 9-bit adders, and three 11-bit adders. It will be used as a reference implementation in this paper.

However, hardware implementations are often optimized to save valuable resources on a chip. In this case, the multiplication by two is implemented by arithmetic shifting. Subtraction is composed of adders v_3 and a set of inverters (neg). The absolute value is obtained by an inversion controlled by the most significant bit representing a negative sign. In other words, the inversion is implemented by an array of XOR gates in which one input of each XOR gate is connected to the most significant bit. The 9-bit and 11-bit adders were replaced by 8-bit adders. Furthermore, the multiplication of one operand is replaced with a division for the other operand, i.e. $(i_0 + i_2) + 2i_1$ was replaced by $(i_0 + i_2)/2 + i_1$. The size of operands of v_2 adders was decreased to 8 bits. For the other adders, the size of the operands was reduced by excluding less significant bits. This optimized version of the Sobel edge detector produces an insignificant error with respect to a solution operating exactly according to formula 2. In total, this edge detector contains sixteen NOT gates, sixteen XOR gates, and eleven 8-bit adders.

2.2   Evolutionary Computing in Edge Detector Design

Evolutionary computing has been utilized to design edge detectors since the nineties [5]. The current research on evolutionary computing for edge detection aims at evolving either edge detectors or edge features, where the features are functions of pixel values used in the process of classifying pixels as edge points or non-edge points [2]. Advanced concepts such as multi-objective methods [27] and Bayesian programs for features definition [2] have been integrated in to the EA-based design approaches. Evolutionary computing was also employed to evolve other computational models that can subsequently be employed for edge detection, for example, cellular automata [15].

Edge detectors and other image operators have been designed by Cartesian genetic programming (CGP). The method and representative case studies are surveyed in [16]. In the case of edge detector evolution, CGP evolves a solution using elementary two-input functions such as minimum, maximum and addition operating over pixel values. The objective is to minimize the error function which is usually defined as a mean absolute error between the image generated by a candidate solution and a "golden image" in which all edges are ideally marked. The golden image is in practice obtained by a suitable conventional edge detector such as Canny or Sobel operator. The evolved circuit, in fact, approximates the conventional solution using hardware-friendly components.

In order to develop low-cost and efficient hardware implementations, edge detectors were evolved using gates and other hardware friendly components as building blocks. For example, an evolvable hardware approach was taken for low level edge detector design in [7] and genetic programming was used to search for digital transfer function of image edge detector [3]. The image filter evolution has been accelerated using specialized hardware such as graphics processing units [4] and field programmable gate arrays [22].

2.3   Approximate Computing in Image Processing

In recent years, approximate computing was established to investigate how computer systems can be made better—more energy efficient, faster, and less complex—by relaxing the requirement that they are exactly correct. One of the approximation techniques is functional approximation whose purpose is to implement a slightly different function to the original one providing that the error is acceptable and power consumption, performance or other parameters are improved. The functional approximation can be conducted at the level of software as well as hardware.

Image operators (including edge detectors) are good candidates for approximations because occasional errors (pixels showing undesired values) are not often recognized by users. On the other hand, the approximate implementations can lead to a reduction in power consumption or processing time.

In [1], a software module implementing the Sobel edge detector was replaced by trained neural network and the module has been later accelerated in a specialized hardware.

Using Axilog, which is a set of language annotations that provide the necessary syntax and semantics for approximate hardware design and reuse in Verilog [26], a conventional implementation of Sobel detector counting 143 lines of code was manually annotated (details not provided in the paper) and an approximate implementation was obtained. In both cases, appropriate papers report the quality of resulting images for a few target errors.

Image operators can also be approximated by approximating selected arithmetic components (adders or multipliers) that are present in conventional implementations, see denoising filters [9,18]. Finally, approximations of median-outputting filters based on simplifying of the compare and swap components can be found in [13].

2.4   Evolutionary Circuit Design

After publishing a seminal paper introducing the field of evolvable hardware [6], new methods based on EAs have been proposed for circuit design. A considerable success have been achieved by Cartesian genetic programming which enabled to improve results of conventional circuit synthesis and optimization algorithms for small combinational circuits and discover new implementations of important circuit components such as filters, classifiers and predictors [11]. More complex circuits were then evolved by means of decomposition techniques [20], functional level evolution, developmental encodings, and advanced fitness evaluation methods utilizing the principles of formal verification [23].

Digital circuits can naturally be approximated by means of CGP. The methods can be classified as

- Error-oriented: In the first phase, CGP tries to evolve a circuit showing a predefined error. In the second phase, the resources are optimized.
- Resources-oriented: Resources (e.g. the number of gates) are constrained and CGP is used to minimize the circuit error with available resources.
- Multi-objective: All criteria are optimized together using a multi-objective EA such as NSGA-II.

Examples of approximate circuits obtained by CGP are approximate medians, 8-bit adders and 8-bit multipliers [24,25]. All these circuits were approximated without any decomposition. In this work, we propose to approximate selected components of the whole circuit and analyze the impact of the approximation on the circuit behavior.

3   Adopting CGP for Circuit Approximation

CGP will be used to approximate selected components of conventional or evolved edge detection circuits. This section deals with the principles of CGP when applied to circuit evolution and approximation.

3.1   Cartesian Genetic Programming

Circuit Representation in the Chromosome: A candidate circuit is modeled as a grid of processing nodes arranged in n_c columns and n_r rows. Each processing node performs specific operation g from the set of functions Γ. In evolutionary circuit design, this function set usually contains logic gates or elementary arithmetic functions. The number of circuit inputs n_i and outputs n_o are fixed. Parameter l-back defines a degree of interconnection between the columns. For example, if $l = 1$ the interconnection is minimal because only neighboring columns may be connected; if $l = n_c$ the circuit interconnection is maximal. Nodes of the same column can not be connected.

The circuit connection is encoded into a chromosome. Each gate (processing node) is represented by a triplet (i_1, i_2, α), where α is a code of operation taken from the function set Γ. Symbols i_1 and i_2 are pointers to nodes (or primary inputs) to which a given gate is connected to, providing that primary inputs are labeled $0 \dots n_i - 1$ and the nodes are labeled $n_i - 1 \dots n_i + n_c n_r - 1$. Finally, the chromosome contains n_o genes determining the nodes or logic constants (0 or 1) where the primary outputs are connected to.

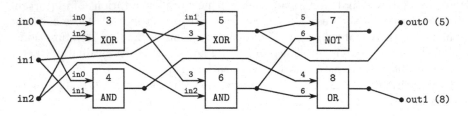

Fig. 2. Full adder represented by CGP with parameters: $n_i = 3, n_o = 2, n_c = 3,$ $n_r = 2, l = 3, \Gamma = \{0^{AND}, 1^{OR}, 2^{XOR}, 3^{NOT}\}$. Chromosome: (0, 2, 2) (0, 1, 0) (1, 3, 2) (3, 2, 0) (5, 6, 3) (4, 6, 1) (5, 8)

Figure 2 provides an example of the CGP encoding. One important feature of CGP is that not all the nodes have to be included in the phenotype. In this case, NOT (node 7) is disconnected.

Fitness Function: As our goal is to approximate arithmetic circuits using the resources-oriented method, the fitness function is defined as a sum of absolute differences (SAD)

$$f_{SAD} = \sum_{j=1}^{K} |y(j) - t(j)|,$$

where y is candidate circuit's response, t is target response and K is the number of fitness cases. Because target circuits are arithmetic circuits, we have to evaluate circuit responses for all possible combinations of operands, i.e. $K = 2^{n_i}$. This definition of the fitness function is preferred over the Hamming distance as suggested in [17].

The proposed approximation method is constructed as a resources-oriented method, in which a good compromise is sought between the number of gates and the error. Resources (gates) can be constrained either by constraining the product ($n_c \times n_r < k$) or by constraining the number active gates in a potentially big array of gates. The first approach was utilized in the literature [17,25]. In this work, the second approach is adopted as it enables CGP to operate with highly redundant arrays of gates which is beneficial for an efficient search [12]. Let the number of available gates be n_n and the number of gates in phenotype be n_{pn}. The fitness function is defined:

$$f = \begin{cases} f_{SAD}, & \text{if } n_{pn} \leq n_n \\ \infty, & \text{if } n_{pn} > n_n \end{cases}$$

Search Algorithm: CGP uses a $(1+\lambda)$ search method consisting of the following steps:

1. An initial population of the size $1 + \lambda$ is created.
2. Each candidate circuit is evaluated by fitness function f.
3. The highest-scored candidate circuit is selected as a new parent. The parent from previous generation is never selected as the new parent if there is another individual with the same fitness value.
4. By applying a point mutation, λ offspring individuals are generated from the parent. Parameter h defines the number of genes (integers) that undergo a mutation.
5. Steps 2–4 are repeated until the termination condition is not satisfied.

Heuristic Population Seeding: In many cases a conventional exact solution (circuit C_0, with z gates) is available and can be used in the initial population of CGP. According to [25], a simple method can be employed in order to obtain the first approximation of C_0. We create $2z$ circuits in such way that every single gate of C_0 is independently replaced by a wire which connects gate's first or second input with its output. The circuit producing the smallest error out of these $2z$ approximations is taken as the first parent of CGP.

3.2   Resources-Oriented Approximation

The proposed method should produce a Pareto front showing the best obtained compromises between the number of gates and the error. As our target circuits are relatively small (tens of gates), it makes sense to execute CGP multiple times and constraint the number of gates in each run to $z - 1, z - 2, \ldots, z_m$, where z is the number of gates in the exact solution C_0 and z_m is the smallest reasonable approximation of C_0. Each CGP run begins with the best approximate circuit obtained from the previous approximation and the objective is to minimize the circuit error for a given amount of gates.

4   Experimental Results

The adders are key components of edge detection circuits. Firstly, results of adder approximations are summarized in this section. Then, performance and cost of various edge detectors utilizing the approximate adders are reported.

4.1   Evolutionary Approximation of Adders

Computational requirements of CGP can significantly be reduced if the initial population is seeded by a conventional solution. In the case of adders, the carry ripple adder and the Kogge-Stone adder have been employed. The exact 8-bit carry ripple adder is composed of one half adder and seven full adders. In total, it consists of 37 two input gates. However, the carry propagates through 15

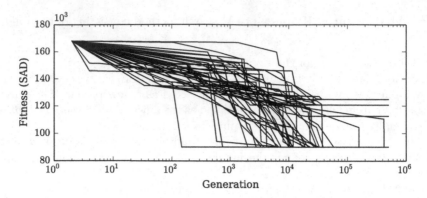

Fig. 3. Convergence curves for the adder composed of 41 gates in all 50 evolutionary runs

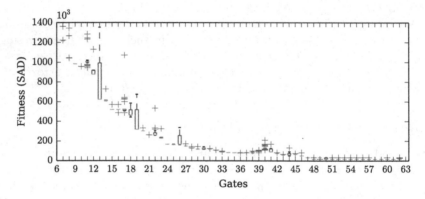

Fig. 4. Statistical evaluation of the evolutionary approximation of 8-bit Kogge-Stone adders using 6–63 gates.

gates and the corresponding delay of 15Δ (where Δ is delay of one gate) is undesirable for many applications. The 8-bit Kogge-Stone adder [8] exploits the carry lookahead logic. As the carry bits are computed in parallel, the resulting delay is only 7Δ. The cost is, however, higher – 73 gates.

The objective is to approximate the 8-bit Kogge-Stone adder. In order to keep the delay less or equal to 7Δ, CGP is used with $n_c = 7$. CGP is executed multiple-times with constrained resources to obtain a Pareto front. The CGP parameters are initialized as follows: $n_r = 13, n_c = 7, l = n_c$, and $\Gamma = \{$BUF, NOT, AND, OR, XOR, NAND, NOR, XNOR$\}$, where BUF stands for an identity function. The first runs are seeded with the circuit obtained by removing one gate from the exact adder, i.e. $n_n = 72$.

The parameters of evolution are set as follows: $h = 5\%, \lambda = 4$, and $n_g = 500000$. After 50 runs, the number of allowed gates n_n is decremented.

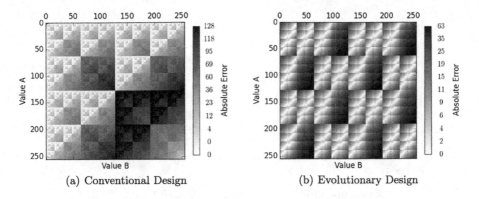

Fig. 5. Error plots of approximate adders containing 9 gates.

Table 1. Parameters of evolved approximate 8-bit adders and one conventional approximate 8-bit adder (9^a).

Gates	Δ	Absolute error					
		Avg.	MIN	Q1	Q2	Q3	MAX
9^a	1	47.87	1	16	47	88	128
9	4	15.02	1	6	12	22	63
13	5	9.51	1	4	9	14	32
24	7	2.56	1	1	3	4	8
37	6	1.23	1	1	2	3	7
43	6	0.85	1	1	1	2	3
62	7	0.00	-	-	-	-	-

In order to demonstrate the progress of evolution, an adder constrained to utilize up to 41 gates is considered. Figure 3 shows convergence curves obtained from 50 independent runs. Every run starts with the error $f = 167 \cdot 10^3$ and ends up with the error $f = 90 \cdot 10^3$ in most cases.

Boxplots summarizing the whole experiment ($n_n = 6, 7, \ldots, 63$ gates enabled) are plotted in Fig. 4. The adders with less than 6 gates were omitted due their large error. It can be seen from the boxplots that the evolution converges to a single value in most cases. Moreover, we discovered a fully functional implementation of the 8-bit adder which contains fewer gates ($n_n = 62$) and features same latency as the 8-bit Kogge-Stone adder. A detailed analysis of the results revealed that a solution composed of 37 gates has delay 6Δ, which is more than two times smaller than the delay of the Carry Ripple Adder. The average error of this approximate adder is only 1.23 (Table 1).

In common conventional approximations, an 8-bit adder is often approximated by a very cheap implementation consisting of 8 OR gates and one AND gate (for the carry bit) [14]. This approximation can be compared with an evolved approximate 8-bit adder of the same implementation cost (i.e. 9 gates).

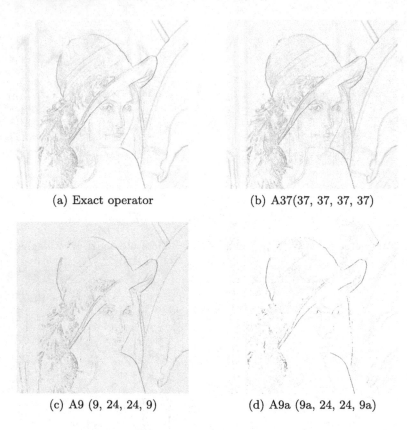

(a) Exact operator   (b) A37(37, 37, 37, 37)

(c) A9 (9, 24, 24, 9)   (d) A9a (9a, 24, 24, 9a)

Fig. 6. Edge detection by approximate Sobel detectors

The conventional adder exhibits bigger maximal, median, and average errors (Table 1). Figure 5 also shows that more significant errors occur in the case of larger operands. On the other hand, in the case of evolved approximate adder, more significant errors are spread out in the space of operand combinations.

4.2   Approximation of Sobel Edge Detector

Several approximations of the reference edge detector implementation from Fig. 1 are proposed in this Section. Firstly, some adders of the reference circuit were replaced by approximate 8-bit adders. The replacement is performed for a set of adders occupying the same depth (v_1, v_2, v_3, and v_4 in Fig. 1) of the adder tree. Let us denote an approximate edge detector by An_1 (n_1, n_2, n_3, n_4), where each element n_i represents the number of gates in adders of a given depth, e.g. A50 $(50, 50, 50, 50)$ is an edge detector containing at all levels approximate adders consisting of 50 gates.

The impact of approximate adders on edge detection is demonstrated in Fig. 6. For example, by using the adders composed of 37 gates, we obtained a

Table 2. MAE and other properties of approximate Sobel edge detectors

Edge Detector			Mean Absolute Error
Name	Gates	Δ	0 3 6 9 12 15 18 21 24 27 30 33 36 39 42 45 48 51 54 57 60
Eq. 2	813	33	
A62	714	30	
A37	439	26	
A13	291	27	
A9	221	24	
A9a	221	18	

Table 3. RMSE and other properties of approximate Sobel edge detectors.

Edge detector			Root Mean Square Error
Name	Gates	Δ	0 3 6 9 12 15 18 21 24 27 30 33 36 39 42 45 48 51 54 57 60
Eq. 2	813	33	
A62	714	30	
A37	439	26	
A13	291	27	
A9	221	24	
A9a	221	18	

low-cost edge detector A37 (37, 37, 37, 37) that produces a very small error which is indiscernible to human eye. This approximate edge detector contains the same amount of gates as the edge detector composed of fully functional 8-bit carry ripple adders. However, the approximate detector is more than two times faster.

Figure 6 compares edge detector A9 (9, 24, 24, 9) with A9a (9, 24, 24, 9). Both solutions utilize the same number of gates. A9 (containing the adders evolved by CGP) shows more precise edge detection than A9a which employs the adders approximated conventionally. On the image of Lenna, A9 produces the mean absolute error (MAE) 19.84 per pixel which is bigger than the error of A9a — 13.53 per pixel. This result is also manifested by darker background of the image produced by A9. If the root mean square error (RMSE) is used as an error metric, the result is 20.84 for A9 which is better result than A9a (RMSE = 22.41).

We tested approximate edge detection circuits on a dataset containing 200 images. Tables 2 and 3 demonstrate the differences in terms of MAE and RMSE between a fully working detector operating exactly according to formula 2, a solution in which the adders containing operands having more than 8 bits are replaced with accurate 8-bit adders (A62), and other approximate detectors – A9, A9a, A13 (13, 24, 34, 43), and A37.

It is difficult to perform a direct comparison with approximate edge detectors published in the literature. The reason is that many circuit parameters and details of experiments are not published. A brief comparison can be done with

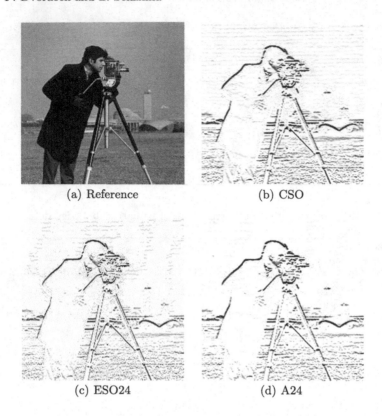

(a) Reference   (b) CSO

(c) ESO24   (d) A24

Fig. 7. Edge detection by approximate Sobel Operators

Axilog. The authors of Axilog reported a 1.82× area reduction for Sobel detector with RMSE = 10 % for a single image. The percentage value of RMSE was computed on the normalized image with pixels in the range $(0, 1)$. Under this metric, A9 detector obtained 3.89× area reduction with maximal RMSE = 6.2 % using the dataset of 200 images [10]. Moreover, A9 detector reduced the area three times in comparison with A62 and almost twice in comparison with A37.

4.3 Approximation of Evolved Edge Operator

Paper [22] presents Sobel operator (CSO) implemented as:

```
uint8 CSO(uint8 kernel[9]) {
  int i;
  i = kernel[0] + 2*kernel[1] + kernel[2];
  i = i - (kernel[6] + 2*kernel[7] + kernel[8]);
  i = max(i, 0);
  i = min(i, 255);
  return i;
}
```

Table 4. Properties of approximate edge operator CSO and approximate evolved operator ESO.

Sobel operator			Root Mean Square Error
Name	Gates	Δ	0 2 4 6 8 10 12 14 16 18 20 22 24 26 28 30 32 34 36 38 40
CSO	411	30	
CSO62	358	28	
CSO37	233	25	
CSO24	188	24	
ESO62	506	73	
ESO37	381	68	
ESO24	336	69	

where kernel is the 3×3-pixel convolution window. Paper [22] also presents an edge detector (ESO), completely evolved by CGP. This operator used the image shown in Fig. 7 as a golden image in the fitness function. The evolved code of ESO is given below.

```
uint8 ESO(uint8 kernel[9]){
  uint i14, i17, i19, i22, i27, i29;
  i14 = min((kernel[1] + kernel[7]) >> 1, kernel[7]);
  i17 = i14 ^ kernel[7];
  i19 = min(i14 + (255 - kernel[1]), 255);
  i22 = 255 - i19;
  i27 = min(i22, (i17 + i19) >> 1);
  i29 = min(i22 + i27, 255);
  return (i27 + i29) & 0xff;
}
```

CSO operator employs just a part of the Sobel edge detector and it thus computes the horizontal derivative. The cost of CSO is five adders and an array of NOT gates. In order to obtain edges as shown in Fig. 7, we saturated the output to 0 or to 255, if the output value is negative or greater than 255, respectively.

Both CSO and ESO can be approximated using the adders presented in Sect. 4.1. As CSO has only three levels of adders, we denote the approximate conventional operator $CSOn_1$ (n_1, n_2, n_3) where n_i represents the number of gates used in approximate adders at the level v_i.

Evolved operator ESO contains five adders which can be replaced by their approximate versions. Approximate detectors will be denoted $ESOn_1$ $(n_1, n_2, n_3, n_4, n_5)$ where n_i represents the number of gates used in a given approximate adders.

ESO contains more gates than CSO mainly because there were no requirements on the area minimization in paper [22]. Table 4 gives RMSE calculated using 200 images for all approximate edge detectors.

It can be seen that RMSE is growing when CSO62 $(62, 62, 62)$ is compared with approximate CSO37 $(37, 37, 37)$ and CSO24 $(24, 24, 34)$. But for example, RMSE boxplots are almost identical in the case of ESO62 $(62, 62, 62, 62, 62)$, ESO37 $(37, 37, 37, 37, 37)$, and ESO24 $(24, 24, 24, 30, 28)$. It seems that the evolved solution is more robust to the approximation than conventional CSO. It also turns out that it is hard to predict the impact of approximations on the overall circuit behavior.

5   Conclusions

In this work, various approaches to the approximation of edge detectors based on the Sobel operator were proposed and evaluated. We replaced exact adders in conventional as well as evolved edge detectors by their approximate versions. The approximate adders were obtained using CGP. Results were reported in terms of the error (MAE and RMSE obtained using 200 test images) and the implementation cost given as the number of gates.

We showed that evolved approximate 8-bit adder composed of 9 gates has smaller average error than a commonly used approximation consisting of the same number of gates. Moreover, evolved inaccurate adder containing 37 gates and producing a very small average absolute error has 3× smaller delay than a fully functional carry ripple adder composed of the same amount of gates. In the case of edge detection, we demonstrated a circuit showing 3.89× area reduction with maximal RMSE $= 6.2\,\%$. It seems that evolved edge detectors are more resilient to approximations than conventional edge detectors.

Our future work will be devoted to a detailed analysis of the approximations not only at the circuit level but also at the whole system level.

Acknowledgements. This work was supported by the Czech science foundation project GA16-17538S.

References

1. Esmaeilzadeh, H., Sampson, A., Ceze, L., Burger, D.: Neural acceleration for general-purpose approximate programs. Commun. ACM **58**(1), 105–115 (2015)
2. Fu, W., Johnston, M., Zhang, M.: Genetic programming for edge detection using multivariate density. In: Proceedings of the 15th Annual Conference on Genetic and Evolutionary Computation, pp. 917–924. ACM (2013)
3. Golonek, T., Grzechca, D., Rutkowski, J.: Application of genetic programming to edge detector design. In: Proceedings of the 2006 IEEE International Symposium on Circuits and Systems, pp. 4683–4686. IEEE (2006)
4. Harding, S., Banzhaf, W.: Genetic programming on GPUs for image processing. Int. J. High Perform. Syst. Archit. **1**(4), 231–240 (2008)
5. Harris, C., Buxton, B.: Evolving edge detectors with genetic programming. In: Proceedings of the First Annual Conference on Genetic Programming, pp. 309–314 (1996)

6. Higuchi, T., Niwa, T., Tanaka, T., Iba, H., de Garis, H., Furuya, T.: Evolving hardware with genetic learning: a first step towards building a Darwin machine. In: Proceedings of the 2nd International Conference on Simulated Adaptive Behaviour, pp. 417–424. MIT Press (1993)
7. Hollingworth, G., Tyrrell, A.M., Smith, S.: Simulation of evolvable hardware to solve low level image processing tasks. In: Poli, R., Voigt, H.-M., Cagnoni, S., Corne, D.W., Smith, G.D., Fogarty, T.C. (eds.) EvoIASP 1999 and EuroEcTel 1999. LNCS, vol. 1596, pp. 46–58. Springer, Heidelberg (1999)
8. Kogge, P.M., Stone, H.S.: A parallel algorithm for the efficient solution of a general class of recurrence equations. IEEE Trans. Comput. **22**, 786–793 (1973)
9. Kulkarni, P., Gupta, P., Ercegovac, M.D.: Trading accuracy for power in a multiplier architecture. J. Low Power Electron. **7**(4), 490–501 (2011)
10. Martin, D., Fowlkes, C., Tal, D., Malik, J.: A database of human segmented natural images and its application to evaluating segmentation algorithms and measuring ecological statistics. In: Proceedings of the 8th International Conference on Computer Vision, vol. 2, pp. 416–423, July 2001
11. Miller, J.F.: Cartesian Genetic Programming. Springer-Verlag, Berlin (2011)
12. Miller, J.F., Smith, S.L.: Redundancy and computational efficiency in cartesian genetic programming. IEEE Trans. Evol. Comput. **10**(2), 167–174 (2006)
13. Monajati, M., Fakhraie, S., Kabir, E.: Approximate arithmetic for low-power image median filtering. Circuits Syst. Signal Process. **34**(10), 3191–3219 (2015)
14. Nepal, K., Li, Y., Bahar, R.I., Reda, S.: Abacus: a technique for automated behavioral synthesis of approximate computing circuits. In: Proceedings of the Conference on Design, Automation and Test in Europe, DATE 2014, pp. 1–6. EDA Consortium (2014)
15. Priego, B., Bellas, F., Souto, D., Lopez-Pena, F., Duro, R.: Evolving cellular automata for detecting edges in hyperspectral images. In: 2012 IEEE International Conference on Fuzzy Systems (FUZZ-IEEE), pp. 1–6. IEEE (2012)
16. Sekanina, L., Harding, L.S., Banzhaf, W., Kowaliw, T.: Image processing and CGP. In: Miller, J.F. (ed.) Cartesian Genetic Programming, pp. 181–215. Springer, Heidelberg (2011)
17. Sekanina, L., Vasicek, Z.: Approximate circuits by means of evolvable hardware. In: 2013 IEEE International Conference on Evolvable Systems. Proceedings of the 2013 IEEE Symposium Series on Computational Intelligence (SSCI), pp. 21–28. IEEE CIS (2013)
18. Shi, K., Boland, D., Stott, E., Bayliss, S., Constantinides, G.: Datapath synthesis for overclocking: online arithmetic for latency-accuracy trade-offs. In: 51st ACM/EDAC/IEEE Design Automation Conference (DAC), pp. 1–6. IEEE (2014)
19. Sonka, M., Hlavac, V., Boyle, R.: Image Processing: Analysis and Machine Vision. Thomson-Engineering, Toronto (1999)
20. Torresen, J.: A scalable approach to evolvable hardware. Genet. Program Evolvable Mach. **3**(3), 259–282 (2002)
21. Ttofis, C., Hadjitheophanous, S., Georghiades, A., Theocharides, T.: Edge-directed hardware architecture for real-time disparity map computation. IEEE Trans. Comput. **62**(4), 690–704 (2013)
22. Vasicek, Z., Sekanina, L.: An evolvable hardware system in Xilinx Virtex II Pro FPGA. Int. J. Innovative Comput. Appl. **1**(1), 63–73 (2007)
23. Vasicek, Z., Sekanina, L.: Formal verification of candidate solutions for postsynthesis evolutionary optimization in evolvable hardware. Genet. Program Evolvable Mach. **12**(3), 305–327 (2011)

24. Vasicek, Z., Sekanina, L.: Circuit approximation using single- and multi-objective cartesian GP. In: Machado, P., et al. (eds.) EuroGP. LNCS, vol. 9025, pp. 217–229. Springer International Publishing, Switzerland (2015)

25. Vasicek, Z., Sekanina, L.: Evolutionary approach to approximate digital circuits design. IEEE Trans. Evol. Comput. **19**(3), 432–444 (2015)

26. Yazdanbakhsh, A., Mahajan, D., Thwaites, B., Park, J., Nagendrakumar, A., Sethuraman, S., Ramkrishnan, K., Ravindran, N., Jariwala, R., Rahimi, A., Esmaeilzadeh, H., Bazargan, K.: Axilog: language support for approximate hardware design. In: Design, Automation Test in Europe Conference Exhibition (DATE 2015), pp. 812–817. IEEE (2015)

27. Zhang, Y., Rockett, P.I.: Evolving optimal feature extraction using multiobjective genetic programming: a methodology and preliminary study on edge detection. In: Proceedings of the 2005 Conference on Genetic and Evolutionary Computation, GECCO 2005, pp. 795–802. ACM (2005)

On the Impact of Class Imbalance in GP Streaming Classification with Label Budgets

Sara Khanchi[✉], Malcolm I. Heywood, and Nur Zincir-Heywood

Faculty of Computer Science, Dalhousie University, B3H 4R2, Halifax, NS, Canada
s.khanchi@gmail.com, {mheywood,zincir}@cs.dal.ca

Abstract. Streaming data scenarios introduce a set of requirements that do not exist under supervised learning paradigms typically employed for classification. Specific examples include, anytime operation, non-stationary processes, and limited label budgets. From the perspective of class imbalance, this implies that it is not even possible to guarantee that all classes are present in the samples of data used to construct a model. Moreover, when decisions are made regarding what subset of data to sample, no label information is available. Only after sampling is label information provided. This represents a more challenging task than encountered under non-streaming (offline) scenarios because the training partition contains label information. In this work, we investigate the utility of different protocols for sampling from the stream under the above constraints. Adopting a uniform sampling protocol was previously shown to be reasonably effective under both evolutionary and non-evolutionary streaming classifiers. In this work, we introduce a scheme for using the current 'champion' classifier to bias the sampling of training instances *during* the course of the stream. The resulting streaming framework for genetic programming is more effective at sampling minor classes and therefore reacting to changes in the underlying process responsible for generating the data stream.

Keywords: Streaming data classification · Non-stationary · Class imbalance · Benchmarking

1 Introduction

The streaming data classification task under label budgets introduces a number of constraints that do not appear under the typical offline supervised learning context [1,5,7,9,14]. Specifically, a streaming data context implies that there is no beginning or end to the data, thus there is no prior partition of the data into training and test sets. Instead, it is necessary to construct a classifier while assuming a limited label budget, i.e. it is too expensive to label all the data, so part of the task of the learning algorithm is to decide which exemplars to request labels for (without exceeding some prior label budget). Naturally, only exemplars for which the streaming data classification algorithm actually requests labels are used for parameterizing candidate classifiers. The remaining data are 'unlabelled' and it is

© Springer International Publishing Switzerland 2016
M. Heywood et al. (Eds.): EuroGP 2016, LNCS 9594, pp. 35–50, 2016.
DOI: 10.1007/978-3-319-30668-1_3

this subset for which a classifier needs to make predictions. Improving the quality of classifier(s) and labelling unlabelled data are therefore not tasks associated with independent prior partitions of the data. Moreover, the data itself might well be generated by a non-stationary process [5, 9]. Thus, the underlying process responsible for creating the data might be subject to sudden shifts or gradual drifts, implying that some form of change detection and / or continuous sampling for labels is necessary in order for classifiers to keep 'up-to-date'.

Naturally, it is not possible to make any guarantees regarding the distribution of class labels within a stream. Moreover, given that streaming classifiers are limited to querying exemplars from a finite 'window' to the stream at any point in time; then, depending on the degree of mixing, the exemplars within the current window location may only be representative of a single class.

In this work, we adopt a generic framework for interfacing genetic programming (GP) to streaming data [9,17]; hereafter referred as streamGP (Fig. 1). The framework identifies four components:

1. the window interface from which new exemplars may be sampled;
2. a sampling policy for deciding which exemplars to request labels for;
3. the data subset against which fitness evaluation is conducted;
4. an archiving policy for deciding which exemplars should be replaced / retained.

In summary, we are interested in considering how decisions that are made regarding the sampling and archiving policies impact on the resulting performance of the classifier. Specifically, we investigate how heuristics for introducing class balance into the data subset can be defined *without* the use of label information.

2   Related Work

Streaming data analysis under label budgets represents a topic of growing interest, with several monographs [1, 7] and journal special issues [12] being devoted to the topic. However, until recently there has been little reference to approaches from evolutionary computation that actively construct models (such as GP, learning classifier systems or neural evolution). Conversely, optimization under 'dynamic environments' represents a distinct topic for evolutionary computation with an emphasis on the accurate 'tracking' of multiple 'peaks' in a multi-modal environment, but without requiring generalization to unseen data. As such, there is no requirement to operate under the constraints of a framework for addressing the issue of label budgets. The survey article of [9] reviews developments from the perspective of both non-evolutionary and evolutionary approaches to model building. Particular highlights include:

Ensemble methods provide the ability to incrementally adapt to changes in the stream. Under a GP context adopting an ensemble approach might imply that multiple individuals from the same population coevolve, as in various frameworks for evolving teams of programs [3,10,13,19]. A recent study demonstrated that supporting coevolutionary teaming under GP is particularly appropriate for streaming data contexts [16]. In a sense, modularity enables greater refinement in the credit assignment process, so that rather than having to replace all of a model, only parts of a model require revision. This is particularly important under tasks that lack a 'complete' definition or undergo change. More generally, the capacity to change is related to representations that are in some way 'elastic', with specific authors making the case for the utility of genotypic-to-phenotypic mappings [4] or neutral networks [18] under dynamic environments.

Anytime operation implies that it must be possible to identify at any point in time a 'champion' individual that will attempt to label the stream. At the same time, the development of new individuals may also be taking place, or alternatively, a change detection process is used to trigger the development of new champion individuals (see below).

Diversity maintenance contributes to the ability to react to change in the minimum amount of time. Both non-evolutionary ensemble methods and evolutionary methods appear to benefit from diversity maintenance, but open questions exist around what 'type' of diversity is most appropriate.

Change detection versus sampling represents a requirement unique to the streaming data task under label budgets. Given that models can only be constructed relative to a very limited subset of exemplars at any point in time *and* there is only a limited label budget, then decisions need to be made regarding which data to request labels for. Change detection might be performed relative to stream content in an attempt to detect such changes. However, this will not detect changes that result from the movement in labels from one concept to another. With this in mind, benefits have been associated with adopting combined approaches in which both the periodic uniform sampling for labels is combined with change driven requests for labels (e.g. [14,20]).

Memory mechanisms are implicit in the use of shared genetic material, support for neutral networks and multi-population models. All forms of memory have a part to play in contributing to solutions to streaming data tasks.

In this work, we will adopt the general framework for applying GP to streaming data from [9] (Fig. 1) and make use of the symbiotic bid based (SBB) framework for coevolving GP programs into teams [10]. The capacity of the latter for task decomposition (or constructing modular solutions) has already been demonstrated to be superior to monolithic GP under non-streaming and streaming tasks (see [11,16] respectively). Moreover, SBB supports multi-class classification from a single run without having to adopt additional heuristics.

3   Methodology

The advancing stream defines a sequential sequence of exemplars, each of dimension d. Without loss of generality, we assume a non-overlapping window interface

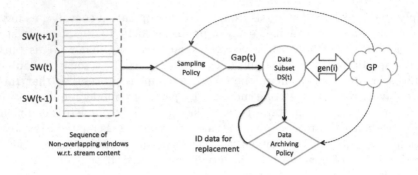

Fig. 1. StreamGP framework with proposed additional feedback paths (dotted).

SW, each consisting of an equal number of exemplars (Fig. 1). Only the exemplars within the current window position, $SW(t)$, are available for sampling. Label information is *not available* when making the decision regarding which exemplars to sample from $SW(t)$. Moreover, for simplicity we assume that each new window location results in Gap label requests, implying that indexes for window instance and the label request are the same.

Only once the sampling policy identifies $Gap(t)$ exemplars for sampling is label information revealed, i.e. $|Gap| \div |SW|$ denotes the label budget. The data subset, DS, represents the finite sized archive of labeled exemplars used for guiding the training process; thus, $|DS| > |Gap|$. A data archiving policy determines which exemplars are replaced each time a new sample of Gap exemplars are taken from the stream. Once data subset content is defined for the current window location, $DS(t)$, one or more generations of GP are performed. It is relative to the content of the data subset that a champion individual is identified for the purposes of anytime operation (Sect. 3.1).

Naturally, the sampling policy might be based on a measure designed to detect change between sequential window locations (for a review see [9]). However, this also limits the circumstances under which model reconstruction is initiated. Conversely, assuming a uniform sampling policy (subject to the label budget) has been empirically shown to be difficult to improve on in practice [14,20]. This was also the approach adopted in the original experiments with the framework of Fig. 1 [16,17].[1]

The question we are interested in explicitly addressing in this work is to what degree the properties of the resulting classifier are biased by decisions made regarding the sampling / archiving policy (Fig. 1). The underlying constraint within which such a sampling policy is required to exist is that label information is *not available* when deciding which exemplars to sample. Two **protocols** will be considered:

---

[1] Earlier work with SBB under streaming data assumed that label information could be used to ensure the data subset was always balanced [15].

1. sample with uniform probability up to the label budget, as per earlier studies [16,17], hereafter the **uniform sampling policy**;
2. make use of the current champion classifier to *suggest labels* and therefore bias the replacement / selection of exemplars, hereafter the **biased sampling policy**.

The first scenario implies that the exemplars within the data subset (against which GP individuals are evolved) will reflect the underlying distribution of exemplars in the stream. The second scenario has the potential to incrementally balance the representation of classes within the data subset. In the following subsections, we develop the framework for anytime operation (champion identification) and then establish the mechanism assumed for reintroducing class-wise sampling of the stream without recourse to any additional label information.

3.1   Anytime Operation

As noted in Sect. 2, streaming data algorithms are required to identify a 'champion' model at any point to label the stream data as it 'passes by' or anytime operation. The only source of information for the purpose of choosing such a champion individual is the current content of the data subset, Fig. 1. Thus, once all GP individuals are evaluated against all DS content (or generation, i), a candidate 'champion classifier' can be identified and deployed. Thereafter, a new champion might be identified on concluding each generation. The process operates entirely within the label budget constraint and results in anytime operation. The metric employed for this purpose is that of multi-class detection rate, as follows: $DR = \frac{1}{C} \sum DR_i$ where $DR_i = \frac{tp_i}{tp_i + fn_i}$; C is the count of classes present in $DS(t)$; tp_i and fn_i are the counts of true positive and false negative for class i, again relative to exemplars present in $DS(t)$.

Naturally, during the course of a streaming data sequence multiple champion individuals might be identified, but only one champion deployed during any segment of the stream. This is distinct from the style of operation assumed for non-streaming data in which models are constructed from a training partition, a champion individual is identified relative to the entire training partition (or an independent validation partition) and test is performed relative to an independent test partition. None of this is possible under streaming data scenarios because access to the data is very limited (with the process creating the data itself potentially being non-stationary) [1,5,7,9].

3.2   Archiving Policy

The modified archiving policy is designed to target overrepresented classes for replacement by the next sample of *Gap* exemplars. This means that exemplars representing the minor class(es) are more likely to be retained in $DS(t)$ between consecutive t, whereas exemplars associated with the major classes are prioritized for replacement. Naturally, this also means that the exemplars representing the major classes are more up to date / turn over at a higher frequency.

The modified archiving policy is detailed as follows:

1. Estimate the current class-wise distribution of exemplars within $DS(t)$, or

$$\forall c \in C' : \forall i \in DS(t) \text{ IF } p_i == c \text{ THEN } w_c = w_c + 1 \qquad (1)$$

where, C' are the number of different classes present within $DS(t)$, p_i is exemplar i from the data subset.

2. Normalize class counts (w_c) so distinguishing between under and over represented classes, or

$$\forall c \in C' : w_c = w_c - \frac{|DS| - |Gap|}{C'} \qquad (2)$$

3. Mark all cases with $w_c > 0$ so identifying the overrepresented classes and identify the corresponding budget of exemplars for replacement, M_c, or

$$\forall c \in C' : M_c = \begin{cases} w_c \times \frac{|Gap|}{\sum w_c} & \text{IF } w_c > 0 \\ 0 & \text{otherwise} \end{cases} \qquad (3)$$

4. For each class, mark M_c instances for replacement with uniform probability, subject to a total budget of Gap. Note that in doing so, older instances are replaced first.

5. IF Step 4 marked less than Gap instances from $DS(t)$ for replacement, THEN the remaining instances $Gap - \sum_{c \in C'} M_c$ are identified uniformly across the overrepresented classes until a total of Gap instances are marked for replacement.

3.3   Sampling Policy

Section 3.2 introduced a bias that resulted in the more overrepresented classes being targeted for replacement. Naturally, this has not done anything to increase the chances of sampling instances from the stream corresponding to less frequently sampled classes. The principle constraint is that we have a limited label budget. Our approach will therefore be to make use of the labels supplied by the champion individual, gp^* (identified in support of the anytime operational requirement, Sect. 3.1) to bias the selection of exemplars for inclusion within Gap relative to the current window location $SW(t)$. Thus, preference will be given to the exemplars that the champion classifier associates with the underrepresented class(es) in the class distribution present in $DT(t-1)$. The resulting sampling policy has the following form:

1. Assume Eq. (1) through (3) from Sect. 3.2 to identify any underrepresented classes and their associated exemplar counts, w_c. Such a process is performed w.r.t. $DS(t-1)$ content, i.e. after the last updating of the Archive Policy.

2. Use the current champion GP classifier, gp^*, to identify any instances under $SW(t)$ that (potentially) correspond to an under represented class, or

$$\forall p_i \in SW(t) : G(p_i, m) = \begin{cases} m & \text{IF } gp^*(p_i) == M_c(m) \\ 0 & \text{otherwise} \end{cases} \qquad (4)$$

where $M_c(m)$ are the subset of classes underrepresented (i.e., those for which $M_c == 0$ from Eq. (3)), $G(p_i, m)$ is a vector of class labels corresponding to the underrepresented instances and m correspond to the class label for an underrepresented class.[2]

3. The non-zero entries of $G(\cdot, \cdot)$ constitute the instances potentially corresponding to the under represented class(es). Sample without replacement until either no instances remain (in $G(\cdot, \cdot)$) or $|Gap|$ instances have been sampled. Such a sampling process is biased to prioritize sampling classes least represented in $DS(t-1)$.

4. If Gap is not yet full, sample from the remaining instances in $SW(t)$ without replacement (uniform p.d.f.).

The above process is enabled *once* some stability is achieved in the identification of champion individuals. For the purposes of the experimental evaluation a generic threshold of the first 2.5 % of the generations is assumed across all data sets.[3] During this initialization period the Gap individuals are selected with uniform p.d.f. alone. No such constraint is employed in the case of the Archiving Policy (Sect. 3.2).

4   Experimental Methodology

An empirical evaluation is performed to investigate the impact of assuming the biased sampling protocol (Sect. 3) versus a uniform sampling of training instances under fixed label budgets.[4] In each case 20 runs are performed over 4 data sets previously employed for benchmarking streaming data algorithms. Two data sets are artificial data sets with specific non-stationary properties present (i.e., explicitly designed in), whereas the other two data sets represent real-world tasks in which general spatial-temporal properties are assumed to be present. Section 4.1 will summarize the properties of these data sets.

Metrics for performance evaluation under streaming data are in itself the subject of active research [2,8]. That is to say, streaming data has a dynamic component on account of the model being under continuous development throughout the stream; thus, the performance metric should be able to characterize performance over the *course* of the stream. In this work, we will adopt a count based metric as it is both robust to different class distributions and capable of expressing the dynamic properties of classifier performance during the course of the stream [9,16]. The specific formulation is presented in Sect. 4.2. Finally, Sect. 4.3 establishes a common parameterization for use throughout the study.

---

[2] Valid class labels appear over the interval $[1, ..., C]$.

[3] Given the later benchmarking parameterization this corresponds to no more than 25 generations.

[4] Previous studies had compared StreamGP under the uniform sampling protocol to non-evolutionary streaming algorithms [16,17].

4.1   Datasets

A total of four data sets will be assumed in which two are artificially constructed in order for specific non-stationary properties to be embedded within the stream: hereafter Shift and Drift.[5] The **Shift** dataset [20] defines a 5-class task in 6-dimensions in which two decision trees are used to define rules for two *separate* 5-class classification tasks: $C1$ and $C2$. The stream is defined in terms of a sequence of 'blocks'. Each block is composed from β % of exemplars defined by decision tree $C1$ and $(100 - \beta)$ % of exemplars defined by decision tree $C2$. The first three blocks assume $\beta = 0$ % thereafter each block results in β incrementally increasing by 10 % until $\beta = 100$ %. The **Drift** dataset [6] is defined by a process of gradual variation in which three classes are described by 10-dimensional hyperplanes. Every 1,000 exemplars half of the parameters may undergo variation. Class labels are defined on the basis of whether the hyperplane exceeds a predefined class threshold.

We also make use of the widely used 'electricity utilization' dataset in which the goal is to predict whether the price of electricity (in a region of Australia) are going to increase or decrease relative to a moving average of the last 24 hours. As such this is an example of a real-world task with implicit temporal properties and has received considerable interest from the perspective of the empirical evaluation of streaming algorithms (e.g., [2]). The final dataset is the 'forest cover type' dataset from the UCI repository, but preprocessed to introduce a sequential ordering in the sequence relative to the elevation attribute [14].[6] Table 1 summarizes the generic properties of each data stream.

Table 1. Properties of the benchmarking datasets. D: number of attributes per exemplar, N: cardinality of the stream, k: number of classes present, 'Class distribution' reflects the overall frequency with which each class is represented over the entire stream. No attempt is made to ensure that this ratio is reflected in the window interface used by the classifier to sample stream content.

Dataset	D	N	k	$\approx$ % Class distribution
Shift	6	6,500,000	5	[37, 25, 24, 9, 4]
Drift	10	150,000	3	[74, 16, 10]
Electricity	8	45,312	2	[58, 42]
Cover	54	581,012	7	[49, 36, 6, 4, 3, 1.5, 0.5]

4.2   Class-Wise Detection Rate

As noted above, given that under streaming data scenarios there are no mechanisms by which stream content can be stratified, then there are no guarantees that

---

[5] Shift and Drift datasets are available from: http://web.cs.dal.ca/~mheywood/Code/ SBB/Stream/StreamData.html.

[6] Electricity and Cover Type are available from: http://moa.cms.waikato.ac.nz/ datasets/.

window content, $SW(t)$, will even contain exemplars from each class. With this in mind, the following definition for the online estimation of multi-class detection rate is assumed [9,17]. A per class detection rate is first defined as follows:

$$DR_c(t) = \frac{tp_c(t)}{tp_c(t) + fn_c(t)} \tag{5}$$

where t is the exemplar index, and $tp_c(t)$, $fn_c(t)$ are the respective online counts for true positive and false negative rates, i.e. up to this point in the stream.

The multi-class detection rate now has the form:

$$DR(t) = \frac{1}{C} \sum_{c=[1,...,C]} DR_c(t) \tag{6}$$

Hence, the multi-class detection rate can also be evaluated at any point in the stream.

4.3 Parameters

GP parameterization follows that adopted in previous work (e.g., [16,17]) and is summarized in Table 2. Moreover, given that for benchmarking purposes the datasets are of a finite length, we enforce label budgets through the use of a fixed number of locations, i_{max}, for the non-overlapping window ($SW(t)$) and knowledge of the dataset cardinality (Table 3). The earlier work also reported that letting GP perform multiple iterations per $DS(t)$ content was beneficial [16,17]. With this in mind, we perform experiments with a maximum total number of generations of i_{max} and $5 \times i_{max}$.[7] The former implies that one generation is performed per DS update, the latter implies that five generations are performed per DS update; hereafter referred to as **single generation** and **multi-generation** respectively. The instruction set takes the form of:

- Single argument operators: $R[x] = \langle op \rangle R[y]$ where $\langle op \rangle \in \{\cos, \exp, \log\}$
- Two argument arithmetic operators: $R[x] = R[x]\langle op \rangle R[y]$ where $\langle op \rangle \in \{+, -, \div, \times\}$
- Two argument conditional operator: IF $R[x] < R[y]$ THEN $R[x] = -R[x]$

Table 2. GP parameters. Mutation rates control the rate of adding / deleting symbionts or changing symbiont action. DS and Gap refer to the data types in Fig. 1. Host population size and gap imply a breeder model of evolution (the worst $Mgap$ hosts are deleted each generation) [10].

Parameter	Value	Parameter	Value
Prob. symbiont deletion (pd)	0.3	Data Subset size (DS)	120
Prob. symbiont addition (pa)	0.3	DS gap size (Gap)	20
Prob. action mutation (μ)	0.1	Host pop	60
Max. symbionts per host (ω)	20	GP gap size ($Mgap$)	20

[7] Any more than five resulted in negligible improvement [16].

Table 3. Stream dataset parameters. Label budget is defined as a function of the number of non-overlapping window locations (i_{max}), DS Gap size (20) and dataset cardinality (N).

Parameter	# unique SW locations (i_{max})	Label budget
Shift (shift)	1,000	0.3 %
Drift (drift)	500	6.7 %
Electricity (elec)	500	22.1 %
Covertype (cover)	1,000	3.4 %

5   Results

Given that the overall detection rate (Eq. (6)) can be decomposed into the contribution from each per class detection rate (Eq. (5)), we can view the independent contributions from each per class detection rate over the course of the stream; hence, providing additional insight into the relative impact of the original uniform sampling protocol versus the proposed biased sampling protocol.

Sections 5.1 and 5.2 review the resulting dynamic multi-class DR as a function of single and multi-generation parameterizations under uniform and biased sampling protocols. Section 5.3 concludes the result section with a static analysis performed in terms of the end-of-stream performance using the overall detection rate (Eq. (6)).

5.1   Single Generation Performance

Figures 2 and 3 reflect the detection rate of each class over the duration of the stream for the two *artificial datasets* (averaged over the 20 runs). Table 1 details the frequency with which each class is represented. Thus, all figures report class 1 as the most frequently occurring and class C as the least frequently occurring. It is readily apparent that the uniform sampling protocol under the Shift dataset explicitly favours the detection of the most frequent classes throughout the stream. Conversely, under the incremental variation of the Drift dataset, the uniform sampling protocol does not reflect this bias, possibly implying that it is more difficult to detect class 2 than 3.

Introducing the biased sampling protocol results in a different preference in class detection rates. Under Shift, the major class (class 1) is still detected most, whereas the second smallest class (class 4) is also detected strongly throughout the stream. Moreover, compared to the uniform protocol, it appears that there is much less difference between the rates at which best and worst classes are detected when using the biased protocol. The Drift dataset resulted in much stronger detection by the biased protocol throughout, albeit with the lest frequent class detected most strongly.

Figures 4 and 5 repeat the dynamic depiction of per class detection rate, this time for the two *real-world* datasets (curves averaged over the 20 runs). Adopting a uniform sampling protocol resulted both classes being detected equally

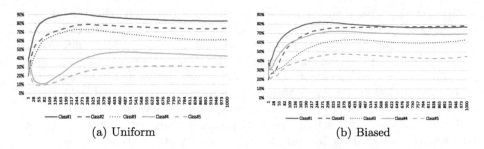

(a) Uniform   (b) Biased

Fig. 2. Shift dataset – Average *per* class detection rate (over 20 runs) under label budget of $i_{max} = 1000$, *single generation*. Curves best viewed in colour

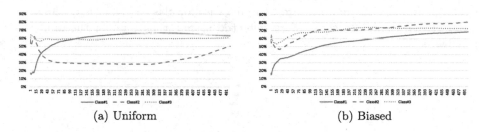

(a) Uniform   (b) Biased

Fig. 3. Drift dataset – Average *per* class detection rate (over 20 runs) under label budget of $i_{max} = 500$, *single generation*

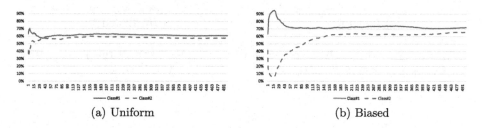

(a) Uniform   (b) Biased

Fig. 4. Electricity dataset – Average *per* class detection rate (over 20 runs) under label budget of $i_{max} = 500$, *single generation*

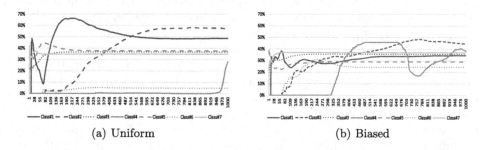

(a) Uniform   (b) Biased

Fig. 5. Cover type dataset – Average *per* class detection rate (over 20 runs) under label budget of $i_{max} = 1000$, *single generation*. Curves best viewed in colour

throughout the stream under Electricity (60%). Conversely, the biased sampling protocol initially resulted in a strong symmetry, with a very distinct notch appearing for the duration of the first 2.5% of the stream. This appears to reflect the parameterization choice assumed for delaying the introduction of the Sampling policy (see comment at the end of Sect. 3.3). That said, the negatively impacted class 2 returns to a detection rate matching that achieved by the uniform framework after $\approx 30\%$ of the stream has passed.

Under the Cover dataset the uniform protocol identified all but two classes with a fixed level of detection rate for the majority of the stream. Class 1 (the major class) is initially detected at a rate of $> 60\%$ before dropping by 10% whereas class 7 is only ever identified right at the end of the stream. Conversely, adopting the biased sampling protocol resulted in class 7 being detected much earlier than under the uniform protocol; likewise for class 6. That said, the two major classes (1 and 2) were always detected more strongly under the uniform framework.

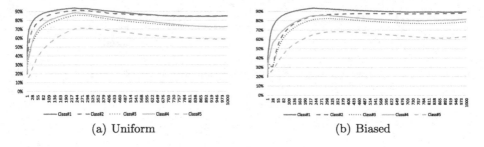

(a) Uniform   (b) Biased

Fig. 6. Shift dataset – Average *per* class detection rate (over 20 runs) under label budget of $i_{max} = 1000$, *multi-generation*. Curves best viewed in colour

5.2   Multi-generation Performance

Adopting a multi-generation parameterization implies that five generations are performed per data subset update $(DS(t))$; thus the label budget is unaffected, but GP is potentially able to react more quickly to change [16,17]. Other than the addition of multiple generations per $DS(t)$, there are no changes relative to the configuration of the uniform and biased protocols.

Figure 6 summarizes per class detection rates for the Shift dataset. Relative to the single generation curves (Fig. 2) all detection rates are improved, i.e., less variation between the detection of best and worst classes. However, it appears that the biased sampling protocol sees most improvement overall. Under the Drift dataset (Fig. 7) all curves are again either improved by the introduction of multi-generation operation or, in the case of the uniform protocol for class 1, negatively impacted. This is interesting, as class 1 is the largest class, thus it might be assumed to see preferential detection by the uniform sampling protocol.

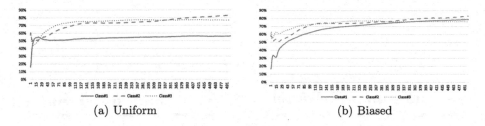

(a) Uniform   (b) Biased

Fig. 7. Drift dataset – Average *per* class detection rate (over 20 runs) under label budget of $i_{max} = 500$, *multi-generation*

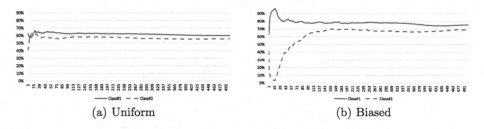

(a) Uniform   (b) Biased

Fig. 8. Electricity dataset – Average *per* class detection rate (over 20 runs) under label budget of $i_{max} = 500$, *multi-generation*

Figure 8 summarizes per class behaviour under the Electricity dataset. Performing multiple generations (per DS update) appears to have very little impact under the uniform sampling protocol, whereas a 5 % improvement appears for the detection of each class under the biased protocol. The notch associated with the delayed introduction of the biased Sample policy is again in evidence.

Finally, the Cover type dataset (Fig. 9) was also generally improved by the addition of multi-generation operation. Note that the uniform sampling protocol tends to result in a wider spread of per class detection rates, whereas the biased protocol allocated it's resources more evenly across the 7 classes. Also evident is a strong preference under uniform sampling to detect the major class, whereas the biased sampling protocol detects the smallest class the strongest. Naturally, attempting to allocate equal numbers of samples to each class implicitly assumes that all classes are equally difficult to classify. Conversely, in practice the difficulty in detecting a class is not related to the number of instances describing it.

5.3   Overall Detection Rates

Overall performance of streaming algorithms is generally characterized in terms of the performance metric at the 'conclusion' of the stream (see for example the widespread use of prequential error as measured at the end of the stream [2,8]). In this case, we can utilize the average class-wise detection rate (Eq. (6)) and then apply a nonparametric Mann-Whitney U test to verify the significance of

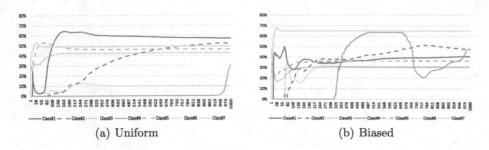

| (a) Uniform | (b) Biased |

Fig. 9. Cover type dataset – Average *per* class detection rate (over 20 runs) under label budget of $i_{max} = 1000$, *multi-generation*. Curves best viewed in colour

Table 4. End-of-stream median multi-class detection rates for uniform and biased sampling protocols and corresponding *p*-value from Mann-Whitney U test

Single generation mode				Multi-generation mode		
Dataset	Uniform	Biased	*p*-value	Uniform	Biased	*p*-value
Shift	56.74 %	67.5 %	1.33×10^{-8}	74.71 %	80.37 %	1.69×10^{-7}
Drift	58.01 %	73.94 %	0.0	72.55 %	79.48 %	0.0
Electricity	59.0 %	69.07 %	0.0	57.95 %	72.6 %	0.0
Cover	35.49 %	34.21 %	0.46	41.9 %	42.9 %	0.063

any difference.[8] Table 4 provides the quantitative summary of this comparison for both the single generation and multi-generation parameterizations under uniform and biased sampling protocols.

In short, under the single generation mode of operation, significant improvements appeared for all but the case of Cover type at the 99 % Confidence interval (with the biased sampling protocol preferred). Under the multi-generation mode Cover type was also improved, thus, both algorithms improved with the inclusion of the biased sampling protocol.

6   Conclusion

Building classifiers for non-stationary streaming data applications with label budgets represents a new challenge for machine learning in general [5,9]. Moreover, only a little research has been conducted to this end using genetic programming. In this work, we benchmark a general framework for applying genetic programming to this task. We show that the current champion from the GP population can be used to provide the basis for defining a biased sampling protocol that more rapidly adapts to dynamical properties in the stream, as well as

---

[8] Violin plots were used to establish that the distributions did not conform to a normal distribution. Space precludes their inclusion.

returning stronger classification performance on the under represented classes. This is achieved without requiring additional label information.

Further investigations will be conducted to determine the relative impact of the 'Archiving Policy' and 'Sampling Policy' independently from each other. We also anticipate characterizing at what points there are changes to the champion classifier during the course of a stream and expand the types of data such algorithms are applied to. Moreover, from the application perspective, we have not sort to explicitly address the issue of how delays in applying an 'oracle' to provide labels when requested impact on the quality of the anytime operation of the classifier.

Acknowledgments. This research is supported by the Canadian Safety and Security Program(CSSP) E-Security grant. The CSSP is led by the Defense Research and Development Canada, Centre for Security Science (CSS) on behalf of the Government of Canada and its partners across all levels of government, response and emergency management organizations, nongovernmental agencies, industry and academia.

References

1. Bifet, A.: Adaptive Stream Mining: Pattern Learning and Mining from Evolving Data Streams. Frontiers in Artificial Intelligence and Applications, vol. 207. IOS Press, Amsterdam (2010)
2. Bifet, A., Read, J., Žliobaitė, I., Pfahringer, B., Holmes, G.: Pitfalls in benchmarking data stream classification and how to avoid them. In: Blockeel, H., Kersting, K., Nijssen, S., Železný, F. (eds.) ECML PKDD 2013, Part I. LNCS, vol. 8188, pp. 465–479. Springer, Heidelberg (2013)
3. Brameier, M., Banzhaf, W.: Evolving teams of predictors with linear genetic programming. Genet. Program. Evolvable Mach. **2**(4), 381–408 (2001)
4. Dempsey, I., O'Neill, M., Brabazon, A.: Grammatical Evolution. In: Dempsey, I., O'Neill, M., Brabazon, A. (eds.) Foundations in Grammatical Evolution for Dynamic Environments. SCI, vol. 194, pp. 9–24. Springer, Heidelberg (2009)
5. Ditzler, G., Roveri, M., Alippi, C., Polikar, R.: Learning in nonstationary environments: a survey. IEEE Comput. Intell. Mag. **10**(4), 12–25 (2015)
6. Fan, W., Huang, Y., Wang, H., Yu, P.S.: Active mining of data streams. In: SIAM International Conference on Data Mining, pp. 457–461 (2004)
7. Gama, J.: Knowledge Discovery from Data Streams. CRC Press, Boca Raton (2010)
8. Gama, J., Sabastiao, R., Rodrigues, P.P.: On evaluating stream learning algorithms. Mach. Learn. **90**, 317–346 (2013)
9. Heywood, M.I.: Evolutionary model building under streaming data for classification tasks: opportunities and challenges. Genet. Program. Evolvable Mach. **16**(3), 283–326 (2015)
10. Lichodzijewski, P., Heywood, M.I.: Managing team-based problem solving with symbiotic bid-based genetic programming. In: ACM Genetic and Evolutionary Computation Conference, pp. 363–370 (2008)
11. Lichodzijewski, P., Heywood, M.I.: Symbiosis, complexification and simplicity under GP. In: ACM Genetic and Evolutionary Computation Conference, pp. 853–860 (2010)

12. Polikar, R., Alippi, C.: Guest editorial: learning in nonstationary and evolving environments. IEEE Trans. Neural Netw. Learn. Syst. **25**(1), 9–11 (2014)
13. Thomason, R., Soule, T.: Novel ways of improving cooperation and performance in ensemble classifiers. In: ACM Genetic and Evolutionary Computation Conference, pp. 1708–1715 (2007)
14. Žliobaitė, I., Bifet, A., Pfahringer, B., Holmes, G.: Active learning with drifting streaming data. IEEE Trans. Neural Netw. Learn. Syst. **25**(1), 27–54 (2014)
15. Vahdat, A., Atwater, A., McIntyre, A.R., Heywood, M.I.: On the application of GP to streaming data classification tasks with label budgets. In: ACM GECCO (Companion), pp. 1287–1294 (2014)
16. Vahdat, A., Morgan, J., McIntyre, A., Heywood, M., Zincir-Heywood, A.: Evolving GP classifiers for streaming data tasks with concept change and label budgets: a benchmarking study. In: Gandomi, A.H., Alavi, A.H., Ryan, C. (eds.) Handbook of Genetic Programming Applications, pp. 451–480. Springer, Switzerland (2015)
17. Vahdat, A., Morgan, J., McIntyre, A., Heywood, M., Zincir-Heywood, A.: Tapped delay lines for GP streaming data classification with label budgets. In: Machado, P., et al. (eds.) Genetic Programming. LNCS, vol. 9025, pp. 126–138. Springer, Switzerland (2015)
18. Wagner, N., Michalewicz, Z., Khouja, M., McGregor, R.R.: Time series forecasting for dynamic environments: the DyFor genetic program model. IEEE Trans. Evol. Comput. **11**(4), 433–452 (2007)
19. Wu, S., Banzhaf, W.: Rethinking multilevel selection in genetic programming. In: ACM Genetic and Evolutionary Computation Conference, pp. 1403–1410 (2011)
20. Zhu, X., Zhang, P., Lin, X., Shi, Y.: Active learning from stream data using optimal weight classifier ensemble. IEEE Trans. Syst. Man Cybern.: Part B **40**(6), 1607–1621 (2010)

Genetic Programming for Region Detection, Feature Extraction, Feature Construction and Classification in Image Data

Andrew Lensen, Harith Al-Sahaf[(⊠)], Mengjie Zhang, and Bing Xue

School of Engineering and Computer Science,
Victoria University of Wellington, PO Box 600, Wellington 6140, New Zealand
{Andrew.Lensen,Harith.Al-Sahaf,Mengjie.Zhang,Bing.Xue}@ecs.vuw.ac.nz

Abstract. Image analysis is a key area in the computer vision domain that has many applications. Genetic Programming (GP) has been successfully applied to this area extensively, with promising results. High-level features extracted from methods such as Speeded Up Robust Features (SURF) and Histogram of Oriented Gradients (HoG) are commonly used for object detection with machine learning techniques. However, GP techniques are not often used with these methods, despite being applied extensively to image analysis problems. Combining the training process of GP with the powerful features extracted by SURF or HoG has the potential to improve the performance by generating high-level, domain-tailored features. This paper proposes a new GP method that automatically detects different regions of an image, extracts HoG features from those regions, and simultaneously evolves a classifier for image classification. By extending an existing GP region selection approach to incorporate the HoG algorithm, we present a novel way of using high-level features with GP for image classification. The ability of GP to explore a large search space in an efficient manner allows all stages of the new method to be optimised simultaneously, unlike in existing approaches. The new approach is applied across a range of datasets, with promising results when compared to a variety of well-known machine learning techniques. Some high-performing GP individuals are analysed to give insight into how GP can effectively be used with high-level features for image classification.

Keywords: Genetic programming · Image classification · Feature extraction · Feature construction

1 Introduction

A common technique used in computer vision is the creation of features which provide a representation of an image that is of a higher level than that of the raw image pixels [12]. Many image classification approaches extract features from an image using a feature extraction algorithm, and then use these features as inputs to a machine learning algorithm to perform classification. A wide range of

M. Heywood et al. (Eds.): EuroGP 2016, LNCS 9594, pp. 51–67, 2016.
DOI: 10.1007/978-3-319-30668-1_4

algorithms for feature extraction have been proposed [17]. One popular approach is the Histogram of Oriented Gradients (HoG) algorithm [8], which produces a histogram of the gradients within an image which can then be used as a feature.

Genetic Programming (GP) has also been applied extensively to image analysis problems [21] since it was introduced in the 1990s. Techniques generally use GP to extract features from raw images by using pixel statistics [24,26], sliding window [22] or filter [3] approaches. GP is able to achieve success on these problems using its evolutionary learning process which allows it to automatically extract and construct high-level features tailored to the dataset it is trained on. This is in contrast to other algorithms such as HoG which do not have a learning process; these algorithms produce general, domain-independent features. The GP approaches tend to extract relatively simple features in comparison to the histograms produced by the HoG algorithm, which might limit their performance. Combining the training process of GP with the powerful features extracted by HoG may improve performance by generating high-level, domain-tailored features automatically. The literature contains many feature extraction methods; we use HoG in this work due to it being one of the most prevalent methods that is simple and efficient enough to implement as a GP function.

Another technique for improving feature quality is to only select regions of an image which are rich in useful features. A two-tier GP (2TGP) [2] method was proposed which automatically selects regions for feature extraction. Using this method in conjunction with more advanced GP feature construction functions would allow region selection and feature construction to be performed simultaneously to improve the image classification performance.

Goals. The goal of this paper is to develop a GP approach to automatically extract and construct high-level features for image classification. To achieve this, we propose a new GP-HoG approach which uses GP with functions based on the HoG method. These new functions are designed to produce more advanced features than the existing GP approaches. In this way, GP will be used for simultaneous region selection, feature extraction and image classification. We aim to achieve this through the following objectives: (1) developing new functions which are inspired by the HoG algorithm. These functions will allow GP to automatically produce high-level features which have the potential to increase classification performance; (2) combining these new functions with a region selection approach 2TGP to allow GP to perform region selection, feature extraction and classification in a single GP tree; (3) analysing the program trees of some good individuals to understand how they are able to generate useful features.

2 Background

Evolutionary Computation (EC) is a large field of artificial intelligence which contains algorithms inspired by biological evolutionary principles [6]. These algorithms are often applied to difficult problems, where the search space is very large. EC algorithms operate iteratively, refining the candidate solutions to a problem in order to gradually improve solutions towards the optimal solution. Evolutionary Algorithms (EAs) are a field of EC algorithms which use Darwinian evolutionary principles to improve solutions by mimicking natural evolution [9].

Genetic Programming (GP) [13] is an EA which models solutions in the form of computer programs. The most common representation is a tree structure, where the root of the tree is the output of the genetic program and the leaves of the tree are inputs or constant values. Non-terminal nodes are functions in the program, which take some inputs (i.e. outputs of other nodes), and then produce an output based on a function applied to those inputs. Terminal nodes are the leaves of a tree.

Feature construction is the process of creating new, high-level features, often by combining multiple existing features [4,14]. Constructed features generally better describe an instance than a single existing feature, reducing the number of features required, which reduces the size of the search space a classifier must train on. GP has been applied extensively to feature construction tasks [10], due to its tree-structure which allows features to be combined using a range of functions to create new features. As GP generally produces a single output value from the root, techniques often use it to produce a single high-level feature.

Al-Sahaf et al. [5] proposed a GP approach to automatically construct an image descriptor that is then used to extract features for multi-class texture classification. Their experiments present the capability of the method to extract important features. The method has significantly outperformed the competitor methods on two texture data sets.

The HoG [8] technique produces a feature vector from an image based on the orientation of gradients within the image. The image is first split into a number of overlapping blocks. Each block produces a histogram of gradients of pixels within that block. For each pixel in a block, both the magnitude and the orientation of its gradient is recorded. The histogram of each block, then, contains bins for various orientations (one bin for a range of orientations), and the height of each bin is the sum of the magnitudes of the gradients falling within that bin. The histogram from each block is then normalised, and all histograms are then combined to give a final feature vector corresponding to the image as a whole. This kind of feature vector has been used for a variety of image analysis problems [8,27].

2.1 Related Work

This subsection briefly surveys typical related work which uses GP to extract and construct features for image classification. The limitations of these works are discussed, showing the motivation behind our proposed GP-HoG approach.

The 2TGP approach [2] used GP to select good regions of an image, extract features from the regions (as simple statistics based on the pixels in the region), and to perform classification. The approach was tested on a variety of datasets which varied in difficulty, with good results across different image domains. The solutions produced were also easy to understand. For example, on a face dataset, solutions were produced using regions which humans would also use for classification, such as the nose, mouth and eyes. On a pedestrian dataset, regions were selected which captured areas, where a standing pedestrian would be expected to appear. By selecting regions, this approach was able to improve classification accuracy. Using more advanced features, i.e., beyond simple pixel statistics, in combination with region selection has the potential to improve the performance even further. This is the direction we take in this study.

In [23], the root of the tree was used as a constructed feature by a Support Vector Machine (SVM) for image classification. This approach created GP trees using a large range of functions which directly considered the pixel values of the given images. By using a multi-objective approach which tried to minimise tree size while maximising classification accuracy, the authors were able to reduce over-fitting and achieved a high classification accuracy. The function set used a range of filtering functions including Gaussian, Laplacian, and Gabor filters, as well as simpler pixel-by-pixel arithmetic operations. While these filtering functions are more advanced than simpler pixel statistic approaches, they are still relatively simpler than the HoG algorithm, as they apply a small filter to each pixel instead of using a more sophisticated histogram technique.

Perez and Olague proposed a GP technique (RDGP) [20] using a function set and a terminal set, which was designed to emulate the Scale-Invariant Feature Transform (SIFT) [15], another widely-used feature extraction algorithm. A range of functions and terminals were used, including arithmetic operators, image derivatives and Gaussian filters. The authors argued that their method would allow GP to automatically synthesise SIFT-like programs by automatically extracting high-level features for object recognition. They claimed that their approach allowed features to be automatically tailored towards the problem being trained on, as the GP programs would be optimised by the evolutionary process. The RDGP approach was shown to produce better features than the standard SIFT approach, with an overall decrease in error in object detection. While this approach performed feature extraction and construction, it did not use GP for classification. As their design broke the SIFT algorithm into its composite parts as GP functions, the evolutionary process must learn to re-construct and optimise these composite parts in order to produce useful features. This may reduce the performance of the method; the approach we propose attempts to mimic HoG within a single function, so that the evolutionary process can instead focus on constructing high-level features and simultaneously evolve a good classifier.

3   The Proposed Method

This section details the proposed method (named GP-HoG) including the program representation, and the fitness function.

3.1   GP Program Representation

The proposed method uses a combination of existing terminals and functions from 2TGP and novel terminals and functions inspired by HoG. In this study, a

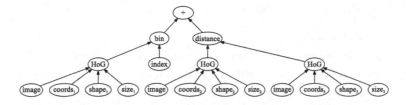

Fig. 1. An example shows an individual structure of the GP-HoG method.

new tree-based program structure is proposed, as presented in Fig. 1. To introduce restrictions on the inputs and outputs of the different nodes in an evolved program, *strongly-typed GP* [18] is used. The full terminal and function sets are listed in Tables 1 and 2 respectively. An individual's tree can be, virtually, divided into three layers. The bottom layer, which includes the HoG and terminal nodes, represents the feature extraction part. The middle layer, which consists of a mix of the *bin* and *distance* nodes, represents the feature construction part. The top layer (including the root node) that is made up of a chain of simple arithmetic operators represents the classification part. As Strongly-Typed GP is used, all three layers appear in every valid program. The feature construction layer represents a main difference between GP-HoG and 2TGP, which has only feature extraction (as pixel statistics across a region) and classification layers. The new feature construction layer aims to further reduce the search space by constructing high-level features from the extracted features from the previous layer, with the expectation of evolving meaningful classifiers and further improving the performance.

The majority of the terminal set is based on the 2TGP method, as the terminals provide the parameters used in the region selection process. The *image* node is the instance, i.e., image being evaluated represented as a 2D array of integer values each of which represents the intensity of a pixel in the image. *rand* is a random value drawn from the half-closed interval $[0, 1)$. The *shape* node defines the shape of a region that can be `rectangle`, `circle`, `row` and `column`. The *coords* node is a pair of (x, y) values that define the location of a region in an image where (x, y) is the centre pixel for the circular shape and the top-left corner pixel for all other shapes. The minimum width and height of all images in the dataset are, respectively, denoted as min_{Width} and min_{Height}. T The *size* node specifies the size of a region. The value of a *size* node is defined to be 1 in `row` and `column` regions. The *size* node is not used in the case of a rectangle region and is replaced by (w, h) which are the *width* and *height* of the region. For a circular region, the *size* node gives the diameter. Restricting the dimensions of regions by using min_{Width} and min_{Height} encourages regions to be created which are valid across the majority of images in a dataset, which improved training and test performance. In order to return a specific value of a histogram, the *index* node is introduced that takes a value between 0 and 7 (inclusive) as each histogram consists of 8 bins (as detailed in the next subsection).

Table 1. The terminal set

Name	Output	Details
image	image	The image being evaluated represented as a 2D array of pixel values
rand	double	Random double in the half-closed interval $[0, 1)$
coords	coords	Provides the location as a region as (x, y). x is the horizontal location randomly generated in $[0, min_{\text{Width}}]$ and y is the vertical location in $[0, min_{\text{Height}}]$
size	size	Random integer value in $[3, \min(min_{\text{Width}}, min_{\text{Height}})]$
index	index	Random integer value in $[0, 7]$, which represents the bin index of a histogram
shape	shape	One of rectangle, circle, row, and column. Rectangle has x in $[0, min_{\text{Width}}]$ and y in $[0, min_{\text{Height}}]$

Table 2. The function set

Name	Input	Output	Details
$+, -, \times, /$	double, double	double	Arithmetic operators
bin	histogram, index	double	Returns the value of the specified index
HoG	image, coords, shape, size	histogram	Performs the HoG algorithm on a region
distance	histogram, histogram	double	Returns the distance between two histograms

The function set comprises of the four arithmetic operators $+$, $-$, $\times$ and protected $/$, *distance*, *bin*, and *HoG* nodes. The arithmetic operators in the function set have their corresponding regular meaning and allow GP to utilise multiple extracted features for classification. The *HoG*, *distance* and *bin* functions and the *index* terminal are used for feature extraction. The design of these functions is discussed in the next subsection.

The most important new function is the HoG function, which is inspired by the HoG algorithm [8]. The HoG function takes the *image*, *coords*, *shape*, and *size* as inputs, and outputs a histogram which represents the distribution of gradients within a region of the image. The standard approach of using a histogram with 8 bins [8] is adopted here, where each bin corresponds to 45° of rotation. By using the provided terminal nodes, a GP tree can construct a histogram across a region of varying shape and size; there are no pre-determined regions as in the normal HoG method. This allows the regions used to be tailored to the dataset.

3.2 Outline of the HoG Function

The region of the image is selected based on the inputs to the *HoG* function. This is done by taking a region of an image defined by the *shape* (how the region is shaped), *coords* (the position of the region) and *size* (size of the region) inputs. Then, the steps presented in Algorithm 1 are applied.

Our HoG approach differs from the standard HoG algorithm [8] in a few ways in order to allow it to be expressed sensibly as a function for GP. The biggest difference is that the *HoG* function is applied only to a single region, given by the function arguments. Normally, the histograms of multiple overlapping blocks across an image are combined to give a more versatile feature vector. As multiple *HoG* functions can be incorporated in a single GP tree, it is not necessary to use multiple blocks to try and analyse the whole image; this will be done automatically as part of the evolutionary process if it gives good performance. As only a single histogram is produced from one run of the proposed GP-HoG algorithm, the normalisation process is only applied to a single histogram, rather than across several as in the original design. This approach also allows for a range of block (i.e. region) sizes and shapes; normally the design of blocks are fixed, such as using an 8×8 square. It is expected that the evolutionary process will be able to automatically find the best block sizes and shapes as individuals with the best block designs, i.e., more representative features, will be rewarded with a better fitness value. The crossover operator allows useful block designs to be exchanged between individuals.

As a number of variations to the original HoG algorithm have been made, we consider Algorithm 1 to be inspired by the HoG algorithm, rather than being a strict implementation of it. As Algorithm 1 outputs a feature vector (histogram) of eight bins, GP can not directly use this vector for classification; therefore, two additional functions were designed which construct high-level features from a histogram. The first function is *distance*, which finds the Euclidean distance between two histograms. This produces a double value which gives a measure of how dissimilar two histograms are. This can be used to compare different regions of an image in order to identify the image's class. For example, on the UIUC dataset (Sect. 4), the regions corresponding to a car's front and back wheels will produce similar histograms. These same regions on a background image are more likely to give different histograms. Hence, the distance between histograms can be used as a feature for classification. The second function is *bin*, which returns the value of a given bin index of a histogram. This function allows GP to select important orientations which have different magnitudes depending on the image class. For example, on the Jaffe dataset, the edges of the mouth have different gradients for the subjects being happy or surprised. The magnitude of a given bin can be used as a feature for distinguishing two classes.

3.3   The Fitness Function

The evolutionary process measures a program's goodness using the fitness function. In this work, the accuracy of a program is used as the fitness value to reflect its ability to discriminate between instances of different classes. The accuracy is the proportion of correctly classified instances to the total number of instances. Hence, an ideal program will have a fitness value of 1 and a fitness value of 0 represents the worst case scenario or performance.

Classification of an instance is performed by feeding it into an evolved GP tree. The *image* terminals of the GP tree are set to contain the image being classified, and then the tree is evaluated from bottom to top, producing a single real number as an output. A threshold 0 is then applied to this value.

Algorithm 1. The procedure used in the *HoG* function

---

1 Find the gradient for each pixel as $G_x = f(x+1,y) - f(x-1,y)$ and $G_y = f(x,y+1) - f(x,y-1)$ where $f(i,j)$ gives the pixel value at (i,j).

2 Find the magnitude at each pixel as $m = \sqrt{G_x^2 + G_y^2}$.

3 Find the orientation of each pixel as $\arctan\left(\frac{G_y}{G_x}\right)$. This is converted to degrees and mapped to be in range $[0°, 360°]$.

4 **for** each pixel **do**

5   Find the two bins of the histogram it lies between based on its orientation. The histogram is divided into 8 bins of size $45°$, so each pixel with an orientation will have a lower bin and an upper bin. For example, a pixel with an orientation of $80°$ would have its lower bin as bin 2 ($45°$), and its upper bin as bin 3 ($90°$).

6   For each bin, find the distance between the bin's orientation and the pixel's orientation. In the previous example, an orientation of $80°$ puts that pixel at $35°$ distance from the lower bin, and $10°$ from the upper bin.

7   For each bin, calculate and add the weighted magnitude as the pixel's magnitude multiplied by how close it is to that bin. As the bin size is $45°$, $m \times \frac{(45-35)}{45}$ is added to the lower bin. The upper bin would have $m \times \frac{(45-10)}{45}$ added to it, as the upper bin's orientation is closer to that of the pixel.

8 **end for**

9 Normalise the histogram by expressing the value of each bin as a fraction of the sum across all bins.

A negative value gives a negative classification, and a non-negative value gives a positive classification. A tree may contain regions that partially fall outside the dimensions of the image. Any such regions are cropped, so only the pixels within the image bounds are used in the computation of the histogram in the *HoG* function.

4   Experiment Design

This section details the datasets, parameter settings, and methods for comparison used in this study.

4.1   Datasets

Three datasets were used to assess the performance of the proposed method. These datasets are for different applications and vary in difficulty. However, each of the datasets is made up of greyscale images and is set for binary classification.

In computer vision, the *Columbia Object Image Library*[1] (COIL-20) [19] dataset is widely used. Two classes of the COIL-20 dataset are used to form the first dataset in this study. Originally, the COIL-20 dataset comprises of 20 classes that each represents a different toy object, e.g., cars, rubber ducks, and boxes. A turntable is used in a scene with a black background to prepare those instances.

---

[1] Available at: http://www.cs.columbia.edu/CAVE/software/softlib/coil-20.php.

For each object, 72 images are provided by taking a snapshot every 5° where the object is rotated through 360°. Then those images are cropped to be 128 × 128 pixels each where the object is centred in the images. Furthermore, those images were normalised by adjusting the brightest pixel to be 255 and scaling the other pixel values accordingly. In this study, only the *cars* and *rubber ducks* classes (Fig. 2(a)) are used as the focus is on performing binary classification.

Meanwhile, the second dataset in this study is formed using the *Japanese Female Facial Expression*[2] (Jaffe) [16] dataset. This dataset is broadly used in the literature for the task of identifying different facial expressions. In total, this dataset consists of 213 images provided by ten Japanese female subjects, which is divided into seven groups: neutral, surprised, angry, sad, happy, disgust, and fear. Each subject provides several images for each facial expression. Following Cheng *et al.* [7], and in order to prevent the classifiers from training on irrelevant features, the images of this dataset were manually cropped in order to remove most of the subject's hair, and the image background leaving only the face. The size of those instances after cropping ranges between 164 and 207 pixels in height, and between 121 and 143 pixels in width. The instances of the happy and surprised classes (Fig. 2(b)) are used in this study to form the second dataset.

To form the third dataset in this study, the *UIUC database for Car Detection*[3] (UIUC) dataset [1] is used. In total, the UIUC dataset consists of 1,050 instances that fall into two classes: cars and background (Fig. 2(c)). The former comprises of 550 instances, whilst there are 500 instances in the latter. The car instances are captured from the same angle and distance (giving the same scale) that show the side view of the vehicle. Each instance in this dataset is 100 × 40 pixels.

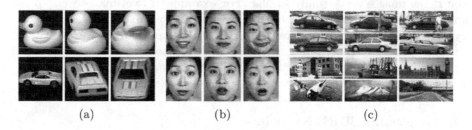

(a)   (b)   (c)

Fig. 2. Samples of the (a) COIL-20, (b) Jaffe, and (c) UIUC cars datasets showing instances of the positive and negative classes in the top and bottom rows respectively.

4.2   Training and Test Sets

The k-fold cross-validation technique was used to evaluate the proposed method and all the baseline methods. The instances of the UIUC and COIL-20 datasets were randomly split into 10 folds. The Jaffe dataset contains only three images of each expression for each human subject, requiring a careful split to ensure that

---

[2] Available at: http://www.kasrl.org/jaffe.html.
[3] Available at: http://cogcomp.cs.illinois.edu/Data/Car/.

the test and training sets are both representative of the dataset, and therefore, this dataset was manually split into 3 folds. Each fold contained one happy and one surprised expression for each subject.

4.3   Baseline Methods

A number of baseline methods are used to compare the new approach to the existing methods in the literature. The Waikato Environment for Knowledge Analysis (WEKA) [11] implementations of the Support Vector Machine (SVM), Decision Trees (J48), Naïve Bayes (NB), Random Forest (RF), and Adaptive Boosting (ABM1) classifications methods are used [25]. As the proposed method is largely based on 2TGP, the 2TGP method is also used as a competitive method in this study. Each of these seven methods (including the proposed method) were evaluated using the k-fold cross-validation scheme described in Sect. 4.2. For each of the non-GP classifiers, an instance is evaluated by giving the classifier a list of concatenated SURF keypoints (as detailed in the next subsection). As a keypoint contains 64 values, there will be $p \times 64$ features provided, where p is the number of keypoints used. For example, if two keypoints called a and b were used, the list would be formatted in the form $[a_0, a_1, .., a_{63}, b_0, b_1, ..., b_{63}]$. As the WEKA implementation of each of these methods is deterministic, they are only run once for a given experiment run. The list of keypoints is ordered by the strength of each keypoint so that the methods are able to learn most effectively. The SURF feature extractor generates keypoints based on the location of the keypoint in the image, which means that a classifier may classify two instances of the same class differently depending on the distribution of keypoints throughout the images, even if the images are actually similar. For example, if two images were of the same person's face but in one the face was shifted 50 pixels to the right, the keypoint corresponding to a "nose feature" could appear in different locations in the keypoint list. By ordering keypoints by how strong they are, the classifier is more likely to classify similar instances to the same class as they will likely have similar strong keypoints at the same index in the list.

4.4   Generating SURF Keypoints

Both of the GP-based (2TGP and GP-HoG) methods are designed to operate directly on the raw pixel values, which is not the case for the other baseline methods. Therefore, SURF image descriptor is used to generate a list of keypoints that can be used as a high-level features by those classifiers. However, SURF generates varying numbers of keypoints based on the number of interest points an image has. For many classifiers, this presents a problem as a static number of features (fixed length feature vector) is expected. Algorithm 2 is developed to address this problem. This method relies on altering the Hessian threshold, which represents a main component of the SURF method, in order to determine the interest points. Binary search is used to adjust this threshold until a predefined number of keypoints are retrieved. A fixed number of keypoints allows more effective training as a solution can perform consistently across a dataset.

Algorithm 2. Selecting top-p keypoints

```
 1 function SELECTKEYPOINTS(image, Lbound, Ubound, p)      ▷ where Lbound and Ubound are the
 2     threshold ← (Ubound − Lbound)/2 + Lbound                lower and upper bounds of the
 3     keypoints ← SURF(threshold)                            Hessian threshold, respectively.
 4     if |keypoints| = p then
 5         return keypoints
 6     else if |keypoints| > p then
 7         Lbound ← threshold
 8     else
 9         Ubound ← threshold
10     end if
11     return SELECTKEYPOINTS(image, Lbound, Ubound, p)
12 end function
```

4.5   Evolutionary Parameters

GP has a number of parameters which can be altered in order to optimise the evolutionary process for a given problem. The GP-HoG and the 2TGP methods were applied to the three datasets (Sect. 4.1). For 2TGP, the same parameters were used as in [2]; namely, 80 % crossover, 20 % mutation and top-10 elitism was used. The population size was $1,024$, and the minimum and maximum tree depth were 2 and 10, respectively. On each dataset, the evolutionary process was independently executed 35 times using different seed values. Each execution was run for 50 generations or until perfect training performance was obtained. GP-HoG used 40 % mutation and 60 % crossover as it was found a higher mutation rate could produce better training performance by allowing a wider exploration of the search space. All other parameters were the same as for 2TGP.

5   Results and Discussion

This section compares the performance of the GP-HoG approach to the 2TGP approach and the five SURF baselines. It also discusses the increase in training time required to train GP using the GP-HoG approach compared to using 2TGP.

5.1   Compared to the 2TGP Approach

The results of the 2TGP and GP-HoG are shown in Table 3. Student's t-test with a 95 % confidence interval was used to evaluate the significance of the performance increase using GP-HoG. A "+" in Table 3 indicates that GP-HoG is significantly better than the 2TGP approach, whereas a "−" indicates it is significantly worse. The GP-HoG approach performs significantly better on the Jaffe and UIUC datasets (the two difficult datasets) while achieving slightly worse mean (but identical maximum) test performance on the COIL-20 dataset. On the most difficult dataset (Jaffe), GP-HoG achieves a 5 % and 11 % increase in mean and maximum test performance respectively over the 2TGP approach.

The average training time has increased notably using the new approach compared to 2TGP, with approximately 3× more computation required on the Jaffe and UIUC datasets, likely due to the larger amount of computation required by the *HoG* function than in the aggregation functions used in the 2TGP approach. The arctan $(\cdot)$ function is the slowest part of the HoG algorithm (from empirical sampling), as it is somewhat expensive to compute even on a modern CPU, and is used once for every pixel in a region. On the UIUC dataset (which has the largest number of images), training never finishes before the maximum number of generations (the maximum training performance is 98 %), and so the training time is much longer than the other datasets. Even with the utilisation of multi-threading, the 600 hours of CPU time across all folds takes about a week of real time. It is important to note that while the increase in training time is a downside of the new approach, the time required to apply the best trained solution to new, unseen images is still minimal. Long training times are common when GP is used, but as long as the evolved programs are not overly complex, they are often quick enough to be used on unseen instances.

5.2   Compared to the Baselines

The results of the five non-GP methods on the three datasets using different numbers of SURF keypoints are presented in Table 4. The values of the last two blocks ($p = 20$ and $p = 50$) of the UIUC dataset are not available as SURF could not generate this many keypoints due to a lack of interest points in the images in this dataset. The GP-HoG approach has similar performance on COIL-20 and improved performance on the Jaffe and UIUC datasets compared to the non-GP baselines. This is unsurprising, as the GP-HoG approach is able to perform region selection and feature construction to give more advanced and dataset-specific features than the domain-independent features produced by SURF.

Table 3. The accuracy and average training time (H:M:S) of the 2TGP and GP-HoG methods on the three datasets.

	2TGP						GP-HoG					
	COIL-20		Jaffe		UIUC		COIL-20		Jaffe		UIUC	
	Train	Test	Train	Test	Train	Test	Train	Test	Train	Test	Train	Test
Max	1.00	1.00	0.98	0.81	0.96	0.94	1.00	1.00	0.99	0.92	0.96	0.95
Mean	1.00	0.99	0.95	0.71	0.94	0.92	1.00	0.97^{-}	0.97^{+}	0.76^{+}	0.95^{+}	0.93^{+}
St.Dev	0.00	0.01	0.02	0.05	0.01	0.01	0.00	0.02	0.02	0.07	0.01	0.01
Training time	00:05:49		01:35:54		16:22:39		03:24:26		04:36:24		52:15:25	

6   Further Analysis

The GP-HoG approach produces programs which can be interpreted and understood by humans. This section analyses three high-performing evolved programs to understand how they can perform classification with high accuracy.

Table 4. The average accuracies of the non-GP baseline methods on the three datasets.

		$p = 5$		$p = 10$		$p = 20$		$p = 50$	
		Train	Test	Train	Test	Train	Test	Train	Test
COIL-20	SVM	1.00	0.99	1.00	0.99	1.00	0.99	1.00	1.00
	J48	0.99	0.86	0.99	0.86	0.99	0.86	0.99	0.85
	NB	0.99	0.97	0.99	0.96	1.00	0.91	1.00	0.94
	RF	1.00	0.99	1.00	0.99	1.00	1.00	1.00	1.00
	ABM1	1.00	0.96	1.00	0.96	1.00	0.93	1.00	0.93
Jaffe	SVM	1.00	0.63	1.00	0.72	1.00	0.74	1.00	0.82
	J48	0.98	0.70	0.98	0.56	0.98	0.59	0.98	0.79
	NB	0.89	0.72	0.92	0.72	0.98	0.75	1.00	0.69
	RF	1.00	0.77	1.00	0.76	1.00	0.77	1.00	0.71
	ABM1	1.00	0.70	1.00	0.77	1.00	0.67	1.00	0.71
UIUC	SVM	0.99	0.92	1.00	0.91	N/A	N/A	N/A	N/A
	J48	0.99	0.84	0.99	0.83	N/A	N/A	N/A	N/A
	NB	0.90	0.89	0.90	0.89	N/A	N/A	N/A	N/A
	RF	1.00	0.94	1.00	0.93	N/A	N/A	N/A	N/A
	ABM1	0.88	0.85	0.88	0.84	N/A	N/A	N/A	N/A

6.1 Example Program 1

An evolved program with high performance on the Jaffe dataset is shown in
Fig. 3. This program is interesting to analyse, as it is very simple, consisting of
two *HoG* operators, and a subtraction operator. The left side of the tree applies
the *HoG* operator to a rectangular region corresponding to the right side of the
face, including the eye, cheek, and part of the nose and lip areas. This region
contains different features of the face in the happy and surprised expressions.
When the subject is happy, the corner of the mouth is narrower than when
surprised, producing a smaller gradient orientation. When they are surprised,
the mouth is widened, creating a right-angle between the chin and where the
subject's ear would be. This edge is at a larger angle than when the subject is
happy, and hence a different histogram is produced. The *HoG* operator produces
a histogram, and the value of the bin corresponding to the 270°–314° orientation
range is selected by the *bin* function. On the right side of the tree, the *HoG*
operator is applied to a circular region of the image, which corresponds to the
upper nose, eyes, and eyebrow areas. The subject's nose appears much narrower
when surprised, due to her mouth being open. There is also more of the nose
included in the surprised image, which introduces an additional edge gradient.
The eyes are also different when surprised; they appear wider and have more
white showing. All of these differences change the histogram that is produced,
allowing GP to extract features that distinguish these two expressions. The value
of the bin corresponding to the 0°–44° range is chosen by the *bin* node on the
right side of the tree. The root of the tree then outputs the difference between
the values from the left and right sides of the tree. This program scores 98 %
and 95 % on training and test sets respectively on the fold it was generated in.

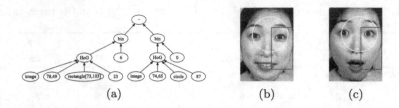

(a)   (b)   (c)

Fig. 3. Example program 1: 98 % training and 95 % test performance on Jaffe dataset (a) GP tree, (b) Happy (+ve) expression, and (c) Surprised (−ve) expression.

6.2 Example Program 2

Another program with very good performance (95 % training and 100 % test accuracy) on the Jaffe dataset is shown in Fig. 4. This program is similar to Program 1 in that it uses only the difference of two histogram values to classify images with high accuracy. The yellow circular region contains the left eye and cheek, and a small part of the nose. The left eye has a different appearance between the happy and surprised expressions; when surprised, a larger amount of the eye is visible. The left nostril is also open in the surprised expression, producing a large gradient around it which is included in the yellow circle. While the blue circle is more difficult to analyse as it covers a large part of the image, one important observation is that it appears to be bounded by the eyes and mouth. The mouth is much darker in the surprised expression, giving a smaller gradient than in the happy expression where there is a distinct gradient between the white teeth and darker lips. By selecting the important mouth, eye and nose regions, the feature selected from the blue circle are used by GP to distinguish between happy and surprised expressions.

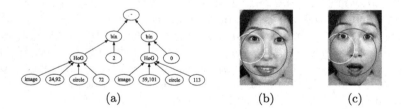

(a)   (b)   (c)

Fig. 4. Example program 2: 95 % training and 100 % test performance on Jaffe dataset (a) GP tree, (b) Happy (+ve) expression, and (c) Surprised (−ve) expression.

6.3 Example Program 3

Unlike the Jaffe and COIL-20 datasets, GP-HoG is unable to obtain perfect training performance on UIUC, and so large programs are produced by the evolutionary training process to maximise the training performance. The program in Fig. 5 (a) is one of the simpler programs that performs well on UIUC, with a tree depth of eight. The regions used by this program are shown in Fig. 5 (b) and (c). While many regions are used, most tend to be small, identifying particular aspects of the image that are useful for classification. Several of these regions enclose specific parts of the car in (b), such as the front wheel, wheel arch and parts of the front door. The front wheel is likely to have a distinctive histogram, as it has a circular tyre surrounding the hubcap. This gives a circular edge which has a consistently changing orientation and a large gradient magnitude (as the tyre is black and the hubcap is grey), producing a histogram with a similar magnitude in each bin. The same region in the background image contains a straight edge, which would produce a histogram with a spike in one bin. This difference in histograms helps the GP tree to distinguish these two classes.

```
(- (* (+ (- (- (bin (HoG image 62,25 rectangle[16,4]
11) 5) 0.3114006771781549) (distance (HoG image 8,4
rectangle[55,15] 17) (HoG image 71,26 rectangle[91,30]
24))) (bin (HoG image 39,39 circle 15) 2)) (distance
(HoG image 50,24 column 12) (HoG image 72,31 circle
4))) (bin (HoG image 56,29 circle 17) 0))
```

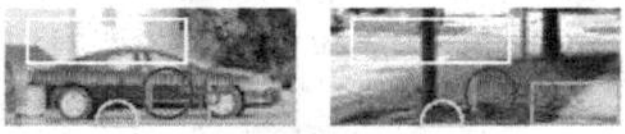

(b)   (c)

(a)

Fig. 5. Example program 3: 96 % training and 97 % test performance on UIUC dataset (a) Lisp expression, (b) Car (+ve) image, and (c) Background (−ve) image.

7   Conclusions

We proposed a new GP method for simultaneous region selection, feature construction and classification, which combines novel functions inspired by the HoG algorithm with the region selection concept proposed in the 2TGP method. Performance evaluation showed good results using the proposed method compared to the performance of 2TGP on three datasets with increasing difficulty. Performance was also promising when compared to other machine learning baselines using SURF features. The analysis of high-performing solutions showed that the GP-HoG approach could perform very well using simple programs. The adaptation of HoG for use as a GP function was shown to be an effective method of performing high-level feature extraction directly within a GP tree.

In the future, we would like to study the use of the extracted and constructed features produced by GP-HoG across different classifiers. This will help in identifying whether GP-HoG can be used for automatic feature extraction and construction, and whether these features are biased towards a specific type of classifiers. Another very important direction is to investigate the possibility of using other algorithms which could also be adapted directly in GP functions. For example, deeper analysis of the SURF or SIFT algorithms could produce functions that could be added to the GP-HoG approach. Other techniques such as edge detection could also be used within a GP tree in order to build a multi-faceted classifier which draws upon a range of techniques for complex classification tasks.

References

1. Agarwal, S., Awan, A., Roth, D.: Learning to detect objects in images via a sparse, part-based representation. IEEE Trans. Pattern Anal. Mach. Intell. **26**(11), 1475–1490 (2004)
2. Al-Sahaf, H., Song, A., Neshatian, K., Zhang, M.: Two-tier genetic programming: towards raw pixel-based image classification. Expert Syst. Appl. **39**(16), 12291–12301 (2012)
3. Al-Sahaf, H., Zhang, M., Johnston, M.: Genetic programming evolved filters from a small number of instances for multiclass texture classification. In: Proceedings of the 29th International Conference on Image and Vision Computing New Zealand, pp. 84–89. ACM (2014)
4. Al-Sahaf, H., Zhang, M., Johnston, M.: Binary image classification: a genetic programming approach to the problem of limited training instances. Evolutionary Computation (Journal, MIT Press) (2015). doi:10.1162/EVCO_a_00146
5. Al-Sahaf, H., Zhang, M., Johnston, M., Verma, B.: Image descriptor: a genetic programming approach to multiclass texture classification. In: Proceedings of 2015 IEEE Congress on Evolutionary Computation, pp. 2460–2467. IEEE (2015)
6. Back, T., Fogel, D.B., Michalewicz, Z.: Handbook of Evolutionary Computation. IOP Publishing Ltd., Bristol (1997)
7. Cheng, F., Yu, J., Xiong, H.: Facial expression recognition in JAFFE dataset based on gaussian process classification. IEEE Trans. Neural Netw. **21**(10), 1685–1690 (2010)
8. Dalal, N., Triggs, B.: Histograms of oriented gradients for human detection. In: Proceedings of the IEEE Conference on Computer Vision and Pattern Recognition, vol. 1, pp. 886–893. IEEE (2005)
9. Eiben, A.E., Smith, J.E.: Introduction to Evolutionary Computing. Springer Science & Business Media, Heidelberg (2003)
10. Espejo, P.G., Ventura, S., Herrera, F.: A survey on the application of genetic programming to classification. IEEE Trans. Syst. Man Cybern. Part C: Appl. Rev. **40**(2), 121–144 (2010)
11. Hall, M., Frank, E., Holmes, G., Pfahringer, B., Reutemann, P., Witten, I.H.: The WEKA data mining software: an update. SIGKDD Explor. Newsl. **11**(1), 10–18 (2009)
12. Huang, Y., Wu, Z., Wang, L., Tan, T.: Feature coding in image classification: a comprehensive study. IEEE Trans. Pattern Anal. Mach. Intell. **36**(3), 493–506 (2014)
13. Koza, J.R.: Genetic Programming: On the Programming of Computers by Means of Natural Selection. MIT Press, Cambridge (1992)
14. Krawiec, K.: Genetic programming-based construction of features for machine learning and knowledge discovery tasks. Genet. Program. Evolvable Mach. **3**(4), 329–343 (2002)
15. Lowe, D.G.: Object recognition from local scale-invariant features. In: Proceedings of the International Conference on Computer Vision, pp. 1150–1157. IEEE (1999)
16. Lyons, M., Akamatsu, S., Kamachi, M., Gyoba, J.: Coding facial expressions with gabor wavelets. In: Proceedings of the 3rd International Conference on Face & Gesture Recognition, pp. 200–205. IEEE (1998)
17. Mikolajczyk, K., Schmid, C.: A performance evaluation of local descriptors. IEEE Trans. Pattern Anal. Mach. Intell. **27**(10), 1615–1630 (2005)

18. Montana, D.J.: Strongly typed genetic programming. Evol. Comput. **3**(2), 199–230 (1995)
19. Nene, S.A., Nayar, S.K., Murase, H.: Columbia Object Image Library (COIL-20). Technical report (1996)
20. Perez, C.B., Olague, G.: Evolutionary learning of local descriptor operators for object recognition. In: Proceedings of the 11th Annual conference on Genetic and evolutionary computation, pp. 1051–1058. ACM (2009)
21. Poli, R.: Genetic programming for feature detection and image segmentation. In: Fogarty, T.C. (ed.) AISB-WS 1996. LNCS, vol. 1143, pp. 110–125. Springer, Heidelberg (1996)
22. Saini, R., Dutta, M.: Image segmentation for uneven lighting images using adaptive thresholding and dynamic window based on incremental window growing approach. Int. J. Comput. Appl. **56**(13), 31–36 (2012)
23. Shao, L., Liu, L., Li, X.: Feature learning for image classification via multiobjective genetic programming. IEEE Trans. Neural Netw. Learn. Syst. **25**(7), 1359–1371 (2014)
24. Winkeler, J.F., Manjunath, B.: Genetic programming for object detection. In: Proceedings of the Second Annual Conference on Genetic Programming, pp. 330–335. Morgan Kaufmann (1997)
25. Witten, I.H., Frank, E.: Data Mining: Practical Machine Learning Tools and Techniques, 2nd edn. Morgan Kaufmann, San Francisco (2005)
26. Zhang, M., Ciesielski, V., Andreae, P.: A domain-independent window approach to multiclass object detection using genetic programming. EURASIP J. Adv. Signal Process. **2003**(8), 841–859 (2003)
27. Zhao, Y., Zhang, Y., Cheng, R., Wei, D., Li, G.: An enhanced histogram of oriented gradients for pedestrian detection. IEEE Intell. Transp. Syst. Mag. **7**(3), 29–38 (2015)

Surrogate Fitness via Factorization of Interaction Matrix

Paweł Liskowski$^{(\boxtimes)}$ and Krzysztof Krawiec

Institute of Computing Science, Poznan University of Technology, Poznań, Poland
{pliskowski,krawiec}@cs.put.poznan.pl

Abstract. We propose SFIMX, a method that reduces the number of required interactions between programs and tests in genetic programming. SFIMX performs factorization of the matrix of the outcomes of interactions between the programs in a working population and the tests. Crucially, that factorization is applied to matrix that is only partially filled with interaction outcomes, i.e., sparse. The reconstructed approximate interaction matrix is then used to calculate the fitness of programs. In empirical comparison to several reference methods in categorical domains, SFIMX attains higher success rate of synthesizing correct programs within a given computational budget.

Keywords: Genetic programming · Test-based problem · Recommender systems · Machine learning · Surrogate fitness

1 Introduction

Conventional fitness evaluation in genetic programming (GP) consists in applying a program to multiple tests (fitness cases) and aggregating the observed differences between the actual and desired program output. Running a program multiple times can be computationally costly, especially when it involves nontrivial computation or requires processing large amount of data. Computational expense becomes particularly high when programs grow large (a common ailment of GP) or engage loops.

Lowering computational cost incurred by evaluation by simply reducing the number of tests is often not a viable option. Few tests implies inaccurate fitness, and consequently a poorly informed search process. Moreover, discarding tests may cause a task to be formally underspecified (underconstrained). For instance, a set of tests for a multiplexer problem that misses even a single test does not technically specify that problem anymore. Also, if the differences between the actual and desired program outputs are discrete, a low number of tests leads to coarse-grained fitness that often fails to differentiate solutions. When, as it is common, the actual and desired output can be compared only for equality, the outcome of a program-test interaction is binary and fitness can assume only $n + 1$ values for n tests.

© Springer International Publishing Switzerland 2016
M. Heywood et al. (Eds.): EuroGP 2016, LNCS 9594, pp. 68–82, 2016.
DOI: 10.1007/978-3-319-30668-1_5

Various means for reducing the number of required program-test interactions other than plain discarding of tests have been proposed in the past. Most of them fall under the category of surrogate fitness and involve measurement of similarity between the inputs of particular tests. In the simplest scenario, an unknown output of a program for a test t is substituted with the known output of that program for a similar test t'. However, designing an appropriate input similarity measure for a given problem requires domain knowledge. And once designed, such a measure may bias the selection of tests to be used as surrogates and lower the likelihood of synthesizing the correct program.

In this paper, we propose a method that builds a Surrogate Function via Factorization of Interaction Matrix (SFIMX) and reduces so the number of interactions. SFIMX, detailed in Sect. 4, is applicable to any domain where interaction outcomes can be encoded as numbers (e.g., symbolic regression, Boolean, integer, etc.) and, unlike the similarity measures mentioned above, does not require additional knowledge. It engages the well-known algebraic concept of matrix factorization, that recently grew in popularity in machine learning and recommender systems. SFIMX is straightforward, performs well in practice (Sect. 6), and has interesting conceptual implications, which we elaborate on in Conclusions.

2 Background

The desired behavior of a program to be synthesized in GP is specified by a set of tests (fitness cases), each being a pair $(x, y) \in T$ of the input x fed into a program and the *desired output* y expected to result from that program execution. T may be sampled from a potentially infinite universe $\mathcal{T} \supset T$.

A GP algorithm solving a program synthesis task maintains a population of programs $P \subset \mathcal{P}$. In every generation, each program $p \in P$ is tested on every test $(x, y) \in T$, in which p is applied to x and returns an output $p(x)$. In other words, p engages in an *interaction* with a test t. The outcome of that interaction can be characterized by a scalar *interaction function* $g(p, t)$. If $p(x) = y$, p is said to *pass* the test and $g(p(x), y) = 1$. Otherwise, we set $g(p(x), y) = 0$ and say that p *fails* (x, y). In this paper, we assume that interaction outcomes are binary, i.e., $g : \mathcal{P} \times \mathcal{T} \to \{0, 1\}$, though in general various *degrees* of passing tests could be considered (for instance by grading them according to the similarity of the actual and desired output).

We gather the outcomes of interactions in an *interaction matrix* G. For a population of m programs and n tests in T, G is an $m \times n$ matrix where g_{ij} is the outcome of interaction between the ith program p_i and jth test t_j. The conventional GP fitness that rewards a program for the number of passed tests can be then written as

$$f(p_i) = \sum_{j=1}^{n} g_{ij}, \tag{1}$$

or alternatively as

$$f(p) = |\{t \in T : g(p, t) = 1\}|. \tag{2}$$

As it follows from the above, *all* elements of G need to be calculated in order to assess fitness values of all programs in P. Therefore, mn program executions are required in every generation of a GP run.

3   Factorization of Interaction Matrix

The motivation behind all methods that aim at reducing the number of program-test interactions is the potential redundancy of interaction matrix. The simplest form of redundancy is test duplication: though we referred above to T as a *set*, it is in practice usually implemented as a *list*, so duplicates are allowed.

Redundancy may also manifest when tests are different but all (or many) programs behave identically on them (in terms of passing or failing). Consider the task of synthesizing a sorting program, where tests are pairs (x, y) of lists and y is the sorted version of x. If the programming language of consideration does not allow any other operation on list elements than comparisons, then once a program passed a specific test of sorting a list of length, say, four, it will pass all other tests with the same permutation of four elements.

The SFIMX method proposed here aims at more subtle type of redundancy, i.e., when the value of the interaction of a program p with a given test t can be reconstructed from the responses of p and other programs in a population to t and other tests. More precisely, reconstructed by means of linear combinations of interaction outcomes. To this aim, we apply the well-known technique of matrix factorization (MF).

Formally, given an non-negative matrix G (interaction matrix in our case) and a desired rank $k \ll min(m, n)$, non-negative matrix factorization (NMF) [1] searches for non-negative matrices (*factors*) W and H that give a lower rank approximation of G as:

$$G \approx WH \quad s.t. \ W, H \geq 0, \tag{3}$$

where $W \in \mathbb{R}^{m \times k}$ is traditionally called *weights matrix* (or *basis matrix*) and $H \in \mathbb{R}^{k \times n}$ is *feature matrix*. Note that each test $t \in T$ is associated with a column in H (a vector $h_t \in \mathbb{R}^k$) and each program $p \in P$ is associated with a row in W (a vector $w_p \in \mathbb{R}^k$). For clarity, we abuse the notation and index the elements, rows, and columns of matrices with programs p and tests t.

The problem given by Eq. 3 is commonly reformulated as the following optimization problem:

$$\min_{W,H} \ f(W, H) \equiv \frac{1}{2} ||G - WH||_F^2 \quad s.t. \ W, H \geq 0, \tag{4}$$

where $|| \cdot ||_F$ is the Frobenius norm.

In the simplest scenario, MF model is trained by fitting to the observed interaction outcomes in G. Notice that if G's rank is $\leq k$, there exists an exact solution to (3). However, as it will become clear in a moment, our goal is to generalize in a way that allows predicting unknown interaction outcomes. Thus, caution should be exercised to avoid overfitting the observed data in G. A common extension of the basic MF formula that addresses this issue is *regularization*, which can be implemented by adding a parameter λ and modifying the squared error objective function:

$$\min_{W,H} \ f(W, H) \equiv \frac{1}{2}\|G - WH\|_F^2 + \lambda(\|W\|_F^2 + \|H\|_F^2) \ \ s.t. \ \ W, H \geq 0. \qquad (5)$$

The minimization problem given by (5) is not convex in both W and H simultaneously, however it is convex in either W or H. Thus, by keeping one matrix constant, the other can be found with a simple least squares computation. This strategy is widely known as alternating least squares [25]. Expression (5) can also be minimized using stochastic gradient descent, however the most popular approach to solve this optimization problem is the multiplicative update algorithm [18], which alternates the following two steps:

$$w_{pk} \leftarrow w_{pk} \frac{(GH^T)_{pk}}{(WHH^T)_{pk}} \qquad (6)$$

$$h_{kt} \leftarrow h_{kt} \frac{(W^T G)_{kt}}{(W^T WH)_{kt}} \qquad (7)$$

In each iteration, the new values of W and H are found by multiplying the current one by a factor that depends on the quality of approximation in (3). The quality of approximation improves monotonically with the application of the above rules [18]. The update rules are applied for a fixed number of iterations or until the error given by the left-hand side of (5) is sufficiently small.

As it follows from (3), to predict an interaction outcome of a program p with a test t from the matrices W and H found by solving (5), we calculate the dot product of two vectors corresponding to p and t:

$$\hat{g}_{pt} = w_p^T h_t = \sum_{k=1}^{k} w_{pk} h_{kt} \qquad (8)$$

Crucially for SFIMX, G can be factorized in the above way *even if some of its elements are unknown*, i.e., when G is *sparse*. This makes matrix factorization a powerful tool in machine learning, where it can be used to fill in the gaps in a large matrix (of, e.g., users' recommendations [12]) even if only small part of that matrix is known for certain. However, the update rules given by (6) and (7) implicitly assume that the input matrix is complete. In order to make them work for sparse matrices, a small modification must be introduced so that the unobserved outcomes in G are masked by zeros and ignored during training of the NMF model. Let M be a binary mask where $m_{pt} = 1$ if g_{pt} is known and $m_{pt} = 0$ otherwise. Then the update rules for so-called *weighted non-negative matrix factorization* (WNMF) [20] become:

$$w_{pk} \leftarrow w_{pk} \frac{((M \odot G)H^T)_{pk}}{((M \odot (WH))H^T)_{pk}}, \qquad (9)$$

$$h_{kt} \leftarrow h_{kt} \frac{(W^T(M \odot G))_{kt}}{(W^T(M \odot (WH)))_{kt}}, \qquad (10)$$

where $\odot$ is the Hadamard (element-wise) product.

4   The SFIMX Algorithm

Based in observations made in the previous section, we propose a method dubbed Surrogate Fitness via Factorization of Interaction Matrix (SFIMX). The method expects two parameters: the factorization rank k and desired density $\alpha \in (0, 1]$ of partial interaction matrix. SFIMX employs the NMF formalisms described in Sect. 3 to replace the conventional fitness evaluation stage of GP algorithm with the following steps:

1. Calculate *in part* the sparse interaction matrix G between the programs from the current population P and the tests from T in the following way:
   (a) For each program p, draw a nonempty random subset of tests $T' \subset T$ of size $\alpha|T|$ to interact with, where $\alpha \in (0, 1]$ is the parameter that controls the fraction of interactions to be calculated.
   (b) Apply p to tests in T', placing the interaction results in the appropriate cells of the corresponding row of G.
   (c) Fill in the remaining (missing) entries in G with zeros.
2. Factorize G in non-negative components W and H using the multiplicative update algorithm ((9) and (10)).
3. Use the obtained matrices to reconstruct the interaction outcomes by calculating $\hat{G} = WH$.
4. Compute from $\hat{G}$ the fitness of each program $p \in P$ using the conventional formula (1), by substituting g_{ij}s with the values taken from $\hat{G}$, i.e., $\hat{g}_{ij}$s.

For the purpose of the above algorithm it is mandatory to redefine the original interaction function $g(p, t)$ defined in Sect. 2, because zero is reserved for missing interaction outcomes. We assume that g returns 1 if p fails (x, y) and 2 if p solves (x, y). Note also that $\alpha \geq \frac{1}{|T|}$ must hold for T' to be nonempty.

Example. Consider population of programs $P = \{p_1, p_2, p_3, p_4\}$ and the population of tests $T = \{t_1, t_2, t_3, t_4, t_5\}$. Assume that SFIMX is run with $\alpha = \frac{3}{5}$ and yields the following sparse matrix of interactions G between P and T:

$$G = \begin{array}{c} \\ p_1 \\ p_2 \\ p_3 \\ p_4 \end{array} \begin{array}{ccccc} t_1 & t_2 & t_3 & t_4 & t_5 \\ \left(\begin{array}{ccccc} 2 & & 1 & 2 & \\ & 2 & 1 & & 1 \\ 1 & & & 2 & 2 \\ 2 & 1 & & & 1 \end{array} \right) \end{array}$$

Let $k = 3$. In step 2 of SFIMX, application of 50 iterations of the multiplicative update algorithm to G results in the following factorization:

$$
W = \begin{array}{c} \\ p_1 \\ p_2 \\ p_3 \\ p_4 \end{array}
\begin{array}{ccc} f_1 & f_2 & f_3 \\ \left(\begin{matrix} 0.46 & 1.96 & 0.6 \\ 1.27 & 0.1 & 0.95 \\ 1.37 & 0.02 & 2.83 \\ 0.4 & 1.86 & 1.60 \end{matrix}\right) \end{array},\quad
H = \begin{array}{c} \\ f_1 \\ f_2 \\ f_3 \end{array}
\begin{array}{ccccc} t_1 & t_2 & t_3 & t_4 & t_5 \\ \left(\begin{matrix} 0.48 & 1.50 & 0.01 & 0.41 & 0.41 \\ 0.87 & 0.14 & 0.19 & 0.77 & 0.01 \\ 0.11 & 0.09 & 1.02 & 0.50 & 0.51 \end{matrix}\right) \end{array}.
$$

When multiplied (step 3 of SFIMX), W and H lead to the following reconstructed interaction matrix:

$$
\hat{G} = WH = \begin{array}{c} \\ p_1 \\ p_2 \\ p_3 \\ p_4 \end{array}
\begin{array}{ccccc} t_1 & t_2 & t_3 & t_4 & t_5 \\ \left(\begin{matrix} 2 & 1.02 & 1 & 2 & 0.52 \\ 0.8 & 2 & 1 & 1.07 & 1 \\ 1 & 2.31 & 2.1 & 2 & 2 \\ 2 & 1 & 2.01 & 2.4 & 1 \end{matrix}\right) \end{array}
$$

Finally, in step 4, we calculate the fitness of particular programs by summing the corresponding rows of the reconstructed interaction matrix, which results in $f(p_1) = 6.54$, $f(p_2) = 5.87$, $f(p_3) = 9.41$, and $f(p_4) = 8.41$. Overall, SFIMX enabled calculating these values using $\alpha|T||P| = 12$ known interaction outcomes, compared to $|T||P| = 20$ interactions required by the conventional method. ■

In the above example, the reconstructed matrix $\hat{G}$ perfectly reproduces the known interaction outcomes, so the square approximation error (5) attains zero. This is guaranteed to happen when $k \geq rank(G)$. In general the approximation error will have the tendency to be greater for smaller values of k and greater values of α.

Properties of SFIMX. Predictions made by the method are based on how similar programs interact with the tests in T. The similarity in behavior of two programs is calculated based on the similarity in the outcomes of interactions with certain tests. Missing interaction outcomes are predicted based on the feedback from other programs and tests in the population.

As a result, evaluation in SFIMX is *contextual*: prediction $\hat{g}_{pt}$ made for a missing outcome depends not only on corresponding p and t but also on other programs in P and other tests in T. All available outcomes of interactions between programs in P and tests in T together determine the MF model and therefore influence how the predictions for missing outcomes are made. As the programs evolve with time, so does the model. Therefore, SFIMX performs NMF anew with each generation to model the missing interaction outcomes.

By factorizing interaction matrix G, the programs and the tests are projected into a reduced latent spaces that capture their most salient abstract features. The weight matrix H has one column for every abstract feature and one row for every program, and maps the features to the programs. The values in W state how much each feature *applies* to each program. Feature matrix H, on the other hand, has a row for each abstract feature and a column for each test. Every value in H indicates the extent to which a test possesses an abstract feature.

Interestingly, NMF with the least squares objective (Eq. 4) is characterized by an inherent clustering property, i.e., it clusters the columns of interaction matrix G. If additional orthogonality constraint on H is added, i.e., $HH^T = I$, then the minimized objective is equivalent to the one of K-means clustering (except for the non-negativity constraint), i.e., the sum of square of distances from clusters' centroids. In such a case, NMF can be viewed as a relaxed form of K-means where the matrix W contains non-negative cluster centroids and the elements of H are cluster membership indicators. This convergence helps understand how the problem of finding similar programs is internally tackled by NMF. It also reveals certain similarities to the recently proposed DOC algorithm, which we touch upon in the following review of related work.

5   Related Work

The values calculated by SFIMX can be treated as a *surrogate fitness*. Also known as *approximate fitness function* or *response surface* [10], a surrogate fitness function provides a computationally cheaper approximation of the original objective function. Surrogates are particularly helpful in domains where evaluation is computationally expensive, e.g., when it involves simulation. They usually rely on simplified models of the process being simulated, hence yet another alternative name: *surrogate models*. In continuous optimization, such models are typically implemented using low-order polynomials, Gaussian processes, or artificial neural networks. Occasionally, surrogate models have been also used in GP. For instance, in [8], Hildebrandt and Branke proposed a surrogate fitness for GP applied to job-shop scheduling problems. A metric was defined that reflected the behavioral similarity between programs, more specifically how the programs rank the jobs. Whenever an individual needed to be evaluated, that metric was used to locate its closest neighbor in a database of historical candidate solutions and neighbor's fitness was used as a surrogate.

Several other studies in GP attempted to reduce the number of programs' evaluations. An arguably simplest approach is to draw a subset of tests $T' \subset T$ and allow the programs interact only with them. This approach was also investigated in the context of evolutionary algorithms, where it is known as Random Subset Selection (RSS) [3]. Apart from speeding up the evolution, the motivation is that programs that perform well on various different subsets might have captured essential knowledge to generalize to all tests in T. Random selection of tests has been shown to improve the success rate and reduce overfitting [6].

SFIMX redefines fitness function. Several other methods proposed in the past in GP do that too, albeit usually not in terms of linear algebra. The arguably oldest approach of this type is implicit fitness sharing introduced by Smith *et al.* [26] and further explored for genetic programming by McKay [21, 22]. IFS lets the evolution assess the difficulty of particular tests and *weighs* the rewards granted for solving them. In this sense, IFS treats tests as limited resources: programs *share* the rewards for solving particular tests, each of which can vary from $\frac{1}{|P|}$ to 1 inclusive. Higher rewards are provided for solving tests that are rarely solved by population members, while importance of tests that are easy is diminished. The assessed difficulties of tests change with evolution, which can help escaping local minima and diversifies population. Diversification maintenance was also the main motivation for the recent lexicase selection algorithm [7], that avoids aggregating interaction outcomes altogether and differentiates programs by comparing them on randomly selected tests.

Another method that aims at scrutinizing the individual outcomes of programs' interactions and leveraging them for better performance is DOC [15]. In every generation, the algorithm identifies the groups of tests on which the programs in the current population behave *similarly* and clusters them together to give rise to new search objectives. Typically, a few such objectives emerge from this process, each of which is intended to capture a subset of 'capabilities' exhibited by the programs in the context of other individuals in population. The newly derived objectives replace then the conventional fitness function are used to drive the selection process. DOC is inspired by previous work in coevolutionary algorithms and test-based problems in [19].

Relying on binary interaction outcomes that only state whether a given test has been passed by a program or not stays in close resemblance to *test-based problems* originating in coevolutionary algorithms [2,5]. In test-based problems, candidate solutions interact with multiple environments – tests. Typically, the number of such environments is very large, making it infeasible to evaluate candidate solutions on all of them. Depending on problem domain, tests may take on the form of, e.g., opponent strategies (when evolving a game-playing strategy) or simulation environments (when evolving a robot controller). Solving a test-based problem requires a learning algorithm to generalize from a sample of tests. Similarly in GP, a synthesized program is expected to generalize beyond the training set and tests often do not enumerate all possible program inputs.

Last but not least, there are certain connections between SFIMX and semantic GP [24] and behavioral [14,16,17] GP methods that define program semantics as the vector of outputs produced by a program for particular tests. From the viewpoint of SFIMX, a single row in an interaction matrix is the outcome of confronting program's semantics with the vector of desired outputs. Recent years have seen a large number of contributions that employ this characterization of program behavior to design new initialization, search, and selection operators [23]. However, those methods are in general not designed to redefine search objectives, which is the primary goal of SFIMX.

6   Experimental Verification

We examine the performance of SFIMX in the domain of tree-based GP. All compared methods implement generational evolutionary algorithm and share the same parameter settings, with initial population of size $|P| = 1000$ filled with the ramped half-and-half operator, subtree-replacing mutation engaged with probability 0.1, subtree-swapping crossover engaged with probability 0.9, and tournament of size 7 in the selection phase. The fitness of each program $p \in P$ is computed using (1). Search lasts up to 200 generations and stops when the assumed number of generation elapses or an ideal program is found; the latter case is considered a success.

Compared Algorithms. We are interested in verifying whether SFIMX is a viable method for reducing the computational cost incurred by evaluation. For that aim, we control the fraction of interactions to be calculated by the parameter $\alpha \in \{0.1, 0.2, \ldots, 1.0\}$ in SFIMX algorithm in Sect. 4. By reducing in each generation the number of interactions by a factor of $(1 - \alpha)$, we spare $(1-\alpha)|P||T|$ interactions per run. We investigate what can be gained by investing these savings in increased population size: we increase the population size by the factor of $(1 - \alpha)$, so that it holds $|P| + (1 - \alpha)|P| = (2 - \alpha)|P|$ individuals. Therefore, population size does not change at all when $\alpha = 1.0$, while for α close to 0 it is almost doubled. Nevertheless, the overall computational budget is the same for all configurations and amounts to $1,000|T|$ interactions per generation and thus $200,000|T|$ interactions per run. This holds for all of the compared algorithms, including the control setups.

We consider three settings of factorization rank k that controls the degree to which the interaction outcomes are being compressed by factorization. The configuration dubbed **SFIMX-full** uses $k = \min(|P|, |T|)$, which is equivalent here to $k = |T|$, because for the considered benchmarks $|P| > |T|$. This value should be considered large, as NMF can then perfectly reproduce the known interaction outcomes, because the rank of G can be at most $\min(|P|, |T|)$.

The **SFIMX-half** configuration uses $k = |T|/2$, which forces the interaction outcomes to be compressed in half the number of weights in matrix W and features in matrix H. However, this number can be still considered quite high, given that we expect the interaction outcomes to be mutually correlated between program and tests.

Finally, the configuration **SFIMX-log** uses the smallest rank $k = \lceil \log_2 |T| \rceil$. In this case, k is in the order of the number of input variables; for instance, for the Mux6 problem $k = \log_2 2^6 = 6$.

The factorization is realized by the WNMF algorithm ((9) and (10)). The regularization factor λ is set to 0.01, as suggested by the common practice. When invoked for a given sparse interaction matrix G, we let WNMF perform up to 50 iterations, each involving both steps, i.e., (9) and (10). If the approximation error (the left-hand size of (5)) drops below 10^{-5}, we stop the optimization earlier. The computational overhead of running WNMF is on average 19 percent of the time spent evaluating programs in the population.

We confront SFIMX with several control setups. The first baseline is the conventional Koza-style **GP** [13]. The second control configuration, dubbed RSS, calculates fitness using $\alpha|T|$ randomly selected tests. The subset of tests is drawn anew in every generation of evolutionary run. We refer to this method as Random

Subset Selection (**RSS**), based on its similarity to an evaluation scheme known in coevolutionary algorithms [3].

Benchmark Problems. SFIMX and the multiplicative update algorithm it involves can in principle factor an arbitrary non-negative interaction matrix G and then reconstruct its approximation $\hat{G}$. However, obtaining good reconstructions for arbitrarily large interaction outcomes might be difficult, and such unconstrained outcomes can be expected for symbolic regression, where they are based on arbitrarily large errors committed by programs on test (not mentioning the possibility of programs returning infinity). Also, the raw interaction outcomes for symbolic regression problems are signed (the difference between the real-valued actual and desired output) and as such would require a well-justified mapping to positive numbers. For these reasons, in this study we limit our interest to problems with discrete interaction outcomes.

The first group are Boolean benchmarks, which employ instruction set {*and, nand, or, nor*} and are defined as follows. For an v-bit comparator $Cmp\,v$, a program is required to return *true* if the $\frac{v}{2}$ least significant input bits encode a number that is smaller than the number represented by the $\frac{v}{2}$ most significant bits. In case of the majority $Maj\,v$ problems, *true* should be returned if more that half of the input variables are *true*. For the multiplexer $Mul\,v$, the state of the addressed input should be returned (6-bit multiplexer uses two inputs to address the remaining four inputs). In the parity $Par\,v$ problems, *true* should be returned only for an odd number of *true* inputs.

The second group of benchmarks are the algebra problems originating from Spector *et al.*'s work on evolving algebraic terms [27]. These problems dwell in a ternary domain: the admissible values of program inputs and outputs are $\{0, 1, 2\}$. The peculiarity of these problems consists of using only one binary instruction in the programming language, which defines the underlying algebra. For instance, for the a_1 algebra, the semantics of that instruction is defined as follows:

a_1	0 1 2
0	2 1 2
1	1 0 0
2	0 0 1

In the following, the employed algebra is indicated by the suffix the name of term to be evolved. See [27] for the definitions of the remaining four algebras. For each of the five algebras considered here, we consider two tasks (of four discussed in [27]). In the *discriminator term* tasks (*Dsc* in the following), the goal is to synthesize an expression that accepts three inputs x, y, z and is semantically equivalent to the one shown below:

$$t^A(x, y, z) = \begin{cases} x & if\ x \neq y \\ z & if\ x = y \end{cases} \tag{11}$$

There are thus $3^3 = 27$ fitness cases in these benchmarks. The second tasks (*Mal*), consists in evolving a so-called *Mal'cev term*, i.e., a ternary term that satisfies the equation:

$$m(x, x, y) = m(y, x, x) = y \tag{12}$$

This condition specifies the desired program output only for some combinations of inputs: the desired value for $m(x, y, z)$, where x, y, and z are all distinct, is not determined. As a result, there are only 15 fitness cases in our *Mal* tasks, the lowest of all considered benchmarks. The motivation for the discriminator and Mal'cev term problems is originally that they're of interest to mathematicians [4]. In this paper, however, we chose them as benchmarks because of their difficulty and formal elegance.

Table 1. Success rate ($\times 100$) of best-of-run individuals, averaged over 30 evolutionary runs. Bold marks the best result for each benchmark.

α	Cmp6	Cmp8	Maj6	Mux6	Par5	Dsc1	Dsc2	Dsc3	Dsc4	Dsc5	Mal1	Mal2	Mal3	Mal4	Mal5
						SFIMX-full ($k = |T|$)									
1.0	50	0	62	**100**	0	7	3	33	0	3	90	63	87	23	93
0.9	37	3	50	**100**	3	0	13	30	0	17	97	63	77	33	93
0.8	50	7	70	**100**	10	3	17	50	0	10	90	77	83	33	**100**
0.7	70	0	73	**100**	7	3	7	57	0	3	87	90	97	53	87
0.6	77	3	80	**100**	**17**	3	13	67	0	17	93	83	97	40	97
0.5	73	3	**83**	**100**	0	10	3	67	0	13	97	77	97	60	93
0.4	**83**	7	80	**100**	3	7	10	**80**	0	17	90	93	77	53	97
0.3	73	0	77	**100**	0	**17**	13	67	0	**23**	90	**100**	**100**	53	**100**
0.2	70	0	57	**100**	0	13	3	73	0	13	73	93	93	53	**100**
0.1	25	0	8	**100**	0	10	7	40	0	0	60	57	77	33	87
						SFIMX-half ($k = |T|/2$)									
1.0	40	0	40	**100**	0	0	10	20	0	10	**100**	50	80	0	90
0.9	53	3	60	**100**	3	3	3	50	0	0	87	63	73	17	87
0.8	57	10	67	**100**	3	0	3	40	0	7	83	63	83	30	90
0.7	77	0	80	**100**	7	3	7	57	0	0	83	70	87	30	90
0.6	80	7	80	**100**	3	0	0	63	0	3	87	90	97	47	**100**
0.5	77	7	77	**100**	0	3	3	70	0	13	83	90	90	**67**	93
0.4	77	10	63	**100**	3	7	13	77	3	13	97	80	93	60	93
0.3	63	0	60	**100**	0	7	**33**	67	0	10	73	80	90	40	93
0.2	50	3	37	**100**	0	7	7	43	0	3	63	77	83	53	97
0.1	17	0	3	73	0	0	0	13	0	0	50	50	63	20	90
						SFIMX-log ($k = log_2 |T|$)									
1.0	27	0	47	**100**	0	0	3	13	0	3	83	47	70	7	73
0.9	47	3	70	**100**	0	0	3	27	0	0	90	40	67	13	83
0.8	50	0	60	**100**	0	0	3	23	0	3	83	53	67	7	77
0.7	70	0	47	**100**	7	7	7	37	0	3	80	50	80	17	93
0.6	57	0	60	**100**	3	7	7	30	0	7	87	67	87	40	97
0.5	70	0	73	**100**	0	0	0	30	0	3	87	63	97	30	90
0.4	73	3	77	**100**	0	3	3	60	0	17	90	80	90	47	97
0.3	67	0	70	**100**	0	7	13	73	0	10	93	83	90	50	**100**
0.2	43	0	43	**100**	0	3	7	57	0	3	70	83	83	57	97
0.1	0	0	0	93	0	0	0	3	0	0	27	20	37	7	73
						RSS									
0.9	65	0	55	95	5	0	0	32	0	0	88	58	62	12	85
0.8	68	0	55	95	0	0	2	35	0	0	82	58	78	15	88
0.7	62	0	65	95	0	0	0	42	0	0	85	55	85	18	88
0.6	68	2	52	92	2	0	0	32	0	5	68	65	82	8	88
0.5	65	0	45	95	0	0	0	45	0	0	68	48	78	8	85
0.4	52	0	55	95	0	0	0	45	0	0	68	48	78	25	85
0.3	58	0	65	92	0	0	0	35	0	0	85	65	82	22	95
0.2	42	0	45	95	0	0	0	35	0	0	62	55	58	22	82
0.1	15	0	0	95	0	0	0	15	0	0	62	45	55	8	85
						GP									
	63	3	63	**100**	3	0	0	30	0	3	93	54	63	27	93

Performance. Table 1 reports the success rates of particular algorithms, resulting from 30 runs of each configuration on every benchmark. To provide an aggregated perspective on performance, we employ the Friedman's test for multiple achievements of multiple subjects [11]. We first determine the best performing configuration within each method. For SFIMX-full and SFIMX-half, the configurations with $\alpha = 0.4$ fare the best, while for SFIMX-log and for RSS $\alpha = 0.3$ is most advantageous. The Friedman test applied to those configurations leads to the following ranking:

SFIMX-half-04	SFIMX-full-04	SFIMX-log-03	GP	RSS-03
2.07	2.13	2.67	**3.90**	**4.23**

The p-value for Friedman test is $\ll 0.001$, which strongly indicates that at least one method performs significantly different from the remaining ones. We conducted post-hoc analysis using symmetry test [9]: bold font marks the methods that are outranked at 0.05 significance level by SFIMX-half-04.

For additional insight, we also ranked all considered configurations for all individual values of α. The best overall average rank of 7.57 was achieved by SFIMX-half-04. Eight out of ten SFIMX-half configurations ranked before any of the control configurations; only SFIMX-half with $\alpha = 0.1$ and $\alpha = 1.0$ ranked behind GP and some RSS setups (average ranks 32.47 and 24.53, respectively). GP attained average rank 21.43.

SFIMX clearly outperforms the other methods. Its average ranks are better than the ranks of control configurations, albeit not so for the logarithmic variant SFIMX-log. That last fact is not surprising, given that SFIMX-log uses roughly an order of magnitude fewer weights and features than SFIMX-full and SFIMX-half. Nevertheless, SFIMX-log still delivers decent performance and for its preferred setting $\alpha = 0.3$ surpasses GP and RSS on most benchmarks. This corroborates our hypothesis that the interaction outcomes are significantly correlated and lend themselves to high compression without affecting the overall performance of the method. This result is particularly appealing, as low k implies low computational overhead of factorization: for SFIMX-log, it amounts only to approximately 6 percent of the total cost of calculating the $1,000|T|$ program-test interactions.

On the other hand, there are no significant differences in performance between SFIMX-full and SFIMX-half. Apparently the relatively high rank of the resulting matrices makes it possible to model the interaction outcomes sufficiently well in both these cases.

The success rates of SFIMX for individual benchmarks are always the best among the considered methods – see the values marked in bold in Table 1. For SFIMX-half, the SFIMX variant that overall fares the best, for three problems (Mux6, Mal1, and Mal5) there is at least one setting of α that makes SFIMX succeed systematically, i.e., in every run (success rate 100). In that respect, it is equaled only by GP and only on the Mal5 problem.

SFIMX performs also well in qualitative terms. It manages to produce solutions for all problems, while GP never solves Cmp8, Dsc1, Dsc2, Dsc4 and Dsc5, and RSS never solves Dsc1 and Dsc4, and hardly ever solves Cmp8, Dsc2 and Dsc5. On those hard problems, SFIMX is in most cases remarkably resistant to the setting of α: for Cmp8 and Dsc1, it succeeds for most values of α in the range $[0.2, 0.9]$, and for Dsc5 for $\alpha \in [0.2, 1.0]$. The only exception is Dsc4 where it managed to solve the problem only for $\alpha = 0.4$, and only once in 30 runs.

As a rule of thumb, we may say that setting α in $[0.3, 0.7]$ is favorable. However, using other values is not very detrimental. For many problems SFIMX maintains decent success rates even for very low setting of this parameter; for instance SFIMX-half is better than or comparable to GP for $\alpha = 0.2$ on Cmp8, Mux6, Dsc1, Dsc2, Dsc3, Dsc5, Mal2, Mal3, Mal4, and Mal5. For some benchmark, it still works quite well even for $\alpha = 0.1$. This is impressive, given that the interaction matrix is reconstructed there from only 10 percent of actual outcomes of program-test interactions.

7   Conclusions and Future Work

In conclusion, we find the idea of reconstructing interaction outcomes via factorization of sparse interaction matrix both conceptually appealing and useful in practice. SFIMX is straightforward, founded on solid mathematical underpinnings, and performs well for a broad range of values of parameters α and k. We assumed here that the algorithm spends the spared evaluation cycles on additional programs in extended population. Obviously, nothing precludes other designs, i.e., extending evolution with additional generations or simply completing a run in a shorter time.

Applicability of SFIMX reaches beyond GP. In general, interaction matrices produced in any test-based problems can be subject to the proposed processing. This applies in particular to interactive domains typically solved with competitive coevolution algorithms. Examples include two-player games, evolution of robot controllers, and abstract problems like density classification task, a classical problem in cellular automata.

In the form presented in this paper, SFIMX deliberately discards certain interactions. However, it might be used in scenarios where G is sparse due to other, more objective and external reasons. The most obvious example are the problems with an infinite or very large number of tests. Many control problems belong to this category. Even in the discrete domains like artificial ant or density classification task, the numbers of possible environments (or initial conditions) are often astronomical, not mentioning the continuous domain with problems like inverted pendulum. In such problems, tests (environments) can be generated on demand, and the interaction function performs agent's simulation in an environment and is thus computationally costly. SFIMX's capability of filling in the missing interaction outcomes can be in such cases invaluable. This preliminary study can be extended in multiple directions. For instance, here we applied SFIMX to discrete domains only; its usefulness in continuous domains typical for symbolic regression is an open question. The required adaptation concerns mapping the – in general arbitrary large – continuous error to interaction outcomes.

We hypothesize that simple transformation with some squeezing function (e.g., sigmoidal function or hyperbolic tangent) may be appropriate for that purpose.

Acknowledgements. P. Liskowski acknowledges support from grant 2014/15/N/ST6/04572 funded by the National Science Centre, Poland.

K. Krawiec acknowledges support from grant 2014/15/B/ST6/05205 funded by the National Science Centre, Poland.

References

1. Berry, M.W., Browne, M., Langville, A.N., Pauca, V.P., Plemmons, R.J.: Algorithms and applications for approximate nonnegative matrix factorization. Comput. Stat. Data Anal. **52**(1), 155–173 (2007)
2. Bucci, A., Pollack, J.B., de Jong, E.: Automated extraction of problem structure. In: Deb, K., Tari, Z. (eds.) GECCO 2004. LNCS, vol. 3102, pp. 501–512. Springer, Heidelberg (2004)
3. Chong, S.Y., Tino, P., Ku, D.C., Xin, Y.: Improving generalization performance in co-evolutionary learning. IEEE Trans. Evol. Comput. **16**(1), 70–85 (2012)
4. Clark, D.M.: Evolution of algebraic terms 1: term to term operation continuity. Int. J. Algebra Comput. **23**(05), 1175–1205 (2013)
5. de Jong, E.D., Pollack, J.B.: Ideal evaluation from coevolution. Evol. Comput. **12**(2), 159–192 (2004)
6. Gonçalves, I., Silva, S., Melo, J.B., Carreiras, J.M.B.: Random sampling technique for overfitting control in genetic programming. In: Moraglio, A., Silva, S., Krawiec, K., Machado, P., Cotta, C. (eds.) EuroGP 2012. LNCS, vol. 7244, pp. 218–229. Springer, Heidelberg (2012)
7. Helmuth, T., Spector, L., Matheson, J.: Solving uncompromising problems with lexicase selection. IEEE Trans. Evol. Comput. **19**(5), 630–643 (2015)
8. Hildebrandt, T., Branke, J.: On using surrogates with genetic programming. Evol. Comput. **23**(3), 343–367 (2015)
9. Hollander, M., Wolfe, D.A., Chicken, E.: Nonparametric Statistical Methods, vol. 751. Wiley, New York (2013)
10. Jin, Y., Olhofer, M., Sendhoff, B.: A framework for evolutionary optimization with approximate fitness functions. IEEE Trans. Evol. Comput. **6**, 481–494 (2002)
11. Kanji, G.K.: 100 Statistical Tests. Sage, London (2006)
12. Koren, Y., Bell, R., Volinsky, C.: Matrix factorization techniques for recommender systems. Computer **8**, 30–37 (2009)
13. Koza, J.R.: Genetic Programming: On the Programming of Computers by Means of Natural Selection. MIT Press, Cambridge (1992)
14. Krawiec, K.: Behavioral Program Synthesis with Genetic Programming. Springer, Switzerland (2015)
15. Krawiec, K., Liskowski, P.: Automatic derivation of search objectives for test-based genetic programming. In: Machado, P., Heywood, M.I., McDermott, J., Castelli, M., García-Sánchez, S., Sim, K. (eds.) EuroGP 2015. LNCS, vol. 9025, pp. 53–65. Springer International Publishing, Switzerland (2015)
16. Krawiec, K., O'Reilly, U.: Behavioral programming: a broader and more detailed take on semantic GP. In: Proceedings of the 2014 Conference on Genetic and Evolutionary Computation, pp. 935–942. ACM (2014)

17. Krawiec, K., Solar-Lezama, A.: Improving genetic programming with behavioral consistency measure. In: Bartz-Beielstein, T., Branke, J., Filipič, B., Smith, J. (eds.) PPSN 2014. LNCS, vol. 8672, pp. 434–443. Springer, Heidelberg (2014)

18. Lee, D.D., Seung, H.S.: Algorithms for non-negative matrix factorization. In: Advances in Neural Information Processing Systems, pp. 556–562 (2001)

19. Liskowski, P., Krawiec, K.: Discovery of implicit objectives by compression of interaction matrix in test-based problems. In: Bartz-Beielstein, T., Branke, J., Filipič, B., Smith, J. (eds.) PPSN 2014. LNCS, vol. 8672, pp. 611–620. Springer, Heidelberg (2014)

20. Mao, Y., Saul, L.K.: Modeling distances in large-scale networks by matrix factorization. In: Proceedings of the 4th ACM SIGCOMM Conference on Internet Measurement, pp. 278–287. ACM (2004)

21. McKay, R.I.B.: Committee learning of partial functions in fitness-shared genetic programming. In: 26th Annual Conference of the IEEE Third Asia-Pacific Conference on Simulated Evolution and Learning 2000, Industrial Electronics Society, IECON, 22–28 October 2000, vol. 4, pp. 2861–2866. IEEE Press, Nagoya, Japan (2000)

22. McKay, R.I.B.: Fitness sharing in genetic programming. In: Whitley, D., Goldberg, D., Cantu-Paz, E., Spector, L., Parmee, I., Beyer, H.G. (eds.) Proceedings of the Genetic and Evolutionary Computation Conference (GECCO-2000), 10–12 July 2000, pp. 435–442. Morgan Kaufmann, Las Vegas (2000)

23. Moraglio, A., Krawiec, K.: Semantic genetic programming. In: Proceedings of the Companion Publication of the 2015 on Genetic and Evolutionary Computation Conference, pp. 603–627. ACM (2015)

24. Moraglio, A., Krawiec, K., Johnson, C.G.: Geometric semantic genetic programming. In: Coello, C.A.C., Cutello, V., Deb, K., Forrest, S., Nicosia, G., Pavone, M. (eds.) PPSN 2012, Part I. LNCS, vol. 7491, pp. 21–31. Springer, Heidelberg (2012)

25. Paatero, P., Tapper, U.: Positive matrix factorization: a non-negative factor model with optimal utilization of error estimates of data values. Environmetrics 5(2), 111–126 (1994)

26. Smith, R.E., Forrest, S., Perelson, A.S.: Searching for diverse, cooperative populations with genetic algorithms. Evol. Comput. 1(2), 127–149 (1993)

27. Spector, L., Clark, D.M., Lindsay, I., Barr, B., Klein, J.: Genetic programming for finite algebras. In: Keijzer, M. (ed.) Proceedings of the 10th Annual Conference on Genetic and Evolutionary Computation, GECCO 2008, 12–16 July 2008, pp. 1291–1298. ACM, Atlanta (2008)

Scheduling in Heterogeneous Networks Using Grammar-Based Genetic Programming

David Lynch[1]($\boxtimes$), Michael Fenton[1], Stepan Kucera[2],
Holger Claussen[2], and Michael O'Neill[1]

[1] Natural Computing Research and Applications Group, UCD, Dublin, Ireland
david.lynch.1@ucdconnect.ie
[2] Bell Laboratories, NOKIA, Dublin, Ireland

Abstract. Effective scheduling in Heterogeneous Networks is key to realising the benefits from enhanced Inter-Cell Interference Coordination. In this paper we address the problem using Grammar-based Genetic Programming. Our solution executes on a millisecond timescale so it can track with changing network conditions. Furthermore, the system is trained using only those measurement statistics that are attainable in real networks. Finally, the solution generalises well with respect to dynamic traffic and variable cell placement. Superior results are achieved relative to a benchmark scheme from the literature, illustrating an opportunity for the further use of Genetic Programming in software-defined autonomic wireless communications networks.

Keywords: Scheduling · Heterogeneous networks · Grammar-based genetic programming

1 Introduction

Traditional cellular infrastructure is under significant strain due to exponentially increasing demand [8]. The number of mobile-connected devices is now greater than the world's population and network traffic will grow tenfold by 2019 [3]. Low-powered antennas called Small Cells (SCs) have been proposed as a means of scaling existing deployments to meet these trends [4].

In traditional networks, high-powered Macro Cells (MCs) are distributed on hexagonal grids to provide blanket coverage to User Equipments (UEs). A UE could be a smartphone, tablet or laptop etc. Heterogeneous Networks (HetNets) are comprised of both SCs and MCs. By offloading UEs in traffic hotspots, SCs alleviate strain on the macro tier. Note that hotspots are regions containing a concentration of UEs. Multi-tiered networks exhibit several desirable properties. Firstly, SCs support ad-hoc deployment by operators. Secondly, they are a cost effective means of densifying networks. Finally, HetNets are spectrally efficient as both tiers share the same channel under the current 3rd Generation Partnership Project–Long Term Evolution (3GPP–LTE) framework [1].

Network operators must offer better quality of service than their competitors to attract and retain customers. In particular, they must maximise the data rates delivered by their networks. The metric that most strongly correlates with user experience is the downlink rate which quantifies the amount of data that can be transferred per unit time. Operators must maximise downlink rates for the least advantaged customers (due to location say), sometimes at the expense of the more privileged. Fairness is vital because dropped calls or slow data speeds are unacceptable from a customer satisfaction standpoint.

HetNets present unique challenges vis-à-vis optimisation because they are highly dynamic. In this paper, we employ a grammar-based form of Genetic Programming (GP) to evolve a HetNet scheduling heuristic. This is a difficult real world problem which to date has not been tackled with GP. Operators currently implement highly suboptimal greedy proportionally fair scheduling. Tailoring such methods to corner cases requires much human effort. These inefficiencies can be alleviated by evolving better software at a cost that is negligible compared to cell densification–deploying a single SC can cost several thousand euros.

The paper is organised as follows. Section 2 describes the problem in detail. Previous work is surveyed in Sect. 3. Our simulation environment is described in Sect. 4. Experiments, results and discussion follow in Sects. 5 and 6. Finally, the paper closes with future directions and conclusions in Sect. 7.

2   Problem Definition

The 3GPP–LTE framework outlines a number of high level protocols for making HetNets viable [1]. SCs are typically underutilised because UEs preferentially attach to stronger MCs. The Cell Range Expansion mechanism is proposed to encourage more efficient offloading from MCs. To achieve this, SCs broadcast a Cell Selection Bias (β_i) such that $\beta_i \geq 0$, $\forall i \in \mathcal{S}$, the set of all SCs. There is no need for MCs to implement bias so $\beta_i = 0$, $\forall i \in \mathcal{M}$, the set of all MCs. A UE (u) attaches to and hence receives data from cell k, where,

$$k = \arg \max_i \left(Signal_{u,i} + \beta_i \right), \ \forall i \in \mathcal{M} \cup \mathcal{S}. \tag{1}$$

If a UE attaches to cell $k \in \mathcal{S}$ but would otherwise attach to a MC $m \in \mathcal{M}$, then we say that the UE resides in the expanded region of k. Since the signal from m in the expanded region is by definition larger than that from k, it follows that interference is significant therein. Edge interference is exacerbated in LTE HetNets because MCs broadcast on the same channel as SCs.

Interference mitigation in the time domain is a defining feature of the 3GPP framework. UEs can receive data in 1 ms intervals referred to as subframes (SFs). A contiguous block of 40 SFs defines a 'frame'. Frames constitute a convenient timespan over which network performance can be analysed. The enhanced Inter-Cell Interference Coordination (eICIC) paradigm introduced the notion of Almost Blank Subframes (ABSs) to mitigate cell-edge interference [11]. MCs mute their data transmissions during an ABS so that only minimal control signals are broadcast, allowing neighbouring SCs to transmit with minimal interference. We refer to the sequence of active and muted SFs at MCs as an 'ABS pattern'.

Clearly, UEs at SC edges experience greatly reduced interference when nearby MCs undergo an ABS. However, UEs that are attached to the muting MC cannot receive any data during an ABS. Intelligent resource interleaving strategies are thus required to realise the benefits from eICIC. A key task in this regard is allocating SFs to SC attached UEs–hence, the scheduling problem.

Scheduling is trivial for UEs served by MCs because they enjoy high signal to interference and noise ratios ($SINRs$) and therefore can be allocated to every non-ABS SF. However, SC attached UEs are subjected to high MC interference. Shannon's formula [27] describes how downlink rates (R) depend on bandwidth and $SINR$:

$$R_{u,f} = \frac{B}{N_f} \times \log_2(1 + SINR_{u,f}), \tag{2}$$

where, B is the available bandwidth, N_f is the number of UEs scheduled in SF f and u denotes a UE. From Eq. 2, observe that $R_{u,f}$ is inversely proportional to N_f, where the downlink rate quantifies how much data can be transferred in a unit of time. Each UE will experience reduced rates in any given SF as it becomes more congested. Consequently, scheduling is a non-trivial problem because we would like to schedule each UE for as many SFs as possible but yet minimise per SF congestion.

SF \ UE	'6'	'4'	'7'	'2'	'8'	'9'
1	T	T	T	T	T	T
2	T	T	T	T	T	T
3	T	T	T	T	T	T
4	T	T	T	F	F	F
5	T	T	T	F	F	F
6	T	T	T	F	F	F
7	T	T	T	F	F	F
8	T	T	T	F	F	F

(Time →)

Fig. 1. Depiction of a SC schedule where rows represent SFs and columns store schedules. UE u receives data in all SFs indexed by 'T' in their schedule.

The non-trivial nature of the problem can be appreciated by visualising schedules in the form of a boolean matrix. Figure 1 describes a feasible set of schedules for a SC with six attached UEs (only 8 out of 40 SFs are displayed for concision). Here 'T' indicates that a UE will receive data in the corresponding SF, and 'F' implies the converse. For instance, UE 9 is scheduled in the first 3 SFs. By construction the illustrative schedules in Fig. 1 exhibit sub-optimal properties. SFs 1–3 are fully congested so the bandwidth is divided six-fold. SFs

4–8 are less congested and so UEs 6, 4 and 7 profit from the liberated bandwidth. However, the reduced congestion is at the expense of UEs 2, 8 and 9 because they receive less airtime. Clearly, this SC could employ a vast number of alternative strategies to allocate SFs, despite the fact that it only serves six UEs. We ask if GP can derive a heuristic to compose synergistic schedules on the fly. Our task is to populate a scheduling matrix like Fig. 1 for all SCs in the network.

HetNet control algorithms are typically evaluated using a proportional fair utility of user experience. The sum log of downlink rates, see for example [12, 22, 26, 29], is given by:

$$PF\ Utility = \sum_{u \in \mathcal{M} \cup \mathcal{S}} \log \overline{R}_u, \tag{3}$$

where,

$$\overline{R}_u = \frac{1}{|\mathcal{F}|} \sum_{f=1}^{|\mathcal{F}|} R_{u,f},$$

is the average downlink rate for UE u over $|\mathcal{F}| = 40$ SFs. Equation 3 rewards individuals that fairly allocate resources. Lifting the downlink rates for poorly performing UEs is heavily rewarded by the logarithm. Conversely, losses for the best performing UEs are not penalised severely. Therefore, the fitness function rewards solutions that 'rob from the rich and give to the poor'.

3   Previous Work

An extensive literature exists on scheduling. Such problems arise in domains of operations research ranging from rostering [14] and job shop scheduling [23] to air traffic control [16]. In general, the feasible solution space is explored directly via search-based techniques. However, heuristic rules that can compute solutions on the fly are often motivated by practical constraints.

Bader-El-Den and Fatima (2010) employed an auction inspired scheme for the exam time-tabling problem [6]. They evolved a 'bidding function' that exams use to bid for time-windows. Auctions are held for each available window until all exams have been allocated. We note that the evolved solution operates within the context of a meta-algorithm, in this case inspired by an auction.

Jakovocić and Marasović (2012) identified the limitations of enumerative and search-based techniques [21]. The combinatorial nature of scheduling problems renders a direct search of the solution space impractical when runtime must be minimised. Following the authors in [6], they manually designed meta-algorithms tailored to specific job-shop scheduling environments. Evolved priority functions operate within these meta-algorithms. Thus, domain knowledge informed the solution structure, lending GP a foothold to search for functional forms.

Sun et al. (2006) instrumented a game theoretic approach to allocate channel resources in a wireless network [28] but our literature review has uncovered no previous work addressing scheduling in HetNets using GP. This paper attempts to fill the gap. Evolutionary methods are indicated in this domain because they are known to yield good solutions in dynamic environments [13,31]. Ho and Claussen (2009) used GP to optimise the coverage of femtocell deployments in enterprise environments [20]. Femtocells are SCs with a range of several meters. Their study represented a proof of concept that it is possible to automatically evolve controllers for wireless networks. Hemberg et al. (2011-13) used Grammatical Evolution to evolve symbolic expressions for setting femtocell powers to optimise coverage [17–19]. The best solutions outperformed human designed heuristics on two of the three objectives.

A number of papers are relevant to our work in the space of eICIC optimisation. Weber and Stanze (2012) compared the performance of two scheduling strategies: strict and dynamic [30]. The former schedules centre UEs in non-ABSs so that ABSs are reserved for expanded region UEs. The latter allows edge UEs to receive both ABS and non-ABS airtime. Experiments showed that the dynamic scheduler achieves a better tradeoff between cell border rates and spectral efficiency.

Pang et al. (2012) proposed a scheduling method based on dynamic programming [26]. Synchronous patterns were assumed so that MCs mute in unison. Exactly two SCs were simulated per MC sector, with UEs uniformly distributed on the map. It is unclear whether their algorithm would perform well under more general conditions. Jiang and Lei (2012) modelled the scheduling problem as a two player Nash bargaining game where protected (ABSs) and normal (non-ABSs) resources at SCs compete for UEs [22]. Each 'player' strives to maximise the total data that it transmits. Performance was improved under the proposed algorithm relative to standard baselines. Edge UEs experienced comparable rates in the proposed and baseline cases. We will demonstrate that a GP evolved heuristic can give considerable gains for edge UEs.

Deb et al. (2014) formulated eICIC optimisation as a non-linear programming instance [12]. Their algorithm computes the airtime UEs should receive from their serving MC and SC during ABS and non-ABS periods. Simulation revealed that cell edge UEs gain the most under eICIC. The authors showed that their algorithm is within 90 % of the optimal but it requires measurement reports from each UE's best SC and MC. In practice UEs only communicate with their serving cell [24].

López-Peréz and Claussen (2013) proposed a heuristic to balance load (number of UEs) between ABSs and non-ABSs at SCs [24]. Load balancing improved the 5th percentile of SC attached UE rates by 55 %, in a scenario with fixed MC ratios and with non-zero biases on SCs. Reduced mean MC throughput under the proposed scheme, relative to the benchmark, was compensated by increased mean SC throughput. In sum, this paper demonstrated the considerable gains achievable with intelligent scheduling. We adopt López-Peréz and Claussen (2013) as a benchmark.

4 Simulation Environment

In order to rapidly evaluate solutions we simulated a HetNet serving $3.61 \, \mathrm{km}^2$ of Dublin City Centre. SCs are typically deployed in an ad-hoc fashion because they serve hotspots, whereas MCs are placed on a grid by network operators. As such, we scattered SCs randomly on the map and arrange MCs in a hexagonal pattern. Figure 3 displays a snapshot of the network used for fitness evaluations. A HetNet with 21 MCs, 50 SCs and 1250 UEs was simulated for training.

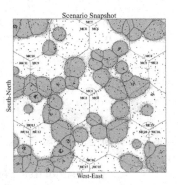

Fig. 2. Environmental encoding.

Fig. 3. SCs are shaded blue, MCs white and UEs are indicated by black dots (Color figure online).

4.1 Generating Inputs

The simulation proceeds sequentially. Firstly, an environmental encoding is generated from a Google Maps [2] image of the serviced region (Fig. 2). This encoding captures the distribution of buildings, bodies of water, open spaces and roads. A signal gain path loss matrix G is then computed for all cells. G models the cell gains, shadow fading and environmental obstacles, so that $G[i, x, y]$ represents the path loss from cell i to location $[x, y]$.

Next, UEs are distributed onto the map. Hotspots, 30 in total, are generated containing between 5 and 25 UEs. With probability 0.1 a hotspot will materialise outside of a SC but mostly they appear within SCs. If a UE is not assigned to a hotspot then it is placed at a random point on the map. A total of 1250 UEs are simulated or about 60 per MC sector.

The signal received by UE u from cell i depends on path loss such that:

$$Signal_{u,i} = P_i^{\mathrm{TX}} + G[i, x, y], \tag{4}$$

where, P_i^{TX} is the transmitting power of i in decibel milliwatts (dBm). SC transmit powers and Cell Selection Biases (β_i) are set by an evolved heuristic devised by the authors [15]. MC powers and biases are constant at $P_i^{\mathrm{TX}} = 37[\mathrm{dBm}]$ and $\beta_i = 0[\mathrm{dBm}], \forall i \in \mathcal{M}$. Hence, u can identify its serving cell using Eq. 1.

MC ABS ratios are set using the simple heuristic from [29]. The ratio of ABSs to non-ABSs for MC $m \in \mathcal{M}$ is established by:

$$ABS_r = \frac{|\mathcal{SC}_{expanded}|}{|\mathcal{SC}_{expanded}| + |\mathcal{A}_m|},\tag{5}$$

where, $|\mathcal{A}_m|$ is the number of UEs served by m and $\mathcal{SC}_{expanded}$ is the set of UEs that would attach to m but instead reside in the expanded regions of SCs within m's coverage area. Recall from Sect. 2 that the 'expanded region' contains those UEs that attach to SC i instead of m because β_i is non-zero. Equation 5 is sensible because if the number of expanded region UEs is large relative to $|\mathcal{A}_m|$, then m should surrender more SFs, thus mitigating cell-edge interference.

Each MC constructs a feasible muting pattern from the ABS ratio (Eq. 5) by combining eight base patterns from the standard [1]. Since five SFs are muted in each of the eight base patterns it follows that the ABS_r must be rounded to an element in the set $\{5/40, ..., 35/40\}$. Intra-frame $SINR$ variance is reduced by 'front-loading' muted SFs so that if $ABS_r = 10/40$ for $m \in \mathcal{M}$, then m will mute in the first two SFs for every block of eight, e.g. in SFs (1, 2, 9, 10, 17, 18, 25, 26, 33, 34). A MC cannot entirely mute or transmit $\forall f \in \mathcal{F}$.

The most important statistic from a scheduling standpoint is the $SINR$ that UEs experience in each SF. $SINR_{u,f}$ is computed by dividing $Signal_{u,serving}$ (the signal in Watts from u's serving cell) in SF f by, the sum of all interfering signals (from all other cells) plus noise. Note that $SINR_{u,f}$ depends on which MCs are muting or transmitting in SF f. The denominator will be reduced during protected SFs because $Signal_{u,m} = 0$ from MC m if it is undergoing an ABS. Therefore, $SINR_{u,f}$ varies over a frame due to the variable number of MCs that mute in different SFs. Our goal is evolve an expression that maps $SINR$ related statistics and attachment information to a binary decision for each UE per SF: schedule or don't schedule. The terminal set for GP is derived from the cell attachment information and $SINRs$ as annotated in Table 1.

4.2   Calculating Fitness

Algorithm 1 delineates the meta-algorithm (in the sense of [6,11]) used to yield schedules from an individual. The GP tree executes independently on each SC as follows. We loop over SFs and UEs, evaluating the tree at each (u, f) tuple. If the tree outputs a positive value then u will receive transmissions in SF f, else u is not scheduled in f. In this sense the tree performs a binary classification task on every execution.

The schedules are implemented in simulation and summary statistics on the realised downlink rates (accounting for congestion) are computed. Performance is expressed as the improvement in sum log downlink rates relative to a baseline strategy whereby u receives data in every SF f if $SINR_{u,f} \geq 1$. This baseline is naive because whilst airtime is maximised for each UE, so too is congestion. Recalling Eq. 3, the fitness function is expressed thusly:

Algorithm 1. Schedule UEs

```
function GETSTATISTICS(u, f)
    return Column 2 of Table 1 for UE u in SF f
procedure DOSCHEDULING(S, Tree)
    for SC ∈ S do                                          ▷ Process each SC independently
        congestion_f ← 0                                   ▷ Track congestion in SF f
        airtime_u ← 0                                      ▷ Track number of SFs received by u
        S ← 0_(|F|×|A|)                                    ▷ Stores SC schedule, see Fig. 1
        for f ∈ F do                                       ▷ F = {1, ..., 40} is the set of SFs
            for u ∈ A do                                   ▷ A stores the attached UEs
                inputs ← GETSTATISTICS(f, u)
                Output = evaluate(Tree(inputs))            ▷ Inputs are listed in Table 1
                if Output > 0 and SINR_{u,f} ≥ 1 then
                    S_{u,f} ← True                         ▷ u will receive data in f
                    congestion_f += 1
                    airtime_u += 1
                else
                    S_{u,f} ← False                        ▷ For the greater good sacrifice u
    return S
```

$$Fitness = \frac{1}{10} \sum_{s=1}^{10} \left(PF\ Utility_s^{tree} - PF\ Utility_s^{baseline} \right). \tag{6}$$

Equation 6 expresses overall fitness as the average performance over ten UE distributions, hereafter scenarios (s). Thus, we provision against overfitting on a single set of UE locations.

5 Experiments

We instrumented a grammar-based form of GP [9,25]. Grammars allow us to incorporate domain knowledge and guarantees that syntactically correct structures are generated. Figure 4 displays the function approximation type grammar used in Backus-Naur Form (BNF). Four non-linear transforms were admitted including 'step' which output -1 if its argument is less than 0, else +1. The logarithm and square root functions were protected via $\log(1 + |x|)$ and $\sqrt{|x|}$ respectively. Random floats in the set $\{-1.0, -0.9, ..., 1.0\}$, statistics on instantaneous rates and memory nodes (airtime and congestion) composed the terminal set. The $SINR$ statistics were mapped to instantaneous downlink rates (assuming no bandwidth splitting) as illustrated in Table 1. Note that num_viable is the number of SFs in which a UE can receive data without packet loss (i.e. when $SINR \geq 1$). By contextualising u relative to all attached UEs, we anticipated that GP would uncover cooperation strategies, whereby u sometimes sacrifices a particular SF f for the global objective.

Thirty independent runs were performed for 75 generations. The Ramped Half-and-Half method was used to initialise the population (pop size = 1000) with an initial max depth of 6. We used fair tournament selection (tournament size was 1 % of pop size) so that all individuals had a chance of getting selected. Subtree Crossover, Subtree Mutation and Point Mutation were used to search the space of derivation trees. Subtree crossover was applied with a probability of 0.5

Table 1. GP Terminal Set

Raw Input	Statistic	Terminal Name
SC attached UEs	$|\mathcal{A}|$	num_att
$SINR_{u,f}$	$\log_2(1 + SINR_{u,f})$	downlink
$SINR_{u,f}, \forall f \in \mathcal{F}$	$\frac{1}{|\mathcal{F}|} \sum_{f \in \mathcal{F}} \log_2(1 + SINR_{u,f})$	avg_downlink_frame
"	$\max_f \left\{ \log_2(1 + SINR_{u,f}) \right\}$	max_downlink_frame
"	$\min_f \left\{ \log_2(1 + SINR_{u,f}) \right\}$	min_downlink_frame
"	$\left| \left\{ SINR_{u,f} \geq 1 \right\} \right|_f$	num_viable
$SINR_{u,f}, \forall u \in \mathcal{A}$	$\frac{1}{|\mathcal{A}|} \sum_{u \in \mathcal{A}} \log_2(1 + SINR_{u,f})$	avg_downlink_SF
"	$\max_u \left\{ \log_2(1 + SINR_{u,f}) \right\}$	max_downlink_SF
"	$\min_u \left\{ \log_2(1 + SINR_{u,f}) \right\}$	min_downlink_SF
$SINR_{u,f}, \forall f \in \mathcal{F}, \forall u \in \mathcal{A}$	$\frac{1}{|\mathcal{A}|} \sum_{u \in \mathcal{A}} \left(\frac{1}{|\mathcal{F}|} \sum_{f \in \mathcal{F}} \log_2(1 + SINR_{u,f}) \right)$	avg_downlink_cell
"	$\max_u \left\{ \max_f \left\{ \log_2(1 + SINR_{u,f}) \right\} \right\}$	max_downlink_cell
"	$\min_u \left\{ \min_f \left\{ \log_2(1 + SINR_{u,f}) \right\} \right\}$	min_downlink_cell
Previous outputs of tree	#SFs received by current UE	airtime
Previous outputs of tree	# other UEs sharing SF	congestion

to each pair of selected parents. Sixty percent of the population was subjected to Subtree Mutation. The remaining forty percent underwent point mutation with probability of 0.05, 0.1, 0.2 or 0.3 per node. We used generational replacement and elitism with elite size equal to 1 % of pop size. A run took ten hours on a twelve core hyperthreaded machine operating at 2.66 GHz.

The search space for this problem admitted many local optima. Trees that output exclusively positive or negative values yielded trivial schedules whereby all UEs were always or never scheduled. To avoid local optimum we assigned

```
<expr> ::= <reg> | <reg> | <reg> | <Terminal>
<reg> ::= <arithmetic>(<expr>,<expr>) | <arithmetic>(<expr>,<expr>) |
          <arithmetic>(<expr>,<expr>) | <arithmetic>(<expr>,<expr>) |
          <non-linear>(<expr>) | <non-linear>(<expr>)
<arithmetic> ::= + | - | * | % (protected division)
<non-linear> ::= sin | log | sqrt | step
<Terminal> ::= U(-1, +1, 0.1) | U(-1, +1, 0.1) |
               U(-1, +1, 0.1) | U(-1, +1, 0.1) |
               num_viable | num_att |
               downlink |
               avg_downlink_frame | max_downlink_frame | min_downlink_frame
               avg_downlink_SF | max_downlink_SF | min_downlink_SF
               avg_downlink_cell | max_downlink_cell | min_downlink_cell
               airtime | congestion
```

Fig. 4. BNF Grammar Definition.

zero fitness to such trees. Runtime was reduced substantially, without degrading solution quality, by terminating the simulation early for trivial trees.

6   Results and Discussion

Figure 5 displays the mean best fitness, with a 95 % confidence interval about the mean (grey shading), across 30 runs for 75 generations. Note we are maximising the improvement in sum log rates–Eq. 6. Clearly, the evolved population has a significantly higher mean fitness than the initial random population. Evidently convergence is not achieved in only 75 generations, so better solutions are likely to emerge from longer runs. Figure 5 suggests that the GP system is stable because the variance about mean best fitness is low across many independent runs.

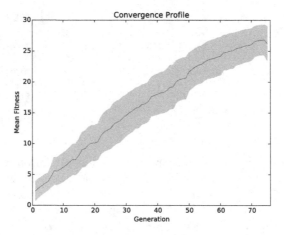

Fig. 5. Mean best fitness on training data including a 95 % confidence interval (Color figure online).

The best overall solution (see Fig. 6[1]) was identified by exposing all 30 best-of-run individuals to unseen test networks. The test fitness of each individual was computed as the average performance across 100 scenarios, in networks with 20, 60 and 100 SCs (each serving 1250 UEs). Most individuals performed well in one or two of the three test networks. A few performed very well across all three topologies. The best individual achieved a fitness of 55.5, 60.4 and 25.8 on the networks containing 20, 60 and 100 SCs respectively. These observations underscore the need for multiple runs despite the fact that we seek only one good solution. That some individuals struggle on one or two of the test networks but perform well on the other(s) may be indicative of overfitting. In future work the use of a validation network will enable intelligent termination if overfitting ensues.

---

[1] Note that the constants have been obfuscated to protect intellectual property.

$$S_{u,f} \leftarrow ((A + D/B) - C + F)$$

$A = (\sin(\log(\sin(\texttt{max\_downlink\_cell}))) + (\log(\texttt{downlink} + \texttt{airtime})/$

$\quad (\texttt{congestion} - (\texttt{congestion}/\texttt{max\_downlink\_SF}))))$

$B = \text{step}(((\text{step}(\texttt{num\_viable}) * \texttt{constant}))/\text{step}(\log(\texttt{max\_downlink\_cell})))$

$C = (\sin(\texttt{num\_viable} - (\text{step}(\texttt{min\_downlink\_frame}/\texttt{min\_downlink\_SF}) -$

$\quad \text{sqrt}(\texttt{avg\_downlink\_SF} + \texttt{avg\_downlink\_frame}))))$

$D = \text{step}(((\text{step}(\texttt{min\_downlink\_frame}) * \texttt{constant})))$

$E = (((\text{step}(\texttt{min\_downlink\_frame})) + ((\texttt{num\_viable} - \texttt{avg\_downlink\_frame})$

$\quad /\texttt{avg\_downlink\_cell})) * \texttt{max\_downlink\_SF})$

$\quad / \log(\texttt{downlink} + \texttt{downlink}) * \log(\texttt{downlink}))$

$F = (\texttt{avg\_downlink\_frame} - (((E/(\texttt{downlink} - \texttt{max\_downlink\_frame}))))$

Fig. 6. Best Evolved Scheduler. Terminals are distinguished by blue text.

The percentage lift in sum log rates in the test network containing 100 SCs was lower relative to that in the less densified networks. One might expect that a larger number of SCs should afford the scheduler greater scope to improve fairness. However on inspection we found that the proportion of UEs in SC expanded regions decreased with increasing SC density. With 20 SCs on the map about 14 % of the SC attached UEs resided in an expanded region compared to just 7 % with 100 SCs. In addition, the average number of UEs per SC decreased as more SCs were added. In combination these factors diminished the marginal impact of scheduling. As expected, the overall network utility was boosted by SC densification. The sum log downlink rates trended 17912→ 18621→18933 as the number of SCs increased 20→60→100. Densification and scheduling are recognised as key requisites for 5G [7].

6.1 Terminal Utilisation

Figure 7 displays each terminal's count in the fittest 150 individuals at each generation. The instantaneous downlink rate (downlink) occurs most frequently. Heavy utilisation of this terminal is unsurprising because it most faithfully predicts realised downlink rates. Antithetically, num_att appears least frequently. This too is unsurprising as the number of attached UEs is a constant statistic and hence it bears no differentiating power. Therefore, num_att represents a baseline with respect to which the importance of other terminals can be inferred. The terminals describing u's performance across all 40 SFs appear more frequently than the other contextual statistics. This may suggest that although u's context within a SF relative to other UEs is important (*_downlink_SF, *_downlink_cell), more relevant for u are the attributes of a SF f relative to other SFs (*_downlink_frame). Of particular interest is the standing of our mem-

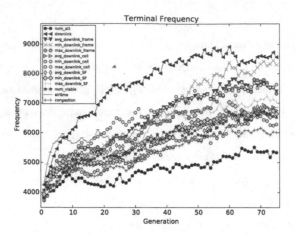

Fig. 7. Count of each terminal in the best 150 individuals over 75 generations.

ory nodes, `airtime` and `congestion`. Their importance is commensurate with that of most SF and cell-wide context nodes. In any case it is clear that they are selected for in the evolving populations. Given the ostensible importance of memory in HetNet scheduling, we are motivated in future work to explicitly factor prior decisions into the current tree output, see Sect. 7.

6.2   Subframe Utilisation

This subsection examines how the scheduler displayed in Fig. 6 behaves seman- tically. To appreciate how MC muting modulates interference at SCs, consider a toy network with three MCs running ABS ratios of 5/40, 10/40 and 15/40. Since ABSs are front-loaded as described in Subsect. 4.1, it follows that all three MCs will mute in SFs 1, 9, 17, 25 and 33. Two of the three MCs will mute in SFs 2, 10, 18, 26, 34, and only one MC will mute in SFs 3, 11, 19, 27, 35. All three will transmit in the remaining SFs. Thus, SFs 1, 9, 17, 25 and 33 are protected from high MC interference. Perhaps GP would learn to award protected SFs to UEs at the SC cell-edge whilst de-allocating those more advantaged cell-centre UEs (in order to relieve congestion). Indeed, simulation of a network with 60 SCs and 21 MCs revealed stark differences in how edge and centre UEs are scheduled. For both groups, we counted the cumulative number of UEs scheduled per SF, for all SCs over ten scenarios.

Columns 2 and 3 of Table 2 display the proportion of expanded region and centre UEs that are scheduled in each SF. For example, column 2 states that on aggregate 85 % of the expanded region UEs are scheduled in the best SFs (1, 9, 17, 25, 33), 58 % in the second best SFs, etc. Column 3 indicates that the scheduler sacrifices centre UEs by denying them the most protected SFs. Consequently, congestion is dramatically reduced for expanded region UEs where their $SINR$ will be greatest. Column 1 confirms that expanded region UEs are awarded more airtime in the better SFs. They can thus leverage both high $SINR$'s and the liberated bandwidth to maximise their achieved downlink rates. Centre UEs make up for the lost premium airtime by dominating less protected SFs where few or no expanded region UEs receive data.

Column 3 shows the cell-wide proportion of UEs scheduled per SF. Clearly, congestion management is of prime importance in SFs 1, 9, 17, 25 and 33 as less that 25 % receive data. In fact quite a few UEs are denied in every SF. The proportion jumps to 91 % in SFs 2, 10, 18, 26 and 34 then decreases monotonically reflecting how the scheduler negotiates trade-offs between airtime and congestion. In sum, non-trivial yet intuitive behaviour is realised by the expression in Fig. 6.

Table 2. Proportion scheduled per SF in various SC regions.

SF	Expanded region UEs	Cell-centre UEs	All attached
1, 9, 17, 25, 33	0.85	0.14	0.23
2, 10, 18, 26, 34	0.58	0.96	0.91
3, 11, 19, 27, 35	0.28	0.94	0.86
4, 12, 20, 28, 36	0.10	0.93	0.82
5, 13, 21, 29, 37	0.02	0.92	0.81
6, 14, 22, 30, 38	0.0	0.92	0.80
7, 15, 23, 31, 39	0.0	0.92	0.80
8, 16, 24, 32, 40	0.0	0.92	0.80

6.3   Benchmarking

The benchmark scheme was proposed by López-Peréz and Claussen (2013) [24]. Based on *SINR*, UEs are split into queues 'overlapping' with host MC ABSs or 'non-overlapping'. The worst UE in both queues is identified. Achievable downlink rates are computed for both worst UEs for both queue types. Next the SC computes target queue lengths based on the expected rates. UEs are transferred iteratively from one queue to the other until convergence. They are then scheduled according to their queue type, i.e. during ABS or non-ABS SFs.

Table 3 compares the evolved GP solution with the benchmark scheme, a Genetic Algorithm (GA) and a Hill Climbing heuristic. The improvements in the 5th and 50th percentile downlink rates and the lifts in sum log rates (Eq. 6) over baseline scheduling are reported. Statistics are generated across 100 scenarios in unseen test networks containing 20, 60 and 100 SCs. The GA was instrumented using a population size of 750 and run for 50 generations for each SC in the test set. The solution space was explored using two-point crossover (probability 1.0) and bit flip mutation on 20 % of the population (probability 0.2 per codon). Similarly, the Hill Climber mutated a randomly generated schedule (bit flip

Table 3. Comparison of Methods

%–tile	GP: Figure 6			Benchmark		
	20 SCs	60 SCs	100 SCs	20 SCs	60 SCs	100 SCs
50th ($\times 10^3$)	52 ± 41	-130 ± 108	-437 ± 133	-92 ± 51	-368 ± 160	-733 ± 276
5th ($\times 10^3$)	159 ± 44	388 ± 69	649 ± 105	118 ± 38	240 ± 56	209 ± 107
Fitness	55.92 ± 0.69	58.10 ± 0.85	24.58 ± 1.08	41.57 ± 1.04	18.84 ± 1.80	-42.63 ± 4.66

%–tile	GA			Hill Climber		
	20 SCs	60 SCs	100 SCs	20 SCs	60 SCs	100 SCs
50th ($\times 10^3$)	43 ± 27	45 ± 69	41 ± 96	46 ± 28	42 ± 68	10 ± 89
5th ($\times 10^3$)	154 ± 43	440 ± 64	570 ± 131	159 ± 44	450 ± 60	563 ± 126
Fitness	84.77 ± 0.42	117.45 ± 0.69	87.09 ± 0.91	87.33 ± 0.51	117.69 ± 0.95	82.36 ± 1.12

with probability 0.025 per codon) for 20,000 iterations and greedily accepted improvements to the current best.

The first panel shows that our evolved solution outperforms the benchmark on all metrics. Two sample t-tests confirm that the differences are significant at a confidence level of 0.99. Of particular interest is row 2 of the first panel which compares the benchmark and evolved heuristics with respect to the 5th percentile of downlink rates. Our fitness function does not incorporate this metric explicitly. We simply maximise sum log rates. Therefore, the improved performance of the worst UEs versus the baseline and benchmark schemes emerges naturally as a by-product of the optimisation.

The second panel shows that there is scope to build on this pilot study. Both the GA and Hill Climbing heuristics significantly outperform the evolved solution and benchmark. Note that these direct search methods are orders of magnitude too slow for online operation.

7 Future Work and Conclusions

We implemented a reinforcement learning approach in this work but some pilot experiments suggest that supervised training will yield far better schedulers. In a follow-up study we will compute near optimal schedules offline using a genetic algorithm. Hence, a more informative fitness function can be devised which respects the distance (e.g. Hamming) to the target semantics. Table 3 reveals that significant gains are achievable.

We observed in Sect. 6 that it is expedient to track intermediate scheduling decisions using counters. Conrads et al. (1998) considered a time series problem where previous tree outputs acted as terminals for the current evaluation [10]. Alfaro et al. showed their recurrent system achieves comparable performance with state of the art methods on real world problems [5]. We propose that a recurrent system like [5, 10] will yield better results on the present problem.

Genetic Programming lends itself well to the task of evolving schedulers for HetNets implementing eICIC. Indeed, the framework proposed in this paper outperforms a state of the art human engineered approach. Future work will build on this pilot study to close the optimality gap.

Acknowledgement. This research is based upon works supported by the Science Foundation Ireland under grant 13/IA/1850.

References

1. 3Gpp, December 2014. http://www.3gpp.org/
2. Google Maps, December 2014
3. Cisco Visual Networking Index: Global Mobile Data Traffic Forecast Update, 2014–2019. Cisco, White Paper (2015)
4. Small Cell Solutions. Alcatel-Lucent (2015). https://www.alcatel-lucent.com/solutions/small-cells
5. Alfaro-Cid, E., Sharman, K., Esparcia-Alcázar, A.I.: Genetic programming and serial processing for time series classification. Evol. Comput. **22**(2), 265–285 (2014)
6. Bader-El-Den, M., Fatima, S.: Genetic programming for auction based scheduling. In: Esparcia-Alcázar, A.I., Ekárt, A., Silva, S., Dignum, S., Uyar, A.Ş. (eds.) EuroGP 2010. LNCS, vol. 6021, pp. 256–267. Springer, Heidelberg (2010)
7. Bhushan, N., Li, J., Malladi, D., Gilmore, R., Brenner, D., Damnjanovic, A., Sukhavasi, R., Patel, C., Geirhofer, S.: Network densification: the dominant theme for wireless evolution into 5G. IEEE Commun. Mag. **52**(2), 82–89 (2014)
8. Bian, Y.Q., Rao, D.: Small Cells Big Opportunities. Global Business Consulting. Huawei Technologies Co., Ltd. (2014)
9. Brabazon, A., O'Neill, M., McGarraghy, S.: Natural Computing Algorithms. Springer, Berlin (2015)
10. Conrads, M., Nordin, P., Banzhaf, W.: Speech sound discrimination with genetic programming. In: Banzhaf, W., Poli, R., Schoenauer, M., Fogarty, T.C. (eds.) EuroGP 1998. LNCS, vol. 1391, pp. 113–129. Springer, Heidelberg (1998)
11. Damnjanovic, A., Montojo, J., Wei, Y., Ji, T., Luo, T., Vajapeyam, M., Yoo, T., Song, O., Malladi, D.: A survey on 3GPP heterogeneous networks. IEEE Wirel. Commun. **18**(3), 10–21 (2011)
12. Deb, S., Monogioudis, P., Miernik, J., Seymour, J.P.: Algorithms for enhanced inter-cell interference coordination (eICIC) in LTE HetNets. IEEE/ACM Trans. Netw. (TON) **22**(1), 137–150 (2014)
13. Dempsey, I., O'Neill, M., Brabazon, A.: Grammatical evolution. In: Dempsey, I., O'Neill, M., Brabazon, A. (eds.) Foundations in Grammatical Evolution for Dynamic Environments. SCI, vol. 194, pp. 9–24. Springer, Heidelberg (2009)
14. Ernst, A.T., Jiang, H., Krishnamoorthy, M., Sier, D.: Staff scheduling and rostering: a review of applications, methods and models. Eur. J. Oper. Res. **153**(1), 3–27 (2004)
15. Fenton, M., Lynch, D., Kucera, S., Claussen, H., O'Neill, M.: Evolving coverage optimisation functions for heterogeneous networks using grammatical genetic programming. In: Proceedings of the 19th International Conference on the Applications of Evolutionary Computation, EvoCOMNET 2016. Springer (2016)
16. Hansen, J.V.: Genetic search methods in air traffic control. Comput. Oper. Res. **31**(3), 445–459 (2004)

17. Hemberg, E., Ho, L., O'Neill, M., Claussen, H.: A symbolic regression approach to manage femtocell coverage using grammatical genetic programming. In: Proceedings of the 13th Annual Conference Companion on Genetic and Evolutionary Computation, pp. 639–646. ACM (2011)
18. Hemberg, E., Ho, L., O'Neill, M., Claussen, H.: Evolving femtocell algorithms with dynamic and stationary training scenarios. In: Coello, C.A.C., Cutello, V., Deb, K., Forrest, S., Nicosia, G., Pavone, M. (eds.) PPSN 2012, Part II. LNCS, vol. 7492, pp. 518–527. Springer, Heidelberg (2012)
19. Hemberg, E., Ho, L., O'Neill, M., Claussen, H.: A comparison of grammatical genetic programming grammars for controlling femtocell network coverage. Genet. Program Evolvable Mach. **14**(1), 65–93 (2013)
20. Ho, L.T., Ashraf, I., Claussen, H.: Evolving femtocell coverage optimization algorithms using genetic programming. In: 2009 IEEE 20th International Symposium on Personal, Indoor and Mobile Radio Communications, pp. 2132–2136. IEEE (2009)
21. Jakobović, D., Marasović, K.: Evolving priority scheduling heuristics with genetic programming. Appl. Soft Comput. **12**(9), 2781–2789 (2012)
22. Jiang, L., Lei, M.: Resource allocation for eICIC scheme in heterogeneous networks. In: 2012 IEEE 23rd International Symposium on Personal Indoor and Mobile Radio Communications (PIMRC), pp. 448–453. IEEE (2012)
23. Jones, A., Rabelo, L.C., Sharawi, A.T.: Survey of job shop scheduling techniques In: Wiley Encyclopedia of Electrical and Electronics Engineering (1999)
24. López-Pérez, D., Claussen, H.: Duty cycles and load balancing in hetnets with eICIC almost blank subframes. In: 2013 IEEE 24th International Symposium on Personal, Indoor and Mobile Radio Communications (PIMRC Workshops), pp. 173–178. IEEE (2013)
25. Mckay, R.I., Hoai, N.X., Whigham, P.A., Shan, Y., O'Neill, M.: Grammar-based genetic programming: a survey. Genet. Program Evolvable Mach. **11**(3–4), 365–396 (2010)
26. Pang, J., Wang, J., Wang, D., Shen, G., Jiang, Q., Liu, J.: Optimized time-domain resource partitioning for enhanced inter-cell interference coordination in heterogeneous networks. In: 2012 IEEE Wireless Communications and Networking Conference (WCNC), pp. 1613–1617. IEEE (2012)
27. Shannon, C.E.: Communication in the presence of noise. Proc. IRE **37**(1), 10–21 (1949)
28. Sun, J., Modiano, E., Zheng, L.: Wireless channel allocation using an auction algorithm. IEEE J. Sel. Areas Commun. **24**(5), 1085–1096 (2006)
29. Tall, A., Altman, Z., Altman, E.: Self organizing strategies for enhanced ICIC (eICIC). In: 2014 12th International Symposium on Modeling and Optimization in Mobile, Ad Hoc, and Wireless Networks (WiOpt), pp. 318–325. IEEE (2014)
30. Weber, A., Stanze, O.: Scheduling strategies for HetNets using eICIC. In: 2012 IEEE International Conference on Communications (ICC), pp. 6787–6791. IEEE (2012)
31. Yang, S., Ong, Y.S., Jin, Y.: Evolutionary Computation in Dynamic and Uncertain Environments. Springer Science & Business Media, New York (2007)

On the Analysis of Simple Genetic Programming for Evolving Boolean Functions

Andrea Mambrini and Pietro S. Oliveto[✉]

University of Sheffield, Sheffield, UK
{a.mambrini,p.oliveto}@sheffield.ac.uk

Abstract. This work presents a first step towards a systematic time and space complexity analysis of genetic programming (GP) for evolving functions with desired input/output behaviour. Two simple GP algorithms, called (1+1) GP and (1+1) GP*, equipped with minimal function (F) and terminal (L) sets are considered for evolving two standard classes of Boolean functions. It is rigorously proved that both algorithms are efficient for the easy problem of evolving conjunctions of Boolean variables with the minimal sets. However, if an extra function (i.e. NOT) is added to F, then the algorithms require at least exponential time to evolve the conjunction of n variables. On the other hand, it is proved that both algorithms fail at evolving the difficult parity function in polynomial time with probability at least exponentially close to 1. Concerning generalisation, it is shown how the quality of the evolved conjunctions depends on the size of the training set s while the evolved exclusive disjunctions generalize equally badly independent of s.

Keywords: Genetic programming · Theory · Runtime analysis

1 Introduction

Genetic programming (GP) was originally proposed by Koza as an evolutionary computation technique for evolving computer programs [4]. Traditionally GP represents programs using *syntax trees* and evaluates their fitness by executing them and then comparing their behaviour against an ideal one (eg., the desired input/output behaviour of the function to be evolved). A population of programs is evolved using typical genetic algorithm (GA) variation and selection operators adapted to work on syntax trees with the goal of eventually identifying a program with the desired functionality. GP has shown great potential by evolving, for example, quantum computing algorithms that outperformed all previous approaches [17], soccer-playing programs [7] and algorithms for the transmembrane segment identification protein problem [5], to name a few.

Despite the wide range of successful applications, there is still very little understanding of GP's behaviour [13,15]. Theoretical work concerning GP has always been undertaken since its early days [6], the majority of which has applied *schema theory* [4,16]. Schema theories are based on the idea of partitioning

© Springer International Publishing Switzerland 2016
M. Heywood et al. (Eds.): EuroGP 2016, LNCS 9594, pp. 99–114, 2016.
DOI: 10.1007/978-3-319-30668-1_7

the search space into subsets, called *schemata*, and modelling the behaviour and dynamics of the population over the schemata. However, such an analysis does not allow any insight towards the understanding of the performance of GP. Chapter 11.1 of [15] concludes that through schema theories *"...we have no way of closing the loop and fully characterising the behaviour of GP systems which is always a result of the interaction between the fitness function and the search biases of the representation and genetic operations used in the system"*. Such characterisations and interactions can, instead, be understood by analysing the time and space complexity of a GP system when attempting to evolve a given class of functions. This approach has been applied to other classes of bio-inspired optimisation heuristics, with remarkable success [1]. Nowadays, the performance quality of various bio-inspired optimisation heuristics is known concerning sophisticated population-based heuristics and even for standard combinatorial optimisation problems with practical applications [11]. These results shed light on which kind of problems a given algorithm works on efficiently and on which it is inefficient and provide a relationship between the size of the problem and the time and space required to solve it. Along the way guidelines towards optimal parameter settings are given.

Some initial runtime analysis results concerning GP systems have already appeared [10]. Such first studies regarded two functions classes called ORDER and MAJORITY where the fitness of a tree (i.e., a candidate solution) depends on the structure of the tree rather than on its execution. Although this is a considerable simplification compared to the problems to which GP is usually applied, these results show that very simple GP systems can optimise both structures efficiently. Furthermore, understanding how and when correct structures are evolved will be necessarily crucial in an analysis of more realistic GP scenarios. Recently, the same simple GP systems have been analysed on the MAX Problem [6] where, given a set of functions, a set of terminals and a bound D on the maximum depth of the solution, the goal is to evolve a tree that returns the maximum value given any combination of functions and terminals [3]. The analysis shows that the simple GP systems can efficiently evolve MAX with function set $F = \{+, *\}$ and one constant as terminal set. Compared to the previous functions, MAX is more similar to those evolved by GP in practical applications since the fitness indeed depends on the behaviour of the computed function on the input. Still, dependence is not very strong, since the space of possible inputs can be partitioned into just two subsets such that for every input in a subset, the optimal solution to the problem is the same.

In this paper we make a further step forward and provide an analysis of GP for typical benchmark functions used in the field of GP [4,6]. Hence, we consider proper learning problems where the fitness depends on the input/output behaviour of the trees. When the initial foundations for a systematic time complexity analysis of EAs were being set, very simple EAs were considered (eg., the (1+1) EA) for simple benchmark problems which are easy for EAs (eg., OneMax and LeadingOnes) and others which are hard (eg., Trap Functions and Needle-In-A-Haystack) [2]. In a similar fashion we will analyse the simple and minimalistic (1+1) GP considered in previous runtime analyses of GP [3,10] for simple Boolean functions with minimal function and terminal sets. Since under the *evolvability* notion in the PAC-learing framework it is well-understood that conjunctions (i.e., AND) are evolvable efficiently while parity problems (i.e., XOR) are not [18], we naturally choose these boolean functions as our starting points for the analysis. In particular, the presented AND problem may be regarded as a GP analogue to OneMax for EAs, while XOR the analogue to Needle. Moreover, we will take into consideration the generalization ability of the solutions found by the algorithms when using incomplete training sets since, as the problem size grows, it is not possible to test the candidate solutions on the complete training set efficiently. We point out that runtime results are available concerning the recently introduced GP variant called Geometric Semantic Genetic Programming (GSGP) [9]. The long term aim of the work presented herein is to understand the behaviour and performance of standard and widely used GP systems.

In the next section we introduce the two simple (1+1) GP and (1+1) GP* systems and formally define the learning problems. In Sect. 3 we present the results for the AND and XOR functions of n variables using the complete training set and minimal function and terminal sets (i.e., respectively $F = \{AND\}, L = \{X_1, \ldots, X_n\}$ and $F = \{XOR\}, L = \{X_1, \ldots, X_n\}$). In particular, we show that both the (1+1) GP and the (1+1) GP* can evolve conjunctions efficiently while they are both inefficient when evolving parity. However, if we add another function to F (i.e., NOT), then the algorithms become inefficient also for evolving conjunctions. In Sect. 4 we present the results when only a training set of polynomial size in the problem size is allowed. We show that the algorithms fit the training set for the AND function in logarithmic time while the XOR function becomes harder than a Needle function for larger than logarithmic training sets because points leading closer to the optimal solution may be rejected. We conclude the section by providing results on how the evolved solutions generalise to the complete training set. In the Conclusions we discuss future work directions.

2 Preliminaries

We consider the (1+1)-GP (Algorithm 1) and (1+1)-GP* (Algorithm 2) from [10], both working with a population of size one and producing at each generation one new offspring using the HVL-Prime mutation operator [14] which chooses, uniformly at random, to either insert, to remove or to replace a node according to the procedures described in Algorithm 3.

Algorithm 1. (1+1) GP	Algorithm 2. (1+1) GP*
1: Initialise an empty tree X	1: Initialise an empty tree X
2: **for** $t := 1$ to ∞ **do**	2: **for** $t := 1$ to ∞ **do**
3: $X' :=$ HVL-Prime(X)	3: $X' :=$ HVL-Prime(X)
4: **if** $f(X') \leq f(X)$ **then**	4: **if** $f(X') < f(X)$ **then**
5: $X := X'$	5: $X := X'$

The only difference between (1+1) GP and (1+1)-GP* is that the former accepts an offspring which is at least as fit as its parent, while the latter accepts only strictly better offspring. An individual X is represented as a binary tree such that each internal node can be an element of the function set F and each leaf can be an element of the terminal set L. The two algorithms do not have a termination criterion since, here, we are interested in the first point of time when the optimal solution is found. For simplicity we initialise the algorithms with empty trees[1]. However, all the presented results can be easily adapted to random tree initialisation with only slightly differing theorem statements.

Let the complete truth table $C = \{(x_1, y_1), ..., (x_N, y_N)\}$ describe the complete input-output behaviour of a Boolean function $\hat{h} : \{0,1\}^n \rightarrow \{0,1\}$ over n variables (i.e., the table has $N = 2^n$ rows). A training set T consisting of $s \leq N$ test cases $T \subset C = \{(x_1, y_1), ..., (x_s, y_s)\}$ is sampled from the truth table uniformly at random with replacement. The black box Boolean learning problem consists of using just the training set T to learn a Boolean function $h : \{0,1\}^n \rightarrow \{0,1\}$ matching as well as possible the input-output behaviour described by C. Given a candidate solution (i.e., a Boolean expression h), the fitness function returns the *training error* $\epsilon_t(h)$ which is the number of rows on which the expression h mismatches the input-output behaviour described by T: $\epsilon_t(h) = \sum_{(x_i, y_i) \in T} I[h(x_i) \neq y_i]$ where $I[\cdot]$ is the indicator function that is 1 if its inner expression is true and 0 otherwise. We impose that the fitness function returns a value of $2^n + 1$ for an empty tree, which is worse than the fitness of any tree.

Algorithm 3. HVL-Prime

1: **procedure** HVL-PRIME(X)
2:     Select uniformly a random an action among INS, DEL, SUB
3:     **if** action is INS **then**
4:         Choose a node $v \in X$ uniformly at random
5:         Select uniformly at random a terminal v' from L
6:         Replace v with a node f selected uniformly at random from F
7:         Set v and v' as children of f, choosing the order of the children uniformly at random.
8:     **if** action is DEL **then**
9:         Choose, uniformly at random, a leaf node v with parent p and sibling s
10:         Replace p with s
11:         Delete p and v
12:     **if** action is SUB **then**
13:         Choose a leaf v uniformly at random
14:         Select uniformly at random a terminal v' from L
15:         Replace v with v'

---

[1] We assume the SUB and DEL of an empty tree return an empty tree.

We will analyse the (1+1) GP and (1+1) GP* on two boolean problems, AND_n with target function $\hat{h} = AND(X_1, \ldots, X_n)$ and XOR_n with target function $\hat{h} = XOR(X_1, \ldots, X_n)$. We say that an algorithm *solves* a boolean problem *efficiently* if it can evolve a solution fitting the training set (i.e., having training error equal to zero) in expected polynomial time, where time is defined as the number of fitness function evaluations.

We will first analyse the situation in which $s = N$, thus the training set encompasses all the possible input-output cases (i.e. complete dataset). In this situation finding an expression that fits the training set T will obviously also lead to an expression that fits the complete set, and thus the original Boolean function. Afterwards we will consider training sets of at most polynomial size, $s = \text{poly}(n) < N$ (i.e. incomplete training set). In this case, minimizing the error on the training set will lead to a generalization error which can be defined as $\epsilon_g(h) = \sum_{(x_i, y_i) \in C} I[h(x_i) \neq y_i]$ where C is the complete truth table.

We define the *generalization ability* of an algorithm $\mathcal{A}$ as $G(\mathcal{A}, s) = 1 - \frac{E[\epsilon_g(\widetilde{X})]}{N}$ where $\widetilde{X}$ is the best individual found by the algorithm (which tries to minimize the error on the training set) after a polynomial number of steps.

3 Analysis for Complete Training Sets

In this section we analyse the (1+1) GP and the (1+1) GP* on AND_n and XOR_n in the case of training sets of size $s = 2^n = N$ (i.e. complete datasets).

3.1 Analysis for AND_n with Complete Training Sets

Theorem 1 shows that both the (1+1)-GP and the (1+1) GP* evolve AND_n efficiently with $L = \{X_1, \ldots, X_n\}$ and $F = \{AND\}$. The theorem also shows that the strict selection of (1+1) GP* enforces solutions of exactly n variables while the (1+1) GP may produce solutions that are asymptotically larger by a logarithmic factor. The following Lemma will be useful.

Lemma 1. *Every conjunction of v distinct variables differs from the target function $AND(X_1, \ldots, X_n)$ on $f_v = 2^{n-v} - 1$ rows.*

Proof. We prove the statement by induction. The base case $f_1 = 2^{n-1} - 1$ follows because the truth table of any conjunction of one variable has $2^n/2$ ones and $2^n/2$ zeros. The target function and the conjunction of one variable will agree on one row, i.e. the one in which all the variables are set to 1. On the rest of the $2^{n-1} - 1$ rows the two functions will not agree.

To prove the inductive step we need to assume $f_i = 2^{n-i} - 1$ and prove that $f_{i+1}(x) = 2^{n-(i+1)} - 1$. When one variable is added to the conjunction, the number of ones will halve (since half of them will be anded with a zero). Since just one of them agrees with the target function, the new conjunction will differ on $f_{i+1} = (f_i + 1)/2 - 1$, which by hypothesis is $2^{n-(i+1)}$. □

Theorem 1. *The (1+1)-GP and the (1+1) GP* using $F = \{AND\}$ and $L = \{X_1, \ldots, X_n\}$ efficiently solve AND_n, using the complete truth table as training set, in time $O(n \log n)$. The size of the final expression is n for (1+1) GP* and $O(n \log n)$ in expectation for (1+1)-GP.*

Proof. We divide the search space into $A_1, \ldots, A_n$ fitness levels such that each level A_i contains all the conjunctions of i distinct variables. By Lemma 1 we have a fitness level partition with increasing fitness and the artificial fitness levels technique [1] is in force. We need to derive a lower bound on the probability p_i that an individual leaves level A_i and reaches level A_{i+1}.

For both (1+1) GPs at level A_i the number of distinct i variables may not be reduced as the fitness would decrease. Hence to reach level A_{i+1} it is sufficient for both algorithms to choose an INS operation and then insert one of the $n - i$ variables that is not already in the conjunction. This probability is $p_i \geq \frac{1}{3}\frac{n-i}{n}$. Thus, by the fitness level method the expected runtime is $E[T] \leq \sum_{i=1}^{n} 1/p_i = \sum_{i=1}^{n} \frac{3n}{n-i} = O(n \log n)$, which proves the first statement.

For the (1+1) GP* (strict selection) an INS operation adding a duplicate of an existing variable will not be accepted since it would have the same fitness of the parent. The same holds for a SUB operation replacing one variable with a duplicate of an existing one. Thus at each iteration the current solution has each variable appearing at most once and, for this reason, a DEL operation will never be accepted, since it would reduce the fitness. Since no duplicates are allowed in the (1+1) GP*, the size of the final expression is exactly n. For the (1+1) GP (weak selection) the situation is different because an INS operation can insert a variable that already exists in the current expression. We observe that the only operation increasing the size of the current expression is INS. Since in expectation $O(n \log n)$ of these operations occur, the expected size of the final expression cannot be more than $O(n \log n)$. □

In the following theorem we show that just adding the negation of all the variables to the terminal set makes both algorithms inefficient.

Theorem 2. *Both the (1+1)-GP and the (1+1)-GP* using $F = \{AND\}$ and $L = \{X_1, \ldots, X_n, \overline{X_1}, \ldots, \overline{X_n}\}$ cannot solve AND_n with probability at least $p > 1 - \left(\frac{1}{4}\right)^{n/3}$ using the complete truth table as training set.*

Proof. Differently from the situation of Theorem 1, the search space contains many local optima from which the algorithms cannot escape. All solutions containing both a variable X_i and its negation $\overline{X_i}$ evaluate to 0, hence have a fitness of 1. We prove that such a solution is found with probability exponentially close to 1. When the current expression is missing just $i \leq n/3$ variables the probability of adding a new variable is bounded from above by the probability of doing so with an INS operation plus the probability that a SUB operation adds a missing variable, which is $p_{add} \leq \frac{1}{3}\frac{n/3}{2n} + \frac{1}{3}\frac{n/3}{2n} = 1/9$. At the same time the probability of inserting the negation of a variable which is already in the current expression is at least $p_{neg} \geq \frac{(2/3)n}{2n} = 1/3$. The probability of the second event happening before the first is $P(neg|neg \cup add) \geq \frac{p_{neg}}{p_{neg}+p_{add}} \geq \frac{1/3}{1/3+1/9} = 3/4$. The probability that neg never happens before all the $n/3$ missing variables are added is $p \leq (1 - 3/4)^{n/3} = \left(\frac{1}{4}\right)^{n/3}$, thus the algorithm ends up in one of these local optima with probability $p > 1 - \left(\frac{1}{4}\right)^{n/3}$.

Since a deletion of $\overline{X_i}$ decreases fitness, the only way to leave such local optima is by adding all the missing variables and then removing $\overline{X_i}$ (plus all the other negated variables that might have been inserted in the process). In the case of strict selection, this is not possible and the (1+1) GP* is stuck forever. For the case of weak selection, the algorithm will have to walk on a plateau of fitness 1 until it reaches a point when all the n variables are in the tree. Only at that point it would be allowed to walk on another plateau by removing all the negated variables. When this happens the optimum would be found. Similarly to the proof of Theorem 4 it is possible to show that this cannot happen in less than exponential time with probability exponentially close to 1. We don't report a complete proof here due to space restrictions. □

3.2 Analysis for XOR_n with Complete Training Sets

The analysis for XOR_n will show a needle-like fitness landscape (Proposition 2). In fact when the training set encompasses all the possible input-output pairs, all the solutions in the search space but the optimum will have the same fitness. As a result, we will show in Theorem 3 that the (1+1) GP* cannot solve XOR_n in finite time because it cannot find strictly improving solutions and in Theorem 4 that the (1+1) GP cannot solve XOR_n in less than exponential time with probability exponentially close to 1. We state the two following facts.

Proposition 1. *Any exclusive disjunction of m variables $(X_1 \oplus \ldots \oplus X_m)$ on a truth table of $n \geq m$ variables outputs 1 for half of its inputs and 0 for the other half.*

Proposition 2. *Any exclusive disjunction of m variables $(X_1 \oplus \ldots \oplus X_m)$ on a truth table of $n \geq m$ variables differs from $X_1 \oplus \ldots \oplus X_n$ on 2^{n-1} inputs, thus has fitness 2^{n-1}.*

Now we are ready to state and prove the theorems.

Theorem 3. *The (1+1)-GP* using $F = \{XOR\}$ as function set and $L = \{X_1, \ldots, X_n\}$ as terminal set cannot solve XOR_n, using the complete truth table as training set.*

Proof. Since by Proposition 2 all the points in the search space have the same fitness, any individual after the first one will not be accepted. Thus XOR_n cannot be solved using strict selection. □

The following Lemma will be useful in the proof of Theorem 4.

Lemma 2. *In a tree containing m leaves, each one sampled uniformly at random from $L = \{X_1, \ldots, X_n\}$ with replacement, with probability $p \geq 1 - e^{-\Omega(n)}$ each variable X_i appears at most $M < \frac{m \log n}{n}$ times for $m \geq \frac{n^2}{e}$, and at most $M < \frac{2n}{\ln\left(\frac{n^2}{em}\right)} < 2n$ times for $m < \frac{n^2}{e}$.*

Proof. We will bound M from above by using a balls and bins argument [8], where the m balls represent the total number of variables in the tree and each of the n bins represents a different variable X_i. The probability that $M > \gamma$ is the probability that after throwing m balls into n bins by selecting each bin uniformly at random, there is at least one bin containing at least γ balls. Let X be a random variable counting the number of balls in a given bin and $P(X > \gamma)$ be the probability that one bin contains more than γ balls. Then by the union bound $P(M > \gamma) \leq n \cdot P(X > \gamma)$. We will calculate $P(X > \gamma)$ for two separate cases.

(1) Let $m \geq \frac{n^2}{e}$. Since the probability of selecting a bin is $1/n$, the expected number of balls in the bin is $\mu = E(X) = m/n$. Applying the Chernoff bound $P(X > (1 + \delta) \cdot \mu) \leq [\frac{1}{1+\delta}(\frac{e}{1+\delta})]^\mu \leq (\frac{1}{1+\delta})^\mu$ for any $\delta > 0$ [8] and by exploiting $m \geq \frac{n^2}{e}$ we get:

$$P\left(X > \frac{m \log n}{n}\right) \leq \left(\frac{1}{\log n}\right)^{\frac{m}{n}} \leq \left(\frac{1}{\log n}\right)^{\frac{n}{e}} \leq \left(\frac{1}{e}\right)^{\frac{n}{e}} = e^{-\Omega(n)}$$

where the inequality before the last one holds for $n > e^e$. Thus $M < \frac{m \log n}{n}$ with probability at least $1 - ne^{-\Omega(n)} = 1 - e^{-\Omega(n)}$.

(2) Let $m < \frac{n^2}{e}$. By applying the Chernoff bound again with $\tau = 1 + \delta$ we get for any $\tau > 1$

$$P(X > \tau \cdot \mu) \leq \left(\frac{e}{\tau}\right)^{\tau\mu}$$

We want τ such that $P(X > \tau \cdot \mu) \leq e^{-n}$ which is equivalent to

$$\left(\frac{e}{\tau}\right)^{\tau\mu} \leq e^{-n} \iff e^n \leq \left(\frac{\tau}{e}\right)^{\tau\mu} \iff e^{\frac{n}{\tau\mu}} \leq \left(\frac{\tau}{e}\right)^{\frac{\tau}{e}} \iff e^{\frac{n^2}{e \cdot m}} \leq \left(\frac{\tau}{e}\right)^{\frac{\tau}{e}}$$

Now, let $K = \tau/e$ and $N = e^{\frac{n^2}{e \cdot m}}$. Since $K > \frac{2 \ln N}{\ln \ln N}$ implies $K^K > N$ provided that $N > e$ which holds since $m < n^2/e$, we get that $\left(\frac{\tau}{e}\right)^{\frac{\tau}{e}} \geq e^{\frac{n^2}{e \cdot m}}$ holds for

$$\frac{\tau}{e} > \frac{2 \ln(e^{\frac{n^2}{em}})}{\ln \ln(e^{\frac{n^2}{em}})} \iff \tau > \frac{2n}{\mu \ln(\frac{n^2}{em})}$$

Hence $\tau > 1$ for $m < 2n^2$ and we can apply the Chernoff bound. Thus,

$$P\left(X \geq \frac{2n}{\ln(\frac{n^2}{em})}\right) \leq e^{-n}$$

So $M \leq \frac{2n}{\ln(\frac{n^2}{em})} < 2n$ with probability at least $1 - ne^{-n}$. $\qquad\square$

Theorem 4. *The (1+1)-GP using $F = \{XOR\}$ and $L = \{X_1, \ldots, X_n\}$ to evolve XOR_n using the complete truth table as training set requires more than $2^{\Omega(\frac{n}{\log n})}$ steps with probability $p > 1 - 2^{-\Omega(\frac{n}{\log n})}$ to reach the optimum.*

Proof. We apply the simplified negative drift theorem [12]. Let k_t denote the number of missing variables in our current solution after simplification[2] (thus e.g. for $X_1 \oplus X_4 \oplus X_1$, $k_t = n - 1$) at step t. Given an expression with i variables missing after simplification we denote with $E[\Delta(i)] = E[(k_{t+1} - k_t)|k_t = i)]$ the negative drift, which is the expected increase in the number of missing variables after simplification in the next step. Since each of the operations (INS, DEL, SUB) happens with equal probability, $E[\Delta(i)] = \frac{1}{3}E[\Delta_{INS}(i)] + \frac{1}{3}E[\Delta_{DEL}(i)] + \frac{1}{3}E[\Delta_{SUB}(i)]$.

When an INS operation occurs $E[\Delta_{INS}(i)] \geq \frac{n-i}{n} - \frac{i}{n}$ since we decrease k_t by one when we insert a variable which was not in the current solution and increase it by one when we add a variable that was already there (because simplification would remove it). For $i \leq \frac{1}{4}n$, $E[\Delta_{INS}(i)] > 1/2$.

When a DEL operation occurs we notice that the number of missing variables after simplification increases if we delete one of the variables appearing an odd number of times in the expression (thus not being simplified out), while we decrease the number of missing variables if we delete a variable appearing an even number of times (because it would not be simplified out anymore). Thus calling m the number of leaves in the tree (i.e. the number of variables before simplification), and $M = \max_i [\text{count}(X_i)]$ the maximum number of occurrences of a variable, we observe pessimistically that $E[\Delta_{DEL}(i)] \geq \frac{m - M \cdot i}{m} - \frac{M \cdot i}{m} = \frac{m - 2 \cdot M \cdot i}{m}$. The conditional drift when $m \geq \frac{n^2}{e}$ (thus $M < \frac{m \log n}{n}$ with probability exponentially close to 1 by Lemma 2) is

$$E[\Delta_{DEL}(i)] \geq \frac{m - 2 \cdot M \cdot i}{m} \geq \frac{m - 2 \cdot \frac{m \log n}{n} \cdot i}{m}$$

which is positive for $i < \frac{n}{2 \log n}$ On the other hand the conditional drift when $m < \frac{n^2}{e}$ (thus $M < 2n$ with probability exponentially close to 1) is

$$E[\Delta_{DEL}(i)] \geq \frac{m - 2 \cdot M \cdot i}{m} \geq \frac{m - 4n \cdot i}{m}$$

which is positive for $i < \frac{m}{4n} < \frac{n}{4e}$ (since $m < n^2/e$).

When a SUB operation occurs we notice that the number of missing variables after simplification increases by two if we replace one of the variables appearing an odd number of times in the expression with another different variable appearing an odd number of times, while it decreases by two if we replace one of the variables appearing an even number of times in the expression with one different variable appearing an even number of times. In all other cases the number of

---

[2] Simplification is a conceptual tool used for the proofs. The actual tree contains all the variables (i.e., the algorithm does not simplify the trees).

missing variables stays the same. Thus $E[\Delta_{\text{SUB}}(i)] \geq 2 \cdot \frac{m-Mi}{m} \cdot \frac{n-i-1}{n} - 2 \cdot \frac{Mi}{m} \cdot \frac{i-1}{n}$.
For $i \leq n/2$ [thus $\frac{n-i-1}{n} \geq \frac{i-1}{n}$] we get

$$E[\Delta_{\text{SUB}}(i)] \geq \frac{2(i-1)}{n} \left[\frac{m-Mi}{m} - \frac{Mi}{m} \right] = \frac{2(i-1)}{n} \left[\frac{m-2Mi}{m} \right]$$

Now we check the two cases of Lemma 2. For $m \geq \frac{n^2}{e}$, we get

$$E[\Delta_{\text{SUB}}(i)] \geq \frac{2(i-1)}{n} \left[\frac{m - 2 \cdot \frac{m \log n}{n} \cdot i}{m} \right]$$

For $2 \leq i \leq \frac{n}{4 \log n}$ we get $E[\Delta_{\text{SUB}}(i)] \geq \frac{1}{n}$. For $m < \frac{n^2}{e}$, we get

$$E[\Delta_{\text{SUB}}(i)] \geq \frac{2(i-1)}{n} \left[\frac{m - 2 \cdot 2n \cdot i}{m} \right]$$

For $2 \leq i \leq \frac{m}{8n} < \frac{n}{8e}$ (since $m < \frac{n^2}{e}$) we get $E[\Delta_{\text{SUB}}(i)] \geq \frac{1}{n}$.

Finally, choosing $a = \frac{n}{8 \log n}$ and $b = \frac{n}{16 \log n}$, the expected negative drift $E[\Delta(i)] = \frac{1}{3}(E[\Delta_{\text{INS}}(i)] + E[\Delta_{\text{DEL}}(i)] + E[\Delta_{\text{SUB}}(i)]) \geq (1/3)[1/2 + 0 + \frac{1}{n}] \geq 1/6 + o(1)$ for $i \in (a, b)$. Since the probability of performing steps greater than 2 is 0 and $p(\Delta(i) = 2) \leq 1/3 \leq (1/2)^{2-r}$ for $r = 1$ the drift theorem is in force. Thus, conditional to the failure probabilities of Lemma 2, the optimum is found in time $T < 2^{\frac{n}{16 \log n}}$ with probability $2^{-\Omega\left(\frac{n}{\log n}\right)}$. By the union bound the probability that the bounds on M do not hold in $2^{\frac{n}{16 \log n}}$ steps is less than $2^{\frac{n}{16 \log n}} \cdot e^{-\Omega(n)} = e^{-\Omega(n)}$. Summing up the failure probabilities completes the proof. □

4 Analysis for Incomplete Training Sets

In this section we consider training sets of at most polynomial size, $s << 2^n$. The algorithm will thus calculate the fitness just on $s = poly(n)$ rows. We say that an algorithm *efficiently solves* a boolean problem if it can find a solution fitting the training set (i.e. with training error equal to zero) in expected polynomial time. We first analyse the time the algorithms takes to get to a solution with fitness zero on the training set and afterwards we will analyse the generalization error.

4.1 Analysis for AND_n with Incomplete Training Set

The analysis for AND_n shows that a polynomial training set is fit in logarithmic time (Theorem 5). Theorem 6 gives an upper bound on the generalization ability and gives a necessary condition on the size of the training set to achieve a generalization ability over a fixed threshold.

Theorem 5. *Let $s = poly(n)$ be the size of a training set chosen from the truth table uniformly at random with replacement. Then both the (1+1) GP and the (1+1) GP* using $F = \{AND\}$ and $L = \{X_1, \ldots, X_n\}$ will solve AND_n in expected time $O(\log s) = O(\log n)$.*

Proof. Since the elements of the truth table are binomially distributed with parameters n and $p = 1/2$, the expected number of 1s in a randomly chosen row of the training set is $n/2$. By a simple application of Chernoff bounds, the probability that more than $Y = n/2 + \epsilon n$ 1s are in the chosen element is bounded from above by $e^{-\Omega(n)}$. By a union bound the probability that any of the s elements of the training set contains more than Y 1s is less than $s \cdot e^{-\Omega(n)} = poly(n) \cdot e^{-\Omega(n)} = e^{-\Omega(n)}$. The algorithm will reach the minimum error (i.e. fitness) of 0 when for each row of the training set there exists a variable X_i in the constructed tree that has a 0 in that row. In this case the AND of X_i with itself or any other variable will return 0. We call a step *successful* if in that step the algorithm adds a new unseen terminal to the tree. For strict selection such a terminal is accepted if it reduces the error in the training set while for non-strict selection it is always accepted as the fitness cannot decrease. The other operators SUB and DEL do not contribute to fitness since DEL may only remove redundant terminals in the non-strict selection version and SUB may only exchange terminals that do not decrease fitness. If a terminal X_i is exchanged for a terminal X_j leading to an improvement (i.e., more rows of the training set have a 0) this may only speed up the algorithm. We consider a phase of k successful steps of the (1+1)-GP and calculate the probability that at the end of the k steps not even one variable with value 0 in a given row of the training set has been added. This probability is the probability that the first k terminals that are selected to be added to the tree have value 1 in the chosen row. Since for each row of the training set there are at most Y 1s, the probability that a terminal has a 1 in that position is at most Y/n and the probability that k consecutive terminals are all 1s in that position is bounded from above by $(Y/n)^k$. By the union bound, the probability that in any of the s rows of the training set all the k terminals have values 1 is less than $p_s = s \cdot (Y/n)^k$. We calculate the value of p_s after a phase of length $k = \log_{\frac{n}{Y}}(2s)$: $p_s \leq s \cdot \left(\frac{Y}{n}\right)^k = s \cdot \left(\frac{Y}{n}\right)^{\log_{\frac{n}{Y}}(2s)} = s \cdot \frac{1}{2s} = \frac{1}{2}$.

Hence with probability at least $1/2$, after $k = \log_{\frac{n}{Y}}(2s)$ successful steps, for each row of the training set the constructed tree contains at least one terminal X_i that has value 0 in that row. This implies that, in expectation, 2 phases of length k each are sufficient to reduce the error in the training set to 0. All that remains to be done is to calculate the expected time for $2k$ successful steps. Given that the tree consists of i different terminals, the probability that the (1+1)-GP adds a new terminal to the tree is $p_i = \frac{1}{3}\frac{n-i}{n}$ where $1/3$ is the probability that an INS operation is chosen. Hence, by a simple coupon collector argument the expected time to collect the AND of $2k = 2\log_{\frac{n}{Y}}(2s)$ different terminals is bounded by:

$$\sum_{i=0}^{2\log_{\frac{n}{Y}}(2s)-1} \frac{3n}{n-i} \leq \sum_{i=0}^{2\log_{\frac{n}{Y}}(2s)-1} \frac{3n}{n - (2\log_{\frac{n}{Y}}(2s) - 1)}$$

$$\leq \sum_{i=0}^{2\log_{\frac{n}{Y}}(2s)-1} \frac{3n}{cn} = O(\log_{\frac{n}{Y}}(2s)) = O(\log n)$$

Finally, we need to remember that the above expected time is conditional to starting with at most Y ones in each row of the training set. This may not happen with probability at most $e^{-\Omega(n)}$. If this is the case we pessimistically assume that a row of the training set consists of its worst case value of $n-1$ 1s and one 0. Then, by the same coupon collector argument used in Theorem 1, the conditional expected runtime is bounded from above by $O(n \log n)$. The statement of the theorem now follows by an application of the law of total expectation:

$$E(T) = p(X) \cdot E(T|X) + p(\overline{X}) \cdot E(T|\overline{X})$$
$$\leq (1 - e^{-\Omega(n)}) \cdot O(\log n) + e^{-\Omega(n)} \cdot O(n \log n) = O(\log n)$$

$\square$

Theorem 6. *Both the (1+1) GP and the (1+1) GP* using $F = \{AND\}$, $L = \{X_1, \ldots, X_n\}$ and a training set of size $s = poly(n)$, have a generalization ability on AND_n of $G \leq 1 - 2^{-\log(s)}$.*

Proof. Recall that the generalization error $\epsilon_g(h)$ of an expression h is the number of rows mismatching the target function. In the case of AND, $\epsilon_g(h) \geq ones(h) - 1$, since all but at most one 1s are mismatched.

Theorem 5 states that the algorithms will stop at an expression $\tilde{h}$ having at most $\log(s)$ variables, thus having at least $2^{n-\log(s)}$ ones. Thus $\epsilon_g(\tilde{h}) \geq 2^{n-\log(s)} - 1$. The generalization ability is then $G = 1 - \frac{\epsilon_g(\tilde{h})}{2^n} \leq 1 - 2^{-\log(s)}$. $\square$

Corollary 1. *A necessary condition to get a generalization ability greater than $1 - \epsilon$, is to have a training set of size $s > \frac{1}{\epsilon}$.*

Proof. From Theorem 6 the generalization ability is $G \leq 1 - 2^{-\log(s)}$. Thus:

$$1 - 2^{-\log(s)} > 1 - \epsilon \Leftrightarrow 2^{-\log(s)} < \epsilon \Leftrightarrow 1/s < \epsilon \Leftrightarrow s > 1/\epsilon$$

$\square$

4.2   Analysis for XOR_n with Incomplete Training Set

The analysis for XOR_n shows that if the training set size is at most $\log(n)$, the training set can be fit efficiently (Theorem 7). On the other hand if the size of the training set grows asymptotically faster than $\log n$, then the algorithms do not fit the training set in polynomial time with probability exponentially close to 1 (Theorem 8). However, in both cases the generalization ability will be equal to $1/2$ with probability exponentially close to 1 in the former case, and with probability 1 in the latter case (Theorem 9).

Theorem 7. *Let $s \leq \ln n$ be the size of the training set chosen from the truth table uniformly at random with repetition. Then the (1+1) GP using $F = \{XOR\}$ as function set and $L = \{X_1, \ldots, X_n\}$ as terminal set will find a solution fitting the training set of XOR_n in time $O(n^2)$ with probability at least $1 - e^{-\Omega(n)}$.*

Proof. Since each row of the training set is binomially distributed with parameters n and $p = 1/2$, also each variable X_i is binomially distributed, but with parameters s and $p = 1/2$ because the training set has s rows. An element j (i.e., $1 \leq j \leq s$) of the optimal solution *opt* fitting the training set has value 1 if the number of 1s in the n variables $X_1, \ldots, X_n$ at position j is odd while it has value 0 if the number of 1s at position j in the n variables is even. Hence, once the training set has been created, the optimal solution is determined and each variable X_i will have the same value as *opt* at each position j with probability $1/2$. By the principle of deferred decisions [8], the same holds for $X_i \oplus, \ldots, \oplus X_m$, $1 \leq i, m \leq n-1$ at each position j. In fact if the first $m-1$ terms $X_i, \oplus, \ldots, \oplus X_{m-1}$ have the "correct" value at position j, then with probability $1/2$ the output will still be correct after the first $m-1$ terms are XOR-ed with X_m (i.e., X_m has a 0 at position j). On the other hand if it was not correct it would become "correct" with probability $1/2$ (i.e., X_m has a 1 at position j). Thus

$$P(\text{j is correct}) = P(X_m(j) = 0 \mid \text{j was correct}) + P(X_m(j) = 1 \mid \text{j was not correct})$$
$$= 1/2 \cdot 1/2 + 1/2 \cdot 1/2 = 1/2$$

As a result the probability that any solution, $X_i \oplus, \ldots, \oplus X_m$, $1 \leq i, m \leq n-1$ is equal to *opt* in all positions is bounded by $p(X_t = opt) = \frac{1}{2^s} \geq \frac{1}{2^{\ln n}} = \frac{1}{n}$. Hence the probability that a solution $X_i \oplus, \ldots, \oplus X_m$, $1 \leq i, m \leq n-1$ is different from *opt* is less than $(1 - 1/n)$ and the probability that cn^2 solutions are all different than *opt* is bounded by

$$\left(1 - \frac{1}{n}\right)^{cn^2} \leq \left(1 - \frac{1}{n}\right)^{n \cdot cn} \leq \left(\frac{1}{e}\right)^{cn}$$

where $c < 1$ is a positive constant. As a result, after visiting cn^2 distinct solutions, with probability at least $1 - e^{-cn}$, the training set has been fitted by the algorithm.

All that remains to be shown is that cn^2 distinct solutions are visited by the (1+1)-GP. We consider a current solution of the algorithm $X_1, X_2, \ldots, X_m$, $\overline{X_{m+1}}, \ldots, \overline{X_n}$ where the $\overline{X_i}$ are the variables that are missing after simplification. By just considering an INS operation, from each such solution it is possible to reach n different neighbours (i.e., $n - i$ neighbours are reached by adding a missing variable and i neighbours are reached by adding one of the m variables, hence effectively removing a variable after simplification). If less than $(9/10)n$ of the neighbours on the current level have not been visited, the probability of visiting one is at least $1/3 \cdot 9/10 = 3/10 > 1/4$. Otherwise we consider the set of neighbours as "full" and look at the probability of moving to a new solution not having any neighbours in common with the current solution. Such a solution can be visited by either: (a) adding two missing variables; (b) removing two of the m variables in the current solution; (c) adding one missing variable and removing one of the m variables. Each of (a), (b), (c) can be achieved by performing two consecutive INS operations of which the first operation must lead to an accepted search point (i.e., which happens with probability at least $1/2$). If, after the first INS operations all the neighbours of $c'n$ different solutions (not having neighbours in common with the previous search point) were all "full" then $c'(9/10)n^2$ distinct solutions would have been visited (recall a neighbourhood is full when $(9/10)n$ neighbours have been visited). We set $c'(9/10) > c$. Hence, unless $(9/10)c'n^2$ distinct solutions have been seen, there must be at least a constant fraction $c''n$ of neighbours reached by the second INS that lead to unseen new solutions by another INS operation. Overall, the probability of reaching a new solution with 'unfilled neighbourhood' is at least $1/3 \cdot 1/3 \cdot 1/2 \cdot (c''n)/n = c^*$. Since each of the distinct solutions is found with constant probability, the expected time to find them all is bounded from above by $(c'n^2 \cdot c^*)$. By another application of Chernoff bounds with success probability c^* and a phase of length $2c'c^*n^2$, we get that $c'n^2$ distinct solutions have not been seen with probability at most $e^{-1/8 \cdot c* \cdot c'n^2}$. Summing up the failure probabilities proves the statement of the theorem.   □

We now consider larger training sets.

Theorem 8. *Let $s = \omega(\log n)$ be the size of the training set chosen from the truth table uniformly at random with repetition. Then the (1+1) GP using $F = \{XOR\}$ and $L = \{X_1, \ldots, X_n\}$ will not find a solution fitting the training set of XOR_n with fitness better than $(1-\delta)s/2$, $\delta > 0$ any constant, in polynomial time with probability at least $1 - e^{-\Omega(s)}$.*

Proof. As shown in the proof of Theorem 7 each column of the training set is binomially distributed with parameters s and $p = 1/2$, and by the principle of deferred decisions the same holds for each candidate solution $X_i \oplus, \ldots, \oplus X_m$, $1 \leq i, m \leq n-1$. As a result the probability that each row of a candidate solution is equal to the value in the respective row in *opt* with probability $p(Y_i) = 1/2$ and the expected number of rows where they are equal is $E(Y) = s/2$. Hence, by a simple application of Chernoff bounds, the probability that a candidate solution has $(1 + \delta)s/2$ rows that agree with the respective rows in *opt* is

$$P\left(Y > \frac{s}{2} + \delta\frac{s}{2}\right) \leq e^{-(\delta/6) \cdot s}$$

By the union bound the probability that after a polynomial number of steps n^c any seen solution agrees in $(1 + \delta)s/2$ rows is less than

$$n^c \cdot e^{-(\delta/6) \cdot s} \leq e^{c \log n - (\delta/6)s} \leq e^{-\Omega(s)}$$

for any $s = \omega(\log n)$, which proves the statement of the theorem.   □

Theorem 9. *Both the (1+1) GP and the (1+1) GP* using $F = \{XOR\}$, $L = \{X_1, \ldots, X_n\}$ and a training set of size $s = poly(n)$ have a generalization ability of $G = 1/2$.*

Proof. By Proposition 2 all the expressions but $X_1 \oplus \ldots \oplus X_n$ have a generalization error of 2^{n-1} thus if the algorithms do not find the expression $Y = X_1 \oplus \ldots \oplus X_n$ in polynomial time their generalization ability is $G = 1 - \frac{2^{n-1}}{2^n} = 1/2$. By Theorem 3 the (1+1) GP* will never move from its first point, hence does not find Y with probability $1 - 2^{-s}$. The (1+1) GP with $s = \omega(\log n)$ cannot fit the training set in polynomial time by Theorem 7, thus cannot find Y efficiently. When $s < \log(n)$ we use a similar argument to that of Theorem 4 to show that, even if the training set can be fit in polynomial time, the expression Y will not be found in polynomial time. In fact since with probability $1/2$ a new solution will have fitness better than its parent, the expected drift will be half of that calculated in Theorem 4 (i.e., in expectation half of the offspring will not be accepted). Thus the time to get to Y is still at least exponential with probability exponentially close to 1, which concludes the proof. $\qquad\square$

5   Conclusions

A further step has been made towards the rigorous computational complexity analysis of GP for evolving functions with desired input/output behaviour. We have analysed the (1+1) GP and (1+1)-GP* for evolving two common boolean functions, XOR_n and AND_n (which may be considered the GP analogues to OneMax and Needle for EAs) both in the case of complete input-output information and with incomplete training sets of polynomial size.

We have rigorously proved that AND_n can be efficiently solved by both the (1+1) GP and the (1+1) GP* in case of complete datasets, with (1+1) GP* producing shorter expressions. We have also shown that both the algorithms can efficiently fit a training set of polynomial size and provided a necessary condition to achieve a given generalization ability. The analysis for XOR_n reveals a needle-like fitness landscape, leading to the (1+1) GP* not being able to solve the problem at all and the (1+1) GP requiring exponential time. The analysis for the incomplete datasets has shown that there is a $\log(n)$ threshold for the size of the training set under which the training set can be fit efficiently and over which asymptotically it cannot be fit in polynomial time with probability exponentially close to 1. The analysis on the generalization ability has shown that, despite the size of the training set, the generalization ability is equal to $1/2$ with probability exponentially close to 1.

Future work will be directed, on one hand, towards extending the results to increasingly deal with more comprehensive terminal and function sets, while on the other hand will focus on more sophisticated GP systems and benchmark functions where typical bloat and overfitting problems can be studied.

Acknowledgements. The authors would like to thank Alberto Moraglio for constructive discussions which initialised this work. Further preliminary discussions occurred at Dagstuhl Seminar N.15211. This work was supported by EPSRC under Grant n. EP/M004252/1.

References

1. Auger, A., Doerr, B.: Theory of Randomized Search Heuristics: Foundations and Recent Developments. World Scientific, Singapore (2011)
2. Droste, S., Jansen, T., Wegener, I.: On the analysis of the (1+1) evolutionary algorithm. Theor. Comput. Sci. **276**(1–2), 51–81 (2002)
3. Kötzing, T., Sutton, A.M., Neumann, F., O'Reilly, U.M.: The max problem revisited: the importance of mutation in genetic programming. Theor. Comp. Sci. **545**, 94–107 (2014)
4. Koza, J.R.: Genetic Programming: On the Programming of Computers by Means of Natural Selection. MIT Press, Cambridge (1992)
5. Koza, J.R.: Genetic Programming II: Automatic Discovery of Reusable Programs. MIT Press, Cambridge (1994)
6. Langdon, W.B., Poli, R.: Foundations of Genetic Programming. Springer, Heidelberg (2002)
7. Luke, S.: Genetic programming produced competitive soccer softbot teams for RoboCup97. In: Proceedings of the Third Annual Conference on Genetic Programming 1998, pp. 214–222. Morgan Kaufmann (1998)
8. Mitzenmacher, M., Upfal, E.: Probability and Computing. Cambridge University Press, Cambridge (2005)
9. Moraglio, A., Mambrini, A., Manzoni, L.: Runtime analysis of mutation-based geometric semantic genetic programming on boolean functions. In: Proceedings of FOGA XII, pp. 119–132. ACM (2013)
10. Neumann, F., O'Reilly, U.M., Wagner, M.: Computational complexity analysis of genetic programming - initial results and future directions. In: Riolo, R., Vladislavleva, E., Moore, J.H. (eds.) Genetic Programming Theory and Practice IX. Genetic and Evolutionary Computation, pp. 113–128. Springer, New York (2011)
11. Neumann, F., Witt, C.: Bioinspired Computation in Combinatorial Optimization: Algorithms and Their Computational Complexity. Natural Computing Series. Springer, Heidelberg (2010)
12. Oliveto, P.S., Witt, C.: Simplified drift analysis for proving lower bounds in evolutionary computation. Algorithmica **59**(3), 369–386 (2011)
13. O'Neill, M., Vanneschi, L., Gustafson, S., Banzhaf, W.: Open issues in genetic programming. Genet. Program. Evolvable Mach. **11**(3–4), 339–363 (2010)
14. O'Reilly, U.-M., Oppacher, F.: Program search with a hierarchical variable length representation: genetic programming, simulated annealing and hill climbing. In: Davidor, Y., Männer, R., Schwefel, H.-P. (eds.) PPSN 1994. LNCS, vol. 866, pp. 397–406. Springer, Heidelberg (1994)
15. Poli, R., Langdon, W.B., McPhee, N.F.: A field guide to genetic programming (2008). http://lulu.com
16. Poli, R., McPhee, N.F., Rowe, J.E.: Exact schema theory and Markov chain models for genetic programming and variable-length genetic algorithms with homologous crossover. Genet. Program. Evolvable Mach. **5**(1), 31–70 (2004)
17. Spector, L., Barnum, H., Bernstein, H.J., Swamy, N.: Quantum computing applications of genetic programming. In: Spector, L., Langdon, W.B., O'Reilly, U.M., Angeline, P.J. (eds.) Advances in Genetic Programming 3, pp. 135–160. MIT Press, Cambridge (1999)
18. Valiant, L.G.: Evolvability. J. ACM **56**(1), 3:1–3:21 (2009)

Genetic Programming Based Hyper-heuristics for Dynamic Job Shop Scheduling: Cooperative Coevolutionary Approaches

John Park[1(✉)], Yi Mei[1], Su Nguyen[1,2], Gang Chen[1], Mark Johnston[1], and Mengjie Zhang[1]

[1] Evolutionary Computation Research Group, Victoria University of Wellington,
PO Box 600, Wellington 6140, New Zealand
{John.Park,Yi.Mei,Su.Nguyen,Aaron.Chen,Mengjie.Zhang}@ecs.vuw.ac.nz,
Mark.Johnston@msor.vuw.ac.nz
[2] Hoa Sen University, Ho Chi Minh City, Vietnam

Abstract. Job shop scheduling (JSS) problems are optimisation problems that have been studied extensively due to their computational complexity and application in manufacturing systems. This paper focuses on a dynamic JSS problem to minimise the total weighted tardiness. In dynamic JSS, attributes of a job are only revealed after it arrives at the shop floor. Dispatching rule heuristics are prominent approaches to dynamic JSS problems, and Genetic Programming based Hyper-heuristic (GP-HH) approaches have been proposed to automatically generate effective dispatching rules for dynamic JSS problems. Research on static JSS problems shows that high quality ensembles of dispatching rules can be evolved by a GP-HH that uses cooperative coevolution. Therefore, we compare two coevolutionary GP approaches to evolve ensembles of dispatching rules for dynamic JSS problems. First, we adapt the Multilevel Genetic Programming (MLGP) approach, which has never been applied to JSS problems. Second, we extend an existing approach for a static JSS problem, called Ensemble Genetic Programming for Job Shop Scheduling (EGP-JSS), by adding "less-myopic" terminals that take job and machine attributes outside of the scope of the attributes commonly used in the literature. The results show that MLGP for JSS evolves ensembles that are significantly better than single "less-myopic" rules evolved using GP with only little difference in computation time. In addition, the rules evolved using EGP-JSS perform better than the MLGP-JSS rules, but MLGP-JSS evolves rules significantly faster than EGP-JSS.

Keywords: Dynamic job shop scheduling · Multilevel GP · Ensemble GP

1 Introduction

Job shop scheduling (JSS) problems are combinatorial optimisation problems that have been studied over the past 50 years [17]. In a JSS problem instance, there are typically a fixed number of machines on the shop floor. Jobs arrive

© Springer International Publishing Switzerland 2016
M. Heywood et al. (Eds.): EuroGP 2016, LNCS 9594, pp. 115–132, 2016.
DOI: 10.1007/978-3-319-30668-1_8

on the shop floor over time, and need to be processed on a specific sequence of machines. However, a machine can only process one job at a time. The sequence of jobs and the times when the jobs are selected to be processed by a machine is called a *schedule*. A JSS problem instance has a measure of "quality" for the final schedule called an *objective function* [14]. Therefore, the goal for solving a JSS problem instance is to find the schedule that will give the best possible outcome according to the objective function.

Various approaches to different JSS problems have been proposed in the literature. Exact mathematical optimisation techniques have been proposed [14,17] to find the *optimal* (best feasible) solution to a static JSS problem instance. In a static JSS problem instance the properties of the jobs and the machines on the shop floor are known *a priori*. On the other hand, in *dynamic* JSS problem instance [10] unforeseen events occur and affect the shop floor, e.g., the breakdown of a machine. Unforeseen events mean that it is not possible to determine whether a schedule is optimal until all arriving jobs have been processed, meaning that mathematical optimisation techniques are not suitable for dynamic JSS problems. Instead, researchers have proposed various heuristic techniques for dynamic JSS problems. Heuristic techniques range from simple dispatching rules [14] to complex meta-heuristic techniques [17]. A dispatching rule is a heuristic that iteratively selects a job to be processed by a machine whenever the machine becomes available. Dispatching rule approaches to dynamic JSS problem are prominent in the literature because in practice they can react quickly and effectively to unforeseen changes on the shop floor [9]. In addition, recent approaches have proposed *Genetic Programming based Hyper-heuristic* (GP-HH) approaches for dynamic JSS problems [1,2]. A GP-HH automatically evolves dispatching rules for a dynamic JSS problem, bypassing the need for human experts and extensive trial-and-error testing required to construct a dispatching rule manually. In addition, GP evolved rules generally perform better than manually designed rules [1]. However, there are many challenges that can arise when evolving dispatching rules using GP-HH. To generate a schedule for a JSS problem instance using a dispatching rule, there can be potential scenarios where complex decisions need to be made. Decisions made early in the schedule can greatly affect the overall quality of the schedule. However, many existing GP-HH approaches evolve "myopic" dispatching rules. These rules only take into account the current state of the job and the machine that the job is waiting at when selecting a job to process [5]. This can result in situations where the evolved rules make good local decisions, but make poor global decisions. In addition, most GP-HH approaches evolve single dispatching rules to handle dynamic JSS problems. In classification, research has shown that an *ensemble* of decision makers (i.e. a diverse set of "experts" that cover for each other's errors) generally performs better than a single decision maker for handling a complex and difficult classification problem [15]. Therefore, it may be more effective to evolve ensembles of dispatching rules instead of evolving single dispatching rules for JSS problems. However, research into evolving ensembles of dispatching rules for JSS problems has been limited.

The goal of this paper is to investigate two GP-HH approaches and determine which of the two approaches will evolve higher quality ensembles of dispatching rules for a dynamic JSS problem. The two GP-HH approaches combine coevolutionary techniques with GP to evolve ensembles, which have not been applied to dynamic JSS problems previously. The first GP-HH approach is an adaptation of the approach called Multilevel Genetic Programming (MLGP) [20]. MLGP has not previously been applied to JSS problems. MLGP is promising due to its ability to automatically find a group of individuals that work together effectively. The adaptation of MLGP for the dynamic JSS problem will be denoted as Multilevel Genetic Programming for Job Shop Scheduling (MLGP-JSS). In addition, a set of "less-myopic" terminals proposed by Hunt et al. [5] will be used in MLGP-JSS. These terminals consist of job and machine attributes outside of the attributes associated with the current state of a job and the machine that the job is waiting at. By combining job and machine attributes which can reduce the myopic nature of dispatching rules [5] with MLGP-JSS, it may be possible to evolve a set of rules that can cover for each other's errors by handling different "locality" of decisions.

The second GP-HH approach is an extension of an existing GP-HH approach called Ensemble Genetic Programming for Job Shop Scheduling (EGP-JSS). EGP-JSS is an approach proposed by Park et al. [12], and uses Potter and De Jong's *cooperative coevolution* [16] to evolve ensembles of dispatching rules for a static JSS problem. This will be updated with the "less-myopic" terminals and will be applied to the dynamic JSS problem for the first time. Therefore, the research objectives in this paper are: (1) Adapting the MLGP framework to the dynamic JSS problem. (2) Incorporate the "less-myopic" terminals to the both MLGP-JSS and EGP-JSS for the dynamic JSS problem. (3) Comparing the evolved rules from MLGP-JSS, EGP-JSS and a benchmark GP-HH that evolves single dispatching rules.

2   Background

This section firstly describes the notation and definitions for JSS, and approaches to dynamic JSS problems. Afterwards, GP-HH approaches to JSS are covered, along with some Cooperative Coevolutionary Algorithms (CCEAs) which have been proposed in the literature.

2.1   Dynamic Job Shop Scheduling

The notation used for JSS problem instances is as follows. In a JSS problem instance, there are M machines on the shop floor. A job j arrives at the shop floor requiring a sequence of N_j operations $\sigma_{1j}, \ldots, \sigma_{N_j j}$ which need to be processed in order for the job to be completed. The operations must be processed sequentially, i.e., operation σ_{2j} for a job j cannot be processed before σ_{1j}, etc. Each operation σ_{ij} of job j needs to be processed by a specific machine $m(\sigma_{ij})$. Job j needs to be processed at the machine m for time $p(\sigma_{ij})$ (shortened to p_{ij}), during which machine m cannot process any other jobs. The time when the last operation of job j is completed is called the completion time, denoted as C_j. The time when job j arrives at a machine to process operation σ_{ij} is called the arrival time, denoted as $r(\sigma_{ij})$. This means that job j arrives at the shop floor at time $r(\sigma_{1j})$, which is abbreviated to r_j. In a JSS problem with total weighted tardiness (TWT) minimisation objective, there are two additional attributes associated with a job j: a due date d_j and a weight w_j. Job j is considered tardy if its completion time C_j is greater than its due date. Job j's tardiness T_j is given by $T_j = \max\{C_j - d_j, 0\}$. The goal of the JSS problem with TWT minimisation is to process all N jobs while minimising $\sum_{j=1}^{N} w_j T_j$.

A dispatching rule is applied to a JSS problem instance as follows. When a machine m becomes available, the dispatching rule first determines the set of jobs that can potentially be selected. The decision to select the job to be processed by machine m is called a *dispatching decision*. If the dispatching rule only considers the jobs already waiting at the machine, then it is said to generate a *non-delay* schedule [14]. On the other hand, a dispatching rule that considers future jobs which arrive earlier than a job which can be completed on the machine before the future job's arrival is said to generate an *active* schedule [14]. From the considered jobs, the dispatching rule selects a job to begin processing based on the job and the machine attributes. This process is repeated until all jobs that have arrived on the shop floor are completed and a schedule is generated. For example, shortest processing time (SPT) is a dispatching rule in which the SPT rule selects the job with the shortest processing time to be processed at a dispatching decision.

Constructing dispatching rules which use a good combination of job and machine attributes has been shown to give better performance than dispatching rules which use single attributes for specific dynamic JSS problems [6]. Therefore, many priority-based dispatching rule approaches have been proposed for dynamic JSS problems [6]. A priority-based dispatching rule consists of a priority function, which takes job, machine and shop floor attributes as inputs. The priority function assigns priority values to jobs waiting at a machine. The job with the highest priority is then selected to be processed by the machine. Examples of well-known priority-based dispatching rules are the apparent tardiness cost (ATC) and cost over time (COVERT) rules [19] for dynamic JSS problems with the TWT minimisation objective.

2.2   Genetic Programming Based Hyper-heuristics for Dynamic JSS

Many GP-HH approaches have been proposed to generate reusable scheduling heuristics. In most cases, the output heuristic is a dispatching rule [1]. Hildebrandt et al. [3] evolved priority-based dispatching rules for a dynamic JSS problem that models a semiconductor manufacturing environment. They showed that the rules evolved using GP perform significantly better than existing man-made dispatching rules. Nguyen et al. [9] explored three different GP representations for static JSS problems to evolve dispatching rules, and applied the evolved rules to a dynamic JSS problem. Although the evolved dispatching rules outperformed benchmark man-made dispatching rules for the static JSS problems, they did not perform as well as man-made dispatching rules for the dynamic JSS problem. Pickardt et al. [13] proposed a two-step procedure, where priority-based dispatching rules are evolved using GP, and combined with existing rules using a genetic algorithm (GA) to assign the rules to specific machines on the shop floor. The two-step approach performed significantly better than evolving rules using GP or allocating rules to specific machines using GA separately, and outperformed the state-of-the-art approaches to a dynamic semiconductor manufacturing problem. Hunt et al. [4] evolved priority-based dispatching rules for a dynamic two-machine JSS problem with TWT minimisation objective. The evolved rules performed significantly better than some benchmark dispatching rules. Hunt et al. [5] then proposed a set of "less-myopic" terminals to the GP process to reduce the myopic nature of dispatching rules. They showed that the rules evolved with the added terminals performed significantly better than the benchmark GP-HH approach which uses the standard job and machine properties.

2.3 Cooperative Coevolution in Genetic Programming

CCEAs are techniques where behaviours from multi-agent systems are incorporated with evolutionary computation [11]. First, the problem is decomposed into smaller subproblems. Afterwards, the agents that make up the multi-agent system are applied to the subproblems. In CCEA approaches, the aim is to allow different individuals of the population to fill different "ecological niches" [16]. The evolved individuals that can best handle the different niches are then combined together as a cohesive solver. A survey of CCEA techniques is provided by Panait and Luke [11].

Nguyen et al. [8] used Potter and De Jong's cooperative coevolution [16] to evolve multiple scheduling policies for a dynamic JSS problem with three objectives. In their approach, they used two subpopulations. The individuals in the first subpopulation were used to assign due dates to the jobs during a dispatching decision. The individuals in the second subpopulation were used as priority-based dispatching rules. Park et al. [12] used cooperative coevolution to evolve ensembles of dispatching rules for a static JSS problem with the makespan minimisation objective. In this approach, individuals in all subpopulations are "voters", and vote on the job to be selected for processing. Other CCEA techniques include Orthogonal Evolution of Teams (OET) [18] and MLGP [20]. OET groups individuals into separate "teams" that compete against other teams during the selection procedure. MLGP is a CCEA which evolves multiple levels of evolution simultaneously. Neither OET and MLGP have previously been applied to dynamic JSS problems. In addition, Potter and De Jong's cooperative coevolution has not been applied to dynamic JSS problems to evolve ensembles of dispatching rules.

3 Coevolutionary GP Approaches to JSS: MLGP-JSS and EGP-JSS

The two coevolutionary GP approaches for evolving ensembles of dispatching rules to the dynamic JSS problem with TWT minimisation objective are given below. The first approach, MLGP-JSS, adapts the MLGP approach of Wu and Banzhaf [20]. The description of MLGP-JSS is broken down into two steps: a high-level overview of the approach, and the evaluation procedure, which includes the fitness functions used. The second approach extends EGP-JSS proposed by Park et al. [12] that uses Potter and De Jong's cooperative coevolution [16]. Finally, the representation of individuals used for MLGP-JSS and EGP-JSS is described.

3.1 MLGP-JSS Process Overview

A key component of MLGP-JSS is *groups*. A group is a set of individuals in the population that cooperate with each other. For our approach, we use groups as ensembles to solve JSS problem instances. This is discussed in further detail in Sect. 3.3. The MLGP-JSS process is broken down into three major steps. The first step is to carry out evolution on the *group level*, where groups are bred, evaluated and added to the GP population. The second step is to carry out evolution on the *individual level*, where GP individuals are bred, evaluated and added to the population. The final step is the selection procedure, where only the elite groups and individuals are retained in the population for the next generation. After the termination criterion, i.e., the maximum number of generations, is reached, the final output is the best group of individuals found so far, an ensemble of dispatching rules that can be applied to dynamic JSS problem instances. The overall MLGP-JSS process is shown in Algorithm 1, where G_B is the number of groups bred at each generation, and I_B is the number of individuals bred at each generation.

Evolution on the Group Level: There are three evolutionary operators which breed new groups from existing individuals and groups in the population. The first operator is *cooperation*. Cooperation combines two *entities* together to form a new group containing all individuals from both entities without duplicates. An entity can be either an individual or a group. This means that two individuals, an individual and a group, or two groups can be merged to form another group. The entities for cooperation are selected using roulette wheel selection over all entities. An example is shown in Fig. 1a, where group G1 is combined with group G2 to form the new group G4, which contains individuals I1, I2 and I4.

The second evolutionary operator on the group level is the group crossover operator. In group crossover, roulette wheel selection selects two groups from the GP population as parents. The two parents randomly exchange one individual to produce the child groups. Individuals in a parent group have equal probabilities of being exchanged. In Fig. 1b, crossover occurs between parents G1 and G3, exchanging individuals I2 and I3 respectively. This generates groups G4 and G5.

Algorithm 1: MLGP-JSS

Initialise Population P with I_R Individuals;
Evaluate Fitnesses of Individuals $idv_1, \ldots, idv_{I_R} \in P$;
while *Maximum generation is not reached* **do**
    for $g \leftarrow 1$ *to* G_B **do**
        Breed a Group grp from Population P;
        Evaluate the Fitness of Group grp;
        Add Group grp to Population P;
    end
    for $i \leftarrow 1$ *to* I_B **do**
        Breed an Individual idv from Population P;
        Evaluate the Fitness of Individual idv;
        Add Individual idv to Population P;
    end
    Retain the G_R Best Groups and the I_R Best Individuals from Population P;
end

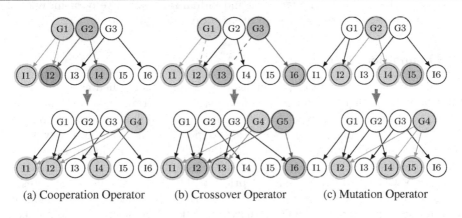

(a) Cooperation Operator   (b) Crossover Operator   (c) Mutation Operator

Fig. 1. Examples of group operators used to breed new groups.

The final evolutionary operator on the group level is the group mutation operator. In the group mutation, a group is selected through roulette wheel selection, and either an individual is added to or removed from the group. If an individual is being added, then an individual is selected through roulette wheel selection over the individuals in the GP population. If an individual is being removed, then an individual in the group is randomly selected with uniform probability. An example of mutation operator adding an individual is shown in Fig. 1c, where individual I5 is added to G2 to produce offspring G4.

Evolution on the Individual Level: Crossover and mutation operators used to breed the individuals are the standard operators for tree-based GP [7]. To select the parent individuals to breed new offspring, a group grp is selected based on a probability proportional to the group's fitness. It is expected that individuals in a group contribute to achieve cooperation despite their fitnesses [20].

Therefore, an appropriate number of parent individuals are selected from group grp with a uniform probability. After crossover or mutation is carried out, the newly bred children do not automatically become part of the group from which the parent individuals were selected from, but are inserted back into the pool of individuals after evaluating the children's fitness.

3.2 Selection

After G_B groups and I_B individuals are breed and added to the population, the selection procedure retains the G_R groups with the highest fitnesses out of all the groups in the population, while discarding the groups with fitnesses lower than the top G_R groups. A similar procedure is then applied to the individuals, where the top I_R fittest individuals in the population are retained and the rest discarded. However, if an individual idv does not belong to the top I_R fittest individuals, but belongs to a group grp, then individual idv is still retained by group grp. This is due to the fact that individual idv may have poor individual fitness, but works well together with other individuals in group grp.

3.3 Evaluation Procedure

To evaluate a group grp in the population, group grp is applied as a non-delay [14] dispatching rule to the set Δ_{train} of JSS training instances. When a machine becomes available in a discrete-event simulation of the job shop on JSS problem instance I, the individuals that are part of group grp act as an ensemble, "voting" on the next job to be processed by the machine. An individual in group grp first assigns priorities to the non-delay jobs waiting at the machine, and then votes on the job with the highest assigned priority. The job with the most votes is then selected by the group to be processed. In the case of a tie in the number of votes between two jobs, the ATC dispatching rule [19] is used as a tie-breaker. The job with the higher priority assigned by the ATC rule is ranked higher than the other job. An example of the voting procedure is shown in Fig. 2 for a group consisting of 3 individuals. The individuals each vote for one of four jobs waiting at the machine to be processed. Rule 1 votes for job 1 which has the priority of 9.787, rule 2 votes for job 2, and rule 3 votes for job 1, resulting in job 1 being selected to be processed.

After group grp is applied to the training instances, we get the fitness Fit_{grp} of group grp as shown in Eq. (1). The group fitness function used as the dynamic JSS problem is modified from the group fitness function proposed by Wu and Banzhaf [20]. In the equation, $Obj(grp, I)$ is the TWT objective value of the solution generated for a JSS problem instance I by group grp. GS_{grp} denotes the size of group grp, and is used as a penalty factor to prevent groups from increasing in size with minimal improvement in performance [20].

$$Fit_{grp} = \frac{1}{|\Delta_{train}|} \sum_{I \in \Delta_{train}} Obj(grp, I) + \sqrt{\frac{2 \times GS_{grp}}{2 + GS_{grp}}} \tag{1}$$

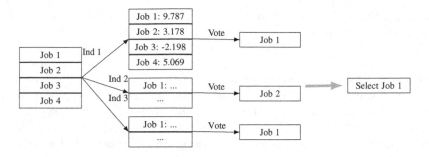

Fig. 2. An example of a voting procedure for individuals in a group.

On the other hand, to evaluate an individual idv, individual idv is applied to the JSS training instances as a non-delay priority-based dispatching rule. The fitness of individual idv, denoted as Fit_{idv}, is the average TWT of the solutions over the training instances as shown in Eq. (2).

$$Fit_{idv} = \frac{1}{|\Delta_{\text{train}}|} \sum_{I \in \Delta_{\text{train}}} Obj(idv, I) \tag{2}$$

3.4 EGP-JSS Process Overview

The EGP-JSS approach incorporates Potter and De Jong's cooperative coevolution [16] to evolve ensembles of dispatching rules. EGP-JSS is an extension of a previous approach by Park et al. [12] for a static JSS problem, where they provide a full description of the EGP-JSS approach. In cooperative coevolution [16], a population of individuals is partitioned into subpopulations, and individuals from each subpopulation only interact with representatives of other subpopulation. A representative in a subpopulation is the best individual of the subpopulation found so far. To evaluate an individual of a subpopulation, the individual is grouped up with the representatives of the other subpopulations to form an ensemble of dispatching rules. The ensemble is then applied to the training instances. In EGP-JSS, an ensemble's voting procedure for selecting jobs is similar to the ensemble's voting procedure in MLGP-JSS, i.e., the job with the majority vote is selected to be processed, and the ATC rule is used as the tie-breaker between jobs with equal number of votes. The fitness of the ensemble over the training instances is used as the fitness of the individual.

3.5 GP Representation, Terminals and Function Sets

For MLGP-JSS and EGP-JSS, we use tree-based GP, where the individuals represent arithmetic function trees. An individual's function tree is used as a priority function in a non-delay dispatching rule. When a machine m^* becomes available at time R_{m^*}, the non-delay jobs waiting in the queue in front of machine m^* are considered for selection. The set of non-delay jobs in the queue in front of machine m^* will be denoted as A_{m^*}. The priority of the operation for a job $j \in A_{m^*}$ waiting at machine m^* is calculated by inputting the attributes of j and the machines on the shop floor into the function tree.

The terminals in the GP terminal set either correspond to base level attributes of jobs and machines or are constructed from multiple base level attributes. Examples of a base level attribute are the due date (d_j) or the processing time $(p(\sigma_{kj}))$ of the kth operation of job j. As an example, suppose that job j is waiting at machine m^* during a dispatching decision, requiring the ith operation σ_{ij} to be processed by the machine. When job j is inputted into an individual's function tree, the DD terminal corresponds to the due date d_j of job j. The RT terminal corresponds to the sum of the remaining total processing times of input job j from the ith operation to the final operation, i.e., $\sum_{k=i}^{N_j} p(\sigma_{kj})$.

The terminal set consists of "basic" terminals and the "less-myopic" terminals. The basic terminals are terminals used by Nguyen et al. [9] to evolve rules using arithmetic representations for individuals in the GP population. The less-myopic terminals are terminals proposed by Hunt et al. which are used to evolve

Table 1. The terminal set used for the GP representation, where job j is one of the jobs waiting at the machine m^* to process operation σ_{ij}.

Terminal		Description	Value
Basic	RJ	Operation ready time	$r(\sigma_{ij})$
	RO	Remaining number of operations of job j	$N_i - m + 1$
	RT	Remaining total processing times of job j	$\sum_{k=i}^{N_j} p(\sigma_{jk})$
	PR	Operation processing time of job j	$p(\sigma_{ij})$
	RM	Machine m^* ready time	R_{m^*}
	NJ	Non-delay jobs waiting at machine m^*	$|A_{m^*}|$
	DD	Due date of job j	d_j
	W	Tardiness penalty of job j	w_j
	#	Constant	Uniform$[0,1]$
Less-myopic	NPR	Next operation processing time	$p(\sigma_{(i+1)j})$
	NNQ	Number of idle jobs waiting at the next machine	$|A_{m(\sigma_{(i+1)j})}|$
	NQW	Average waiting time of last 5 jobs at the next machine	$\frac{1}{5}\sum_{k=1}^{5} q_{(N_{m_j^{**}}+1-k)m_j^{**}}$
	AQW	Average waiting time of last 5 jobs at all machines	$\frac{1}{5M}\sum_{m=1}^{M}\sum_{k=1}^{5} q_{(N_m+1-k)m}$
Function		$+, -, \times, /,$ if, max, min	

"wider-looking" rules [5]. The less-myopic terminals are used because they allow individuals that are part of the ensemble to focus on different "locality" of the decisions, with some individuals focusing on very local decisions and other individuals focusing on decisions further down the line. An example of Hunt et al.'s [5] less-myopic terminals is the average wait time of last five jobs processed at the next machine (NQW). Let $q_{1m}, \ldots, q_{N_m m}$ be the waiting times of the 1st, $\ldots$, N_mth jobs processed by machine m, and let m_j^{**} be the machine for which the job j's next operation is to be processed on. Then the NQW terminal corresponds to $\frac{1}{5} \sum_{k=1}^{5} q_{(N_{m_j^{**}}+1-k), m_j^{**}}$ for j. On the other hand, the function set consists of arithmetic operators $+, -, \times$, protected $/$, if, max and min. Protected division returns a value of 1 if the denominator is zero. if is a ternary operator which returns the value of the second argument if the value of the first argument is ≥ 0, and returns the value of the third argument otherwise. max and min are binary operators which return the maximum and minimum of the arguments respectively. The full list of terminals and functions are given in Table 1.

4   Experimental Design

This section covers the dataset that is used for training and testing. A description of the benchmark GP-HH approach used to compare with MLGP-JSS and EGP-JSS will also be presented. In addition, this section gives the GP parameter settings used to evolve the rules.

4.1   Dataset

The effectiveness of evolved rules can be significantly affected by the number and the frequency of job arrivals [3,9]. In addition, long-running discrete-event simulations are commonly used to evaluate rules for dynamic JSS problems [1]. On the other hand, applying a dispatching rule within a simulation is expensive, and a large training set may not necessarily lead to significantly higher quality heuristics [1]. Therefore, the dataset that is used for evaluating the GP-HH approaches are discrete-event simulations that were used by Hunt et al. [5], where there are a small number of training and test problem instances, each with large number of job arrival. Each problem instance in the dataset has 10 machines, and job attributes are randomly generated. A job's operations have random processing times that follow a discrete uniform distribution with mean μ. Jobs arrive according to a Poisson process with mean λ, where λ is configured so that it meets a desired utilisation rate ρ. Therefore, $\lambda = \frac{\rho \times p_M}{(1/\mu)}$, where p_M is the expected number of operations per job divided by the number of machines. If all jobs arriving on the shop floor require 4 operations in a 10 machine environment shop, then $p_M = 0.4$. Due date d_j of job j is given by $d_j = r_j + h \sum_{i=1}^{N_j} p(\sigma_{ji})$, where h is the due date tightness parameter. For each problem instance, a list of possible due date tightness values is provided. A job's due date tightness h is randomly selected from the list with an equal probability. Weight w_j for job j is selected from 1, 2, or 4 with probabilities of 0.2, 0.6 and 0.2 respectively [14].

Table 2. Configurations used for the discrete-event simulation representing dynamic JSS problem instances.

Training, 4op	Mean processing time, μ	25
	Expected utilisation rate, ρ	0.85, 0.95
	Due date tightness, h	$\{3, 5, 7\}$
	# of operations per job	4
	# of configurations	2
Training, 8op	Mean processing time, μ	25
	Expected utilisation rate, ρ	0.85, 0.95
	Due date tightness, h	$\{3, 5, 7\}$
	# of operations per job	8
	# of configurations	2
Testing	Mean processing time, μ	25, 50
	Expected utilisation rate, ρ	0.90, 0.97
	Due date tightness, h	$\{2, 4, 6\}$
	# of operations per job	$4, 6, 8, 10, X \sim \text{Unif}(2, 10)$
	# of configurations	20

There is a warm up period of 500 jobs. Thereafter, the weighted tardiness values of subsequent jobs are added to the objective function value. The simulation ends once $N = 2500$ jobs have been processed.

To evolve the rules, there are two training sets which have a different number of operations for each job. A problem instance in the first training set, 4op, has 4 operations per job. A problem instance in the second training set, 8op, has 8 operations per job. Both training sets have 2 problem instances, where the first problem instance in both sets have $\rho = 0.85$, and the second problem instance in both sets have $\rho = 0.95$. On the other hand, the test set has 20 problem instances, divided into four subsets of five problem instances. In a subset, the first four problem instances have $4, 6, 8, 10$ operations per job. The last problem instance in the subset has anywhere between 2 and 10 operations per job, where the number of operations per job is random with an equal probability. Therefore, $p_M = 0.4, 0.6, 0.8, 1.0, 0.6$ for each problem instance in the subset respectively. The five instances in each of the four subsets have $\mu = 25, 50, 25, 50$ and $\rho = 0.90, 0.90, 0.97, 0.97$. For all problem instances the machine which the operation is carried out is random, and there is no re-entry for the jobs, i.e., no two operations for a single job occur on the same machine. The overall parameter settings for the dataset are shown in Table 2.

4.2   GP-HH Benchmark Methods for Comparison

As MLGP-JSS has not been applied to JSS problems, it will first be compared against a benchmark GP-HH approach. As the most prominent method for evolving dispatching rules [1], the benchmark GP-HH approach in this paper evolves single priority-based dispatching rules. The benchmark GP-HH will be abbreviated to GP-JSS, and uses the same terminals as both MLGP-JSS and EGP-JSS (Table 1), which includes the less-myopic terminals.

4.3   Parameter Settings

The following parameters are used for GP-JSS, EGP-JSS and MLGP-JSS. The total population size for all three GP-HH is set to 1024. Similar to parameter configurations used by Park et al. [12], the population for EGP-JSS is divided to 4 subpopulations of size 256. The rest of the GP parameter configuration is the same as the configuration suggested by Koza [7]. However, tournament selection is used in the MLGP process when there are no groups in the population. The additional parameters for MLGP-JSS are the number of individuals bred (I_B), the number of groups bred (G_B), the number of groups retained (G_R), and the group cooperation/crossover/mutation rate. The number of individuals bred is equal to the number of individuals in the population, i.e., 1024. MLGP-JSS starts off with no groups in the population, and the number of groups bred and retained are set to 200 and 100 respectively. The probability of using a specific group operator is based off the probability used by Wu and Banzhaf [20] for their MLGP approach. Finally, the k value for the tie-breaker ATC heuristic is set to k [19]. All these parameter settings are listed in Table 3.

Table 3. GP parameters used by the GP-HH approaches for evolving ensembles

Approaches		GP-JSS	EGP-JSS	MLGP-JSS
Population related	Number of subpopulations	1	4	1
	Subpopulation size	1024	256	1024
Common Parameters	Generations	51		
	Crossover rate	80%		
	Mutation rate	10%		
	Reproduction rate	10%		
	Max-depth	8		
	Selection method	Tournament Selection		
	Selection size	7		
MLGP Specific	Number of groups breed G_B	-	-	200
	Number of groups retained G_R	-	-	100
	Group cooperation rate	-	-	31.25%
	Group crossover rate	-	-	50%
	Group mutation rate	-	-	18.75%
Tie-breaker Specific	ATC k value	-	3.0	3.0
Function set		$if, +, -, \times, /, max, min$		
Terminal set		as shown in Table 1		

5   Results

This section covers the results of the evaluation between the different GP-HH approaches, which is broken down into two steps. First, MLGP-JSS will be evaluated against GP-JSS. Second, MLGP-JSS will be evaluated against EGP-JSS in order to determine which CCEA approach out of the two performs better on dynamic JSS problems. For each GP-HH approach, rules are evolved from the training sets 4op and 8op over 30 independent runs. The computation times taken to evolve a rule for the GP-HH are also measured and analysed.

5.1   MLGP-JSS vs GP-JSS

The evolved rules from MLGP-JSS and GP-JSS are applied to the 20 JSS problem instances in the test set. From this, we get sets of TWT solution values for each JSS problem instance that are used to compare the performance of the rules. For a fair comparison, the MLGP-JSS rules evolved over a specific training set are compared against the GP-JSS rules evolved over the same training set, e.g., rules evolved over training set 4op are compared against each other. One set of rules is significantly better than another set of rules for a particular JSS problem instance if the obtained p-value under the one-sided Z-test is less than 0.05. The results are shown in Table 4. In the table, $\langle x, y \rangle$ denotes that $\mu = x$ and $\rho = y$ for the JSS problem instance. The problem instance for which the MLGP-JSS rules are significantly better than the GP-JSS rules are highlighted, and vice versa.

Table 4. The evolution times and TWTs for the MLGP-JSS rules and the GP-JSS rules over the training and the test JSS problem instances.

Approach		4op		8op	
		MLGP-JSS	GP-JSS	MLGP-JSS	GP-JSS
Evolution Time (s)		7079 ± 1532	6701 ± 3337	6202 ± 1456	5418 ± 1932
	Training Set	4.74 ± 1.53	4.69 ± 1.54	4.74 ± 1.55	4.82 ± 1.85
TWT ($\times 10^5$)	$\langle 25, 0.90 \rangle$	4.50 ± 0.33	5.17 ± 1.17	4.52 ± 0.33	5.37 ± 1.35
		4.61 ± 0.51	5.34 ± 1.88	4.70 ± 0.66	5.26 ± 1.81
		11.88 ± 0.88	14.27 ± 4.83	11.93 ± 0.92	14.70 ± 4.87
		8.68 ± 0.64	9.83 ± 2.20	8.67 ± 0.64	10.31 ± 2.79
		17.46 ± 1.66	20.08 ± 3.31	17.44 ± 1.56	20.59 ± 4.19
	$\langle 50, 0.90 \rangle$	12.66 ± 0.98	14.33 ± 3.06	12.54 ± 0.98	14.95 ± 3.90
		31.89 ± 3.03	38.30 ± 6.34	31.76 ± 3.15	39.43 ± 6.99
		35.32 ± 5.77	43.18 ± 7.45	35.09 ± 5.88	44.37 ± 8.71
		3.56 ± 0.33	4.61 ± 1.79	3.62 ± 0.38	4.84 ± 2.03
		25.92 ± 3.01	34.27 ± 9.20	25.69 ± 2.33	35.64 ± 11.20
	$\langle 25, 0.97 \rangle$	40.13 ± 4.18	45.96 ± 6.14	40.03 ± 4.31	46.20 ± 6.09
		64.07 ± 14.23	92.04 ± 58.84	63.98 ± 13.77	88.68 ± 38.84
		11.23 ± 1.01	13.54 ± 3.73	11.43 ± 1.00	13.99 ± 4.45
		15.53 ± 1.78	21.71 ± 9.24	15.46 ± 1.71	20.74 ± 7.30
		73.13 ± 3.90	84.67 ± 11.92	75.45 ± 8.97	84.82 ± 10.85
	$\langle 50, 0.97 \rangle$	79.52 ± 5.96	93.83 ± 22.25	79.96 ± 6.85	96.82 ± 18.52
		6.22 ± 0.58	10.29 ± 17.80	6.09 ± 0.46	7.13 ± 2.10
		11.61 ± 1.01	14.00 ± 3.56	11.44 ± 0.86	14.45 ± 3.87
		21.01 ± 1.90	26.61 ± 5.95	20.91 ± 1.83	26.59 ± 4.96
		56.86 ± 4.46	72.84 ± 11.19	57.54 ± 5.05	72.57 ± 13.03

From the results, the MLGP-JSS rules perform significantly better than the GP-JSS rules for all problem instances. In addition, although the time required to evolve the MLGP-JSS rules is slightly greater than the time required to evolve GP-JSS rules, the difference is not statistically significant. Therefore, the MLGP-JSS is more effective than the standard GP-HH approach for the dynamic JSS problems with TWT minimisation objective. This supports the hypothesis that an ensemble of dispatching rules is more capable of handling complex decisions in a dynamic JSS problem instance than a single rule. Hence, it would be promising to carry out further research into evolving new and effective ensembles of rules using coevolutionary GP for dynamic JSS problems.

5.2   MLGP-JSS vs EGP-JSS

The MLGP-JSS and EGP-JSS rules evolved from the same training set Δ_{train} are compared against each other over the test set in terms of the qualities of the solutions generated, i.e., the TWT of the solutions for the JSS problem instances. The results are shown in Table 5, where a set of rules that is significantly better than the other set of rules for a particular JSS problem instance is highlighted.

Table 5. The evolution times and TWTs for the MLGP-JSS rules and the EGP-JSS rules over the training and the test JSS problem instances.

Approach		4op		8op	
		MLGP-JSS	EGP-JSS	MLGP-JSS	EGP-JSS
Evolution Time (s)		7079 ± 1532	16555 ± 5468	6202 ± 1456	21128 ± 7243
	Training Set	4.74 ± 1.53	3.83 ± 1.00	4.74 ± 1.55	3.81 ± 0.94
	⟨25, 0.90⟩	4.50 ± 0.33	3.82 ± 0.62	4.52 ± 0.33	3.82 ± 0.66
		4.61 ± 0.51	4.50 ± 0.92	4.70 ± 0.66	4.57 ± 1.09
		11.88 ± 0.88	12.88 ± 3.45	11.93 ± 0.92	12.51 ± 2.35
		17.46 ± 1.66	15.95 ± 2.54	17.44 ± 1.56	16.46 ± 2.94
		8.68 ± 0.64	7.45 ± 1.30	8.67 ± 0.64	7.47 ± 1.30
	⟨50, 0.90⟩	12.66 ± 0.98	10.92 ± 1.70	12.54 ± 0.98	10.91 ± 1.76
		31.89 ± 3.03	29.09 ± 6.01	31.76 ± 3.15	28.70 ± 5.55
		35.32 ± 5.77	33.06 ± 6.16	35.09 ± 5.88	33.27 ± 6.61
TWT		3.56 ± 0.33	3.29 ± 0.73	3.62 ± 0.38	3.27 ± 0.68
(×10^5)		25.92 ± 3.01	21.99 ± 5.08	25.69 ± 2.33	21.65 ± 5.01
	⟨25, 0.97⟩	40.13 ± 4.18	44.84 ± 7.78	40.03 ± 4.31	45.07 ± 8.11
		64.07 ± 14.23	80.52 ± 36.05	63.98 ± 13.77	78.94 ± 27.27
		11.23 ± 1.01	10.20 ± 1.99	11.43 ± 1.00	9.98 ± 1.77
		15.53 ± 1.78	13.59 ± 2.95	15.46 ± 1.71	13.52 ± 2.71
		73.13 ± 3.90	71.60 ± 15.63	75.45 ± 8.97	70.27 ± 14.08
	⟨50, 0.97⟩	79.52 ± 5.96	70.90 ± 13.03	79.96 ± 6.85	68.32 ± 13.05
		6.22 ± 0.58	5.41 ± 0.99	6.09 ± 0.46	5.42 ± 0.90
		11.61 ± 1.01	10.19 ± 1.79	11.44 ± 0.86	10.15 ± 1.79
		21.01 ± 1.90	19.61 ± 3.21	20.91 ± 1.83	19.33 ± 2.96
		56.86 ± 4.46	56.65 ± 9.00	57.54 ± 5.05	55.00 ± 10.03

From the results, the EGP-JSS rules have significantly lower TWT values than the MLGP-JSS rules for most of the JSS problem instances in the test set. A possible reason for this observation may be due to the MLGP-JSS's breeding and selection procedure. As described in Sect. 3.1, the MLGP uses roulette wheel selection for choosing different groups to breed new groups and individuals. During the training procedure, the groups generally have similar fitnesses to each other. This means that roulette wheel selection may not be biased enough towards a potentially good group, resulting in a lack of exploitation. Both exploitation, the idea of greedily pursuing the local optima, and exploration, the idea of moving towards areas of the search space to potentially find better points, need to be finely balanced for a GP-HH to find good heuristics in the heuristic space [7].

Another possible reason why the EGP-JSS rules perform better than the MLGP-JSS rules may be linked to the difference in time taken to evolve the rules using EGP-JSS and MLGP-JSS and the performance of the evolved rules on the training sets. MLGP-JSS requires much lower computation time to evolve the rules during the training procedure than EGP-JSS. On average, MLGP-JSS took less than half of the amount of time to evolve the rules over 4op compared to EGP-JSS, and slightly more than a third the amount of time to evolve the rules over 8op. During an evaluation procedure, the number of times the individuals in the GP population is applied to the training instances is significantly lower for MLGP-JSS than EGP-JSS. Therefore, it might be the case that EGP-JSS takes a longer time to explore for good rules in a single generation than MLGP-JSS. This is supported by the differences in the performance over the training sets of the MLGP-JSS rules and the EGP-JSS rules, the EGP-JSS rules perform slightly better than the MLGP-JSS rules. However, the differences in the performances over the training sets are not statistically significant.

In summary, two major findings were made: (1) MLGP-JSS evolves significantly better rules than GP-JSS for the dynamic JSS problem, supporting the hypothesis that ensembles of dispatching rules are better than single dispatching rules. The differences in the evolution times is insignificant. (2) EGP-JSS evolves significantly better rules than MLGP-JSS for the dynamic JSS problem, but EGP-JSS takes longer to evolve the rules than MLGP-JSS.

6   Conclusions

In this paper, we proposed the MLGP-JSS approach for the dynamic JSS problem with the TWT minimisation objective. MLGP-JSS is an adaptation of the MLGP approach proposed by Wu and Banzhaf [20] that has not been applied to dynamic JSS problems. In addition, we extended the EGP-JSS approach by Park et al. for a static JSS problem [12] by incorporating new "less-myopic" terminals [5] and applying it to a dynamic JSS problem for the first time. The experimental results show that MLGP-JSS evolves rules with insignificant increase in time compared to a benchmark GP-HH approach but with significant improvements in the performance. On the other hand, EGP-JSS produces higher quality evolved rules than MLGP-JSS, but MLGP-JSS evolves rules significantly faster than EGP-JSS.

For future work, it may be promising to investigate alternative selection methods during the breeding and selection procedure of MLGP, such as replacing the roulette wheel selection with tournament selection. In addition, it is likely that MLGP-JSS and EGP-JSS can evolve better rules by adding niching or diversity measures, which have been shown in classification to improve the quality of the ensembles [15]. In addition, the evolved rules from both EGP-JSS and MLGP-JSS will also be investigated for further analysis.

References

1. Branke, J., Nguyen, S., Pickardt, C.W., Zhang, M.: Automated design of production scheduling heuristics: a review. IEEE Trans. Evol. Comput. **20**(1), 110–124 (2016). doi:10.1109/TEVC.2015.2429314
2. Burke, E.K., Gendreau, M., Hyde, M., Kendall, G., Ochoa, G., Ozcan, E., Qu, R.: Hyper-heuristics: a survey of the state of the art. J. Oper. Res. Soc. **64**(12), 1695–1724 (2013)
3. Hildebrandt, T., Heger, J., Scholz-Reiter, B.: Towards improved dispatching rules for complex shop floor scenarios: a genetic programming approach. In: Proceedings of the 12th Annual Conference on Genetic and Evolutionary Computation, pp. 257–264 (2010)
4. Hunt, R., Johnston, M., Zhang, M.: Evolving machine-specific dispatching rules for a two-machine job shop using genetic programming. In: Proceedings of the IEEE Congress on Evolutionary Computation, pp. 618–625 (2014)
5. Hunt, R., Johnston, M., Zhang, M.: Evolving "less-myopic" scheduling rules for dynamic job shop scheduling with genetic programming. In: Proceedings of the 2014 Conference on Genetic and Evolutionary Computation, pp. 927–934 (2014)
6. Jayamohan, M.S., Rajendran, C.: Development and analysis of cost-based dispatching rules for job shop scheduling. Eur. J. Oper. Res. **157**(2), 307–321 (2004)
7. Koza, J.R.: Genetic Programming: On the Programming of Computers by Means of Natural Selection. MIT Press, Cambridge (1992)
8. Nguyen, S., Zhang, M., Johnston, M., Tan, K.C.: A coevolution genetic programming method to evolve scheduling policies for dynamic multi-objective job shop scheduling problems. In: Proceedings of the IEEE Congress on Evolutionary Computation, pp. 1–8 (2012)
9. Nguyen, S., Zhang, M., Johnston, M., Tan, K.C.: A computational study of representations in genetic programming to evolve dispatching rules for the job shop scheduling problem. IEEE Trans. Evol. Comput. **17**(5), 621–639 (2013)
10. Ouelhadj, D., Petrovic, S.: A survey of dynamic scheduling in manufacturing systems. J. Sched. **12**(4), 417–431 (2009)
11. Panait, L., Luke, S.: Cooperative multi-agent learning: the state of the art. Auton. Agent. Multi-Agent Syst. **11**(3), 387–434 (2005)
12. Park, J., Nguyen, S., Zhang, M., Johnston, M.: Evolving ensembles of dispatching rules using genetic programming for job shop scheduling. In: Machado, P., et al. (eds.) EuroGP 2015. LNCS, vol. 9025, pp. 92–104. Springer, Heidelberg (2015)
13. Pickardt, C.W., Hildebrandt, T., Branke, J., Heger, J., Scholz-Reiter, B.: Evolutionary generation of dispatching rule sets for complex dynamic scheduling problems. Int. J. Prod. Econ. **145**(1), 67–77 (2013)
14. Pinedo, M.L.: Scheduling: theory, algorithms, and systems. In: Gaul, W., Bachem, A., Habenicht, W., Runge, W., Stahl, W.W. (eds.) Operations Research Proceedings 1991, 4th edn. Springer, Heidelberg (2012)
15. Polikar, R.: Ensemble based systems in decision making. IEEE Circ. Syst. Mag. **6**(3), 21–45 (2006)

16. Potter, M.A., De Jong, K.A.: Cooperative coevolution: an architecture for evolving coadapted subcomponents. Evol. Comput. **8**(1), 1–29 (2000)
17. Potts, C.N., Strusevich, V.A.: Fifty years of scheduling: a survey of milestones. J. Oper. Res. Soc. **60**, S41–S68 (2009)
18. Soule, T., Komireddy, P.: Orthogonal evolution of teams: a class of algorithms for evolving teams with inversely correlated errors. In: Riolo, R., Soule, T., Worzel, B. (eds.) Genetic Programming Theory and Practice IV. Genetic and Evolutionary Computation, vol. 5, pp. 79–95. Springer, Heidelberg (2007)
19. Vepsalainen, A.P.J., Morton, T.E.: Priority rules for job shops with weighted tardiness costs. Manage. Sci. **33**(8), 1035–1047 (1987)
20. Wu, S.X., Banzhaf, W.: Rethinking multilevel selection in genetic programming. In: Proceedings of the 13th Annual Conference on Genetic and Evolutionary Computation, pp. 1403–1410 (2011)

A Genetic Programming Approach for the Traffic Signal Control Problem with Epigenetic Modifications

Esteban Ricalde[✉] and Wolfgang Banzhaf

Memorial University of Newfoundland, St. John's, NL A1B 3X5, Canada
{ergl45,banzhaf}@mun.ca

Abstract. This paper presents a proof-of-concept for an Epigenetics-based modification of Genetic Programming (GP). The modification is tested with a traffic signal control problem under dynamic traffic conditions.

We describe the new algorithm and show first results. Experiments reveal that GP benefits from properties such as phenotype differentiation, memory consolidation within generations and environmentally-induced change in behavior provided by the epigenetic mechanism. The method can be extended to other dynamic environments.

Keywords: Genetic Programming · Epigenetic modification · Dynamic environments · Traffic signal control

1 Introduction

Because of the flexibility of its representation and its context independent methodology, GP can be used to generate solutions to problems in different areas of application in science and technology. However, in real world problems, the goal is often not fixed and can change during the evolutionary process. In a dynamic environment GP needs to be able to adapt to constant changes of the goal and fitness evaluation criteria. One approach to face these challenges is to generate variable locally adaptable solutions.

Biological evolution has different mechanisms to deal with environmental perturbations. Recently, Epigenetics, defined as phenotypic modifications without requiring changes in the nucleotide sequence (DNA), has been discovered to have important influences on the development of adaptation mechanisms at cellular, individual and species levels [13,16]. These imply a more active role for epigenetic mechanisms on the cellular, individual and species development.

In this paper an Epigenetics-based mechanism is presented and integrated into the Genetic Programming algorithm using a decision tree forest representation. A proof-of-concept in a dynamic environment is presented and future experiments are described.

© Springer International Publishing Switzerland 2016
M. Heywood et al. (Eds.): EuroGP 2016, LNCS 9594, pp. 133–148, 2016.
DOI: 10.1007/978-3-319-30668-1_9

In this paper, the term decision tree is used in a loose sense. By decision tree we mean a tree that evaluates to an integer value with a conditional statement as the first node.

We use a traffic signal control problem as our testbed. Urban traffic network control is a complex nonlinear problem and traffic congestion affects daily life of millions of citizens. Furthermore, the rapid increment of metropolitan populations makes control of traffic signals a challenging task. Most of the traffic controller systems currently in use are pre-timed and cannot handle the dynamic nature of the problem. However, in the last decades, different adaptive methods have been implemented in simulated environments, reducing the delay during rush hours.

This paper is organized as follows: Sect. 2 describes one of the biological epigenetic mechanisms named DNA methylation and gives an overview of the different approaches followed to integrate epigenetic mechanisms into Evolutionary Computation. Section 3 introduces the Traffic Signal Control Problem and explores different Evolutionary Algorithm methods implemented for its solution. Section 4 describes the simulator and the traffic network used in this paper. Section 5 defines the chromosome representation and genetic operators used in the GP environment. Section 6 describes the epigenetic mechanism introduced in this paper. Section 7 provides details on the experimental configuration used. Results are presented and discussed in Sect. 8. Section 9 presents conclusions.

2 Epigenetics

Epigenetics is the study of cellular and physiological phenotypic trait variations that are caused by external or environmental factors affecting how cells read genes. This could be seen in contrast to the modifications caused by changes in the DNA sequence.

One of the clearly heritable mechanisms of Epigenetics is DNA methylation [16]. Methylated DNA has a methyl group (CH_3) attached to some of its bases. It is found in vertebrates, plants, and even in many invertebrates, fungi and bacteria [11]. A methyl group is normally attached to the cytosine (C) nucleotide. Methylated cytosine doesn't change its role in the genetic code. It is still paired with guanine. However, the methyl group affects protein transcription by binding with special proteins and preventing Ribonucleic acid (RNA) polymerase to work on it, or by interfering with the binding of regulatory factors to the gene control region. In other words, cytosine methylation is a mechanism to silence DNA sections. During development, methylation marks can change and the modified (methylated) DNA sequence is transferred from cell to cell during cell division.

Even though the importance of epigenetic inheritance in cell differentiation and memory processes has been recognized, its influence on macroscopic phenomena has been discovered only recently. Some examples are environmentally induced epigenetic modification of behavior [10], the influence of Epigenetics on memory consolidation within generations [5], the inherited propensity for learning [2], the role of Epigenetics in morphological differentiation (Honeybee reproductive queen differentiation mediated by royal jelly consumption [8]) and even species differentiation through morphological specializations (for instance phenotypic changes in the modern human brain and behavior compared to other hominids [13]).

The Evolutionary Algorithm (EA) community has recently started to consider the discoveries in the area of Epigenetics. Different approaches have been used to represent the phenotypic mechanism, but it has normally been implemented as extra optimization to accelerate the adaptation of the EA.

Tanev and Yuta [22] worked with a modification of the predator-prey pursuit problem. GP is used to define a set of stimulus-response rules to model the reactive behavior of predator agents. The implementation includes active and inactive histones in the representation and uses age-based predators moving through different life stages (birth, development, survival and death). An extra step called Epigenetic Learning (EL) is included in the fitness evaluation. EL is basically a hill climber acting through epi-mutations of the histone activation signals.

It was found that the probability of success is larger when the Epigenetic Learning mechanism is included. The authors ascribe the difference to the robustness gained with the representation by preserving the individuals from the destructive effects of crossover by silencing certain genetic combinations and explicitly activating them only when they are most likely to be expressed in corresponding beneficial phenotypic traits.

Fontana [6] used other multi-cellular morphogenic models for development with an integer number genetic representation controlled by a regulatory network with epigenetic activation and deactivation signals in different development phases. A two-dimensional cellular grid and a Genetic Algorithm running on the genome allow the model to generate predefined 2-dimensional shapes.

In [21], Sousa and Costa present an epi-genetically controlled agent system for Artificial Life. The agents wander around a 2D environment with walls and different attributes -temperature, light and food- that can vary over time. The goal of the agents is to survive and to reproduce.

The behavior of the agents is coded on binary strings. Activation of genes is controlled by methylation marks. An Evolutionary Algorithm controls the survival and reproduction of the different organisms. Several experiments were performed with different levels of epigenetic transfer between parents and offspring. The results show a significant improvement: Non epigenetic populations found it hard to thrive in dynamic environments, while epigenetic populations were able to regulate themselves under dynamic conditions.

Chikumbo et al. [3] proposed a Multi-Objective Evolutionary Algorithm with epigenetic silencing for the land use management problem. The goal of the farm was to reduce the environmental footprint whilst maintaining a viable farming business through land use and/or management option changes.

The chromosome encoded each paddock land use and the system emulated gene regulation with epigenetic silencing based in histone modification and RNA editing mechanisms. A Pareto front visualization tool was developed composing the 14 fitness criteria into 3 super-objectives. However, the approach was not compared against a classical Multi-Objective Evolutionary Algorithm. Therefore, the improvement of the epigenetic variation could not be estimated.

In 2014, the same authors [4] extended their previous work using a similar epigenetic based modification. The main modification is the use of Hyper Radial Visualization, 3D Modeling and Virtual Reality to reduce the 14 fitness functions and display the solutions in a understandable way to a group of experts. Again, the approach is not compared with a classical EA.

Turner et al. [23] used an Artificial Gene Regulation model with an epigenetic mechanism based on DNA methylation and chromatin modifications. The inclusion of epigenetic information gave the network the ability to allocate different genes to different tasks, effectively regulating gene expression according to the environment in which it was operating.

The goal of the model was to follow specific trajectories in a chaotic system (Chirikov's standard map). The network was evolved using a Genetic Algorithm. The epigenetic mechanism improved performance of the model in a dynamic system. With the ability to inactivate genes came the ability to increase the efficiency of the network. Hence, with each inactive gene for an objective, there was less computational effort required to complete a single iteration of the network simulation.

La Cava et al. [15] included an Epigenetic Hill Climber into the Linear Genetic Programming algorithm by the addition of a binary array equivalent in length to the genotype of each individual. This array, referred to as an epiline, indicated the active genes. The algorithm was used to solve different symbolic regression problems and performed better than the non-epigenetic one. Even when there was no statistically significant improvement in Mean Best Error, the authors reported improvements in effective program size and beneficial genetics (genetic operations that resulted in fitter offspring).

The same research group used a similar epigenetic mechanism in [14] to solve symbolic regression and program synthesis problems. Stack-based GP representations are used for both types of problems. The binary epiline is used to deactivate nodes. Epigenetic hill climber and epigenetic mutation variations are compared against a GP method were all the nodes are active. The epigenetic methods outperformed the GP baseline implementation in terms of fitness minimization, exact solutions, and program sizes.

3   Traffic Signal Control

Urban traffic network control is a complex nonlinear problem and traffic congestion affects daily life of millions of citizens. Furthermore, the rapid increment of metropolitan populations makes the control of the traffic signals a challenging task. Different traffic signal control methods have been implemented over time to try to reduce the negative effects of traffic congestion.

A basic fixed traffic signal has static phase lengths based on historical information for each intersection. However, traffic doesn't behave in the same way during different hours of the day. An engineer can analyze the behavior of the traffic during the day and define different phase lengths for specific intervals. This method is called pre-timed control. It presents an improvement over the fixed control depending on human expertise and the correct modeling of traffic conditions, but requires constant surveillance and constant update, but cannot adapt to sudden modifications in traffic behavior.

Actuated control or traffic-responsive control consists of phase length sets that are extended in response to vehicle detectors. Detection is used to provide information about traffic demand to the controller. Each phase length is determined by a detector input and corresponding controller parameters.

Different traffic units have been used in the literature to measure traffic: average car speed, average intersection delay, average queue length, total system delay, etc. Total system delay is defined as the sum of the stop time of all vehicles in the system for a defined interval of time.

Traffic-control systems are affected by many factors: infrastructure, vehicles, drivers, pedestrians, weather, seasonal effects, etc. Each factor has its own characteristics, which makes the entire traffic system a large complex nonlinear stochastic system which poses many interesting problems and challenges for researchers and engineers.

Wang [24] proposed a general "parallel" control model, where parallel implies parallel interactions between a real transportation system and its corresponding artificial or simulated counterpart. The approach consists of three steps: (1) generation of a simulated model; (2) analysis and evaluation by computational experiments; (3) control and management through parallel execution of the real and artificial system.

This general framework can be used with different simulation, control and learning algorithms, with the constant feedback of differences between real world events and simulated environments as one of its main benefits.

In [25], Zhang et al. proposed a real-time online urban traffic signal control approach using a multi-objective discrete differential evolution modification to optimize the light phase periods of a three-lane, single intersection road including left-turn phases. The authors compared their algorithm with a pre-timed controller using a Poisson distribution to regulate the traffic flow. The proposed approach behaved better in the single intersection problem.

Sánchez-Medina et al. [20] used a Cellular Automaton based traffic simulator and a Genetic Algorithm to simulate and optimize the traffic light phase periods of a section of Saragossa city. The section has seven intersections, 16 input nodes, 18 output nodes and 17 traffic signals. Individuals were represented as an array containing light phase periods of all traffic signals. Four different parameters were used as fitness function. The algorithm was tested with different traffic situations and limited results were obtained. The methodology does not provide a significant improvement for regular traffic conditions of the network; however, it increases the performance for more congested scenarios.

Nie et al. [18] used a two-dimensional Cellular Automaton and a $1 + \lambda$ Evolutionary Strategy to update the time parameters of CA rules in a 20×20 cell network. The authors performed experiments with different traffic densities and the results demonstrated a better performance of the evolutionary approach compared to previous work done with the same Cellular Automaton. However, the simulated environment was too rigid and was not able to represent all conditions of a real environment.

In [1], Braum and Kemper modified an open source area-wide traffic light signal optimizer, called BALANCE [7]. They replaced the hill-climbing algorithm used on the tactical level of BALANCE with a Genetic Algorithm. The chromosome representation used is similar to the one used in [20]; however, the optimization was done online with a real system. The architecture used is similar to the parallel control model defined in [24].

Several experiments were performed with the traffic network of Ingolstadt, Germany. The results demonstrated a better performance of the GA over the Hill Climber (HC) in almost all (different) traffic density tests. The authors conclude that as the network becomes larger and more complex, the evolutionary algorithm provides larger advantages. Once the system started operating in the real world, daily average delays were reduced by 21 % compared to the standard 10 % expected using the traditional HC algorithm of BALANCE.

In [19], Padmasiri and Ranasinghe used a GP and fuzzy logic hybrid approach to define a single fine-tuned fuzzy rule for a single intersection using a Poisson distribution to control the vehicle arrival rate under different traffic volume scenarios. The set of evolved rules use traffic parameters as input and decide to extend or terminate the current green lapse. The results present an improvement compared to previous work. However, solutions lack adaptability to changes in the traffic conditions and the method was tested only with a single intersection.

4   The Traffic Simulator

Microscopic traffic simulation models study individual elements of transportation systems, such as individual vehicle dynamics and individual traveler behavior. The model depends on random numbers to generate vehicles, to select routes and to determine the behavior of the system. In a microscopic simulator the dynamic variables of the model represent microscopic properties like the position and velocity of single vehicles.

Even though several commercial and open source simulators are available, we decided to create a microscopic model simulator in order to have full control of the environment. It allows the parallel execution of experiments in a multi-processor environment, and to simulate different dynamic traffic conditions by the hour.

The simulator works in a similar way to the Cellular Automaton described in [12,20], but operates in a two-dimensional environment. It can represent roads with multiple-lanes and two directions. An extra Object-Oriented layer was incorporated to update only the cells containing vehicles and to reduce simulation time. Instead of using a toroidally closed environment, the network entries are controlled by a Poisson distribution described in Sect. 4.2.

4.1   Traffic Network

The size of the network, its number of connections, geometry, number of lanes and type of intersections can be modified before running the simulator. For this paper, the experiments were performed in a 10 intersections network with 9 input/output nodes and 31 traffic signals. All the nodes are connected by two-lane bi-directional roads. The network is presented in Fig. 1.

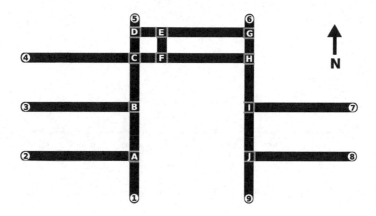

Fig. 1. Traffic network used for the experiment

4.2   Vehicle Insertion

The Poisson distribution correctly models arrival of vehicles, on one or multiple lanes [17]. The flexibility of the Poisson distribution allows the simulation of changes in the traffic densities. In order to simulate real-world similar conditions, the scenario simulates 16.5 h of traffic. The traffic densities change during the simulated day and each entry point to the network follows a different distribution.

The first hour is considered a training step where all the entries follow a standard Poisson distribution going from zero traffic conditions to the maximum saturation peak and declining again to zero traffic. During the remaining 15.5 h of the scenario, two traffic waves are executed. The first one initiates from south-west entries between 7 and 11 am. The second one from north-east entries between 4 and 7 pm. Figure 2 presents the probability distributions generated corresponding to the defined behavior for the network presented in Fig. 1.

Even when the complete scenario covers more than 16 h of traffic, each simulation runs only for one hour of traffic. A time window is used during the experiments. The window moves 5 min after each execution. Using this approach, the full scenario is covered with 200 simulations.

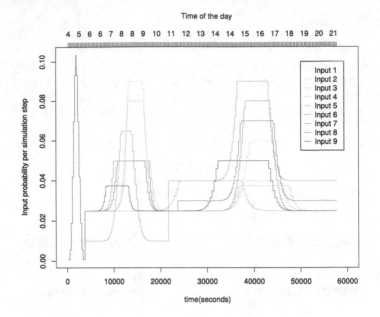

Fig. 2. Traffic input probability distributions

5   Representation

We used a forest of decision trees as the GP representation. Each decision tree is employed to evaluate a set of intersections with similar characteristics; i.e., same number of intersecting roads and equivalent proximity to entry points. For example, the network in Fig. 1 requires a forest of 4 decision trees: (E, F, H), (B, D, G, I), (A, J) and (C).

The terminal set is formed by integer numbers, between −10 and 10, and traffic parameters listed in Appendix A. The function set is formed by mathematical operators (addition, subtraction, multiplication and protected division), logical operators (conjunction, disjunction and negation), comparison operators (equal to, bigger than and smaller than), and a conditional operator.

During the simulation, a decision tree is executed for each intersection twice in every light cycle with current traffic parameters. The resulting integer number is added to the vertical green phase period, subtracted from the vertical red phase period, added to the horizontal red phase period and subtracted from the horizontal green phase period.

Figure 3 presents a single decision tree. This tree represents a human designed solution. The idea is to increase the mobility (increase the green phase in 1 s and reduce the red phase in 1 s) of the vertical or horizontal directions if the corresponding queue is larger than the opposite direction queue for more than 5 vehicles.

Two different crossover operations are available: Tree exchange and sub-tree exchange. The former occurs in 10 % of all crossover operations and exchanges

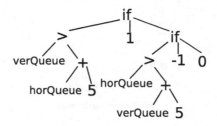

Fig. 3. Decision tree

one tree between two chromosomes. The trees exchanged are in the same position of the two different forests. The latter operator selects a random crossover point of a specific decision tree in both chromosomes and exchanges the two sub-trees selected only if both are of the same type; otherwise, it selects a new crossover point.

Two different mutation operations are available: New tree mutation and node mutation. The former occurs with a probability of 0.1 %, selects a random tree of the forest and replaces it with a newly generated tree. The latter replaces a single node with a node of the same type.

Strong typing is performed through evaluation of the selected points before the application of the reproduction operators. These GP parameters and those presented in Table 1 were selected based on a set of preliminary experiments.

6   The Epigenetic Mechanism

The epigenetic mechanism proposed is based in DNA methylation. Each conditional node is associated with an activation index in analogy to the concentration of methyl groups attached to cytosine nucleotides along the DNA structure. As in the biological counterpart, the evolutionary process of chromosomes is not affected by the activation index. However, during the evaluation step, if the activation index is smaller than an activation threshold, defined as 50 % for this experiment, the conditional node is ignored and the *else sub-tree* is executed, deactivating with that action the *conditional sub-tree* and the *then sub-tree*.

The activation indices are initialized randomly between 0 % and 100 % for the first generation. However, they are transferred to the offspring as part of the crossover operation in the same way methylated DNA is transferred between generations. The collection of activation indices is stored in an epigenetic vector for easy manipulation. The epigenetic vector is included as part of the chromosome, but it is not affected by the genetic operators.

Figure 4 presents the effect of the activation thresholds in a forest of decision trees. The branches in gray are inactive. This change modifies the behavior of the decision tree without modifying the chromosome.

Since methylation marks change during development, it was decided to modify the epigenetic vector during the simulation process using the following procedure: For each intersection that uses a tree expression the traffic balance, defined

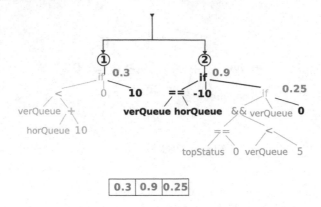

Fig. 4. Decision trees under influence of activation thresholds and epigenetic vector

as the difference between the traffic congestion in vertical directions and the traffic congestion in horizontal directions, is calculated using (1), where i represents an intersection evaluated through the tree expression e, and t represents the current time step.

$$B_{e_i}(t) = \text{verticalQueue}_{e_i}(t) - \text{horizontalQueue}_{e_i}(t) \qquad (1)$$

Every 5 light cycles, a mean traffic balance of the interval is calculated per intersection with (2), where T is the number of time steps of the interval.

$$\overline{B}_{e_i} = \frac{\sum\limits_{t=1}^{T} B_{e_i}(t)}{T} \qquad (2)$$

The interval mean is then compared to the last element of the time interval in order to get the adaptive factor of the expression as it is defined by (3).

$$\Lambda_{e_i} = |B_{e_i}(T) - \overline{B}_{e_i}| \qquad (3)$$

The goal of the adaptive factor is to identify differences between the interval mean congestion levels and the current state for each intersection in the system. A large difference between the current congestion level and the mean behavior of the intersection indicates a change in the environment. In that case, a modification in the behavior of the intersection could help the system to adapt to this environmental change.

Therefore, the adaptive factor of the expression is used as a mutation probability to modify the activation indices of the expression tree. A mutation is performed on the local activation indices. This step is performed as an internal mutation during the simulation process for each intersection. The mechanism works in a similar way to the epigenetic mutation variations presented in [14] and has the purpose to adapt the intersection behavior to environmental changes.

The conceptual idea behind this process is to keep the system behavior stable under environmental perturbations, one of the roles of Epigenetics at the cellular level in Nature. At the end of the simulation, the final activation indices are stored in the epigenetic vector of the chromosome and transferred to the next generation.

7   Experiments

Five different algorithms were tested with the traffic network of Sect. 4: (1) a fixed static control, (2) an actuated control using a human designed fixed decision tree, (3) a pre-timed control evolved using a Genetic Algorithm (GA), (4) an actuated control using the GP representation described in Sect. 5 and (5) an actuated control using the GP representation including the epigenetic mechanism described in the previous section.

The baseline is a fixed control with synchronization of all the intersections. For this method, all the lights are synchronized and the lapses are fixed (15 s for the green light, 5 s for yellow light, 10 s for red light and 10 s for a left turn). The system behavior keeps static for the 16 h of traffic.

The decision tree of Fig. 3 is the human designed actuated control used for the second algorithm. In each simulation, the lights start with the fixed configuration used in the static method, but the decision tree is executed at each intersection twice every light cycle. Therefore, the lapses of each intersection can be modified depending on traffic conditions. The same decision tree is used for all intersections in every simulation.

A pre-timed control is evolved using a GA similar to those presented in [1,20]. The length of the lapses for each intersection in the system is stored as an integer chromosome. An online optimization approach is used with the GA for 200 generations to approximate an optimal pre-timed configuration for the 16.5 h of traffic as it is described in Sect. 4.2.

The GP actuated control and the GP actuated control including the epigenetic mechanism evolve a forest of decision trees (see Sects. 5 and 6) using an online approach.

During the evolution process each individual is evaluated with 20 independent simulation runs. Total system delay, defined in Sect. 3, is used as objective function.

For the fixed control and fixed tree actuated control, 20 independent simulation runs are effectuated for each of the 200 simulation configurations. The total system delay is calculated for each of them and the objective function is defined as the average total system delay of the 20 simulations.

The parameters employed for the Genetic Programming are presented in Table 1. A similar configuration in terms of population size, number of generations, selection method, mutation probability and crossover probability is used for the Genetic Algorithm.

Table 1. Summary of the configuration parameters for the Genetic Programming Model

Configuration parameters	Selected values
Population size	50 individuals
Number of generations	200
Mutation probability rate per node	5%
Crossover probability	80%
Initial size limit	5 levels
Maximum size limit	7 levels
Selection operator	Tournament selection with group size of 7 individuals
Elitism	1 individual

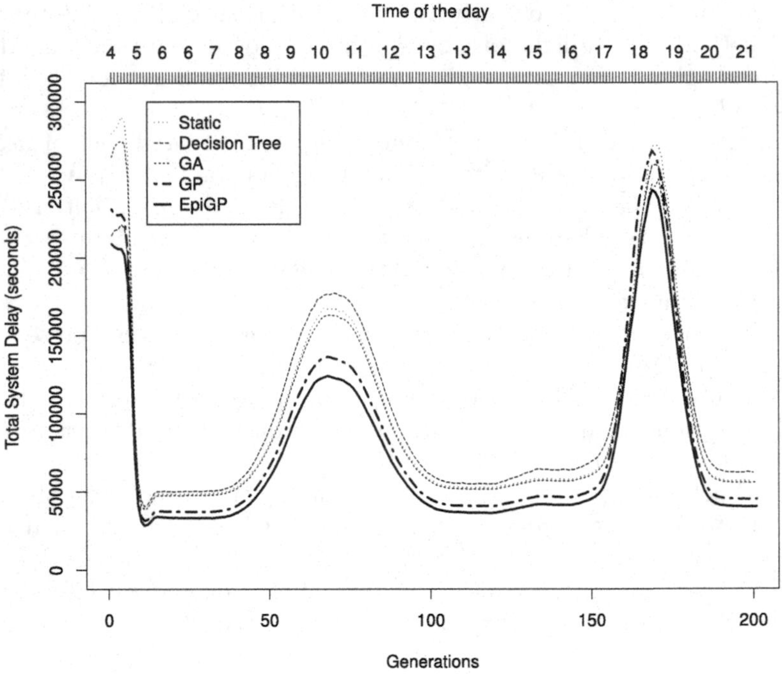

Fig. 5. Fitness curves of fixed control, tree actuated control, GA pre-timed control, GP and epigenetic modification of GP for the experiment

8   Results and Discussion

15 independent runs were performed for each algorithm. Figure 5 presents the comparison of the fitness obtained by the five methods. For GA, GP and the epigenetic modification of GP the fitness value of the best individual per generation is displayed.

Table 2. Vehicle waiting time differences of the four methods

Compared methods	Vehicle waiting time difference (seconds)	Relative difference with static method
Static - GA	13.85	3.33 %
Static - GP	61.82	14.91 %
Static - EpiGP	98.29	23.71 %
GP - GA	47.58	11.57 %
GA - EpiGP	84.45	20.37 %
GP - EpiGP	36.47	8.80 %

From the first generation, the learning curve of the evolutionary actuated control methods starts to provide better solutions than the fixed control approach for almost all evaluation steps. This behavior can be caused by the high variability of traffic densities used in the experiment. Further experiments should be performed with lower variability to analyze the behavior of the methods in more detail.

The epigenetic modification of GP has a lower delay than the standard GP algorithm for almost all the evaluation points. The difference between both methods is more drastic during rush hours. A possible explanation is the adaptive ability provided by the activation-deactivation of code of the epigenetic method during the simulation.

Table 2 presents pairwise comparisons of the vehicle waiting time for combinations of the methods. The second column indicates the difference of the average waiting time per vehicle for the different algorithms. The third column is that difference divided by the average vehicle waiting time of the pre-timed experiment.

It is noteworthy that the epigenetic modification outperformed the other four methods used in the experiments, providing an improvement of more than 20 % compared to the fixed control and the pre-timed control. However, an evaluation of the methods with different variability in traffic conditions needs to be conducted to provide a better understanding of the behavior of the methods. For now, this set of experiments is a proof-of-concept.

9   Conclusions and Future Work

The GP modification described in this paper is an epigenetic approach specifically designed to work on traffic signal control problems. A basic set of experiments was performed and the results demonstrate an increase in the performance compared to the basic GP method and other methods previously used.

Extensive experimentation is required to give statistical significance to the results. To achieve that, statistical tools should be used to perform analysis of the data generated by the independent runs. Scenarios of different sizes should be evaluated to analyze the behavior of the method under different circumstances. It would be ideal to acquire data from a real world network.

Furthermore, the modification needs to be compared against traditional methods used in Traffic Signal Control. An example of these methods is the green wave algorithm described in [9]. Because the architecture developed can be easily transformed into an online real-virtual parallel system as the one described in [24], it can be used in real world traffic optimization.

Moreover, the epigenetic modification can be used to solve other problems. Problems were some elements of the domain vary with the progression of time (dynamic environments) can benefit of the short term memory mechanism presented in this paper. The key elements to implement the epigenetic modification are: the identification of a variable independent to the objective function (traffic balance in our experiment) to calculate the adaptive factor, the insertion of activation indices in nodes of a specific type and the code activation-deactivation process described in Sect. 6.

A   Traffic Parameteres

Traffic parameters included in the terminal set:

- **topStatus**: Status of the north-south direction light of the current intersection (returns 0 if the light is red, 1 if the light is yellow, 2 if the light is green and 3 if the turn left right is on).
- **bottomStatus**: Status of south-north direction light of the current intersection (same output configuration that topStatus).
- **leftStatus**: Status of west-east direction light of the current intersection (same output configuration that topStatus).
- **rightStatus**: Status of east-west direction light of the current intersection (same output configuration that topStatus).
- **verQueue**: Sum of the number of vehicles stopped in the north-south direction and the number of vehicles stopped in the south-north direction in the current intersection.
- **horQueue**: Sum of the number of vehicles stopped in the west-east direction and the number of vehicles stopped in the east-west direction of the current intersection.
- **1stTopNeighborQueue**: Number of vehicles stopped in the north-south direction of the first intersection in the north direction of the current crossing.
- **1stBottomNeighborQueue**: Number of vehicles stopped in the south-north direction of the first intersection in the south direction of the current crossing.
- **1stLeftNeighborQueue**: Number of vehicles stopped in the west-east direction of the first intersection in the west direction of the current crossing.
- **1stRightNeighborQueue**: Number of vehicles stopped in the east-west direction of the first intersection in the east direction of the current crossing.
- **2ndTopNeighborQueue**: Number of vehicles stopped in the north-south direction of the second intersection in the north direction of the current crossing.

- **2ndBottomNeighborQueue**: Number of vehicles stopped in the south-north direction of the second intersection in the south direction of the current crossing.
- **2ndLeftNeighborQueue**: Number of vehicles stopped in the west-east direction of the second intersection in the west direction of the current crossing.
- **2ndRightNeighborQueue**: Number of vehicles stopped in the east-west direction of the second intersection in the east direction of the current crossing.

References

1. Braun, R., Kemper, C.: An evolutionary algorithm for network-wide real-time optimization of traffic signal control. In: 2011 IEEE Forum on Integrated and Sustainable Transportation System (FISTS), pp. 207–214, June 2011
2. Champagne, D.L., Bagot, R.C., van Hasselt, F., Ramakers, G., Meaney, M.J., de Kloet, E.R., Joels, M., Krugers, H.: Maternal care and hippocampal plasticity: evidence for experience-dependent structural plasticity, altered synaptic functioning, and differential responsiveness to glucocorticoids and stress. J. Neurosci. **28**(23), 6037–6045 (2008)
3. Chikumbo, O., Goodman, E., Deb, K.: Approximating a multi-dimensional pareto front for a land use management problem: A modified moea with an epigenetic silencing metaphor. In: 2012 IEEE Congress on Evolutionary Computation (CEC), pp. 1–9, June 2012
4. Chikumbo, O., Goodman, E., Deb, K.: Triple bottomline many-objective-based decision making for a land use management problem. J. Multi-Criteria Decis. Anal. **22**(3–4), 133–159 (2015). http://dx.org/10.1002/mcda.1536
5. Day, J.J., Sweatt, J.D.: Epigenetic modifications in neurons are essential for formation and storage of behavioral memory. Neuropsychopharmacology **36**(1), 357–358 (2011). http://dx.org/10.1038/npp.2010.125
6. Fontana, A.: Epigenetic tracking: biological implications. In: Kampis, G., Karsai, I., Szathmáry, E. (eds.) ECAL 2009, Part I. LNCS, vol. 5777, pp. 10–17. Springer, Heidelberg (2011)
7. Friedrich, B.: Balance and control: Methods for traffic adaptive control. In: World Congress on Intelligent Transport Systems (2nd: 1995: Yokohama-shi, Japan). Steps forward, vol. 5 (1995)
8. Gabor Miklos, G.L., Maleszka, R.: Epigenomic communication systems in humans and honey bees: from molecules to behavior. Horm. Behav. **59**(3), 399–406 (2011)
9. Gershenson, C., Rosenblueth, D.A.: Adaptive selforganization vs static optimization. Kybernetes **41**(3/4), 386–403 (2012)
10. Herrera, C.M., Pozo, M.I., Bazaga, P.: Jack of all nectars, master of most: DNA methylation and the epigenetic basis of niche width in a flower-living yeast. Mol. Ecol. **21**(11), 2602–2616 (2012)
11. Jablonka, E., Lamb, M.: Evolution in Four Dimensions: Genetic, Epigenetic, Behavioral, and Symbolic Variation in the History of Life. Life and Mind. MIT Press, Cambridge (2005). http://books.google.ca/books?id=EaCiHFq3MWsC
12. Nagel, K., Schreckenberg, M.: A cellular automaton model for freeway traffic. J. Phys. I France **2**(12), 2221–2229 (1992)
13. Krubitzer, L., Stolzenberg, D.S.: The evolutionary masquerade: genetic and epigenetic contributions to the neocortex. Curr. Opin. Neurobiol. **24**, 157–165 (2014). http://www.sciencedirect.com/science/article/pii/S0959438813002213

14. La Cava, W., Helmuth, T., Spector, L., Danai, K.: Genetic programming with epigenetic local search. In: Proceedings of the 2015 Annual Conference on Genetic and Evolutionary Computation, GECCO 2015, NY, USA, pp. 1055–1062 (2015). http://doi.acm.org/10.1145/2739480.2754763

15. La Cava, W., Spector, L., Danai, K., Lackner, M.: Evolving differential equations with developmental linear genetic programming and epigenetic hill climbing. In: Proceedings of the 2014 Conference Companion on Genetic and Evolutionary Computation Companion, GECCO Comp 2014, pp. 141–142. ACM, New York (2014)

16. Ledon-Rettig, C.C., Richards, C.L., Martin, L.B.: Epigenetics for behavioral ecologists. Behav. Ecol. **24**, 211–324 (2012)

17. Mauro, R., Branco, F.: Update on the statistical analysis of traffic countings on two-lane rural highways. Modern Appl. Sci. **7**(6), 67–80 (2013)

18. Nie, X., Li, Y., Wei, X.: Based on evolutionary algorithm and cellular automata combined traffic signal control. In: 2010 3rd International Symposium on Knowledge Acquisition and Modeling (KAM), pp. 285–288, October 2010

19. Padmasiri, T., Ranasinghe, D.: Genetic programming tuned fuzzy controlled trafficlight system. In: 2014 InternationalConference on Advances in ICT for Emerging Regions (ICTer), pp. 91-95, Dec 2014

20. Sanchez-Medina, J., Galan-Moreno, M., Rubio-Royo, E.: Traffic signal optimization in la almozara district in saragossa under congestion conditions, using genetic algorithms, traffic microsimulation, and cluster computing. IEEE Trans. Intell. Transp. Syst. **11**(1), 132–141 (2010)

21. Sousa, J., Costa, E.: Epial - an epigenetic approach for an artificial life model. In: International Conference on Agents and Artificial Intelligence (2010)

22. Tanev, I., Yuta, K.: Implications of epigenetic learning via modification of histones on performance of genetic programming. In: Deb, K., Tari, Z. (eds.) GECCO 2004. LNCS, vol. 3102, pp. 213–224. Springer, Heidelberg (2004)

23. Turner, A.P., Lones, M.A., Fuente, L.A., Stepney, S., Caves, L.S., Tyrrell, A.M.: The incorporation of epigenetics in artificial gene regulatory networks. BioSystems **112**(2), 56–62 (2013)

24. Wang, F.Y.: Parallel control and management for intelligent transportation systems: concepts, architectures, and applications. IEEE Trans. Intell. Transp. Syst. **11**, 630–638 (2010)

25. Zhang, M., Zhao, S., Lv, J., Qian, Y.: Multi-phase urban traffic signal real-time control with multi-objective discrete differential evolution. In: 2009 International Conference on Electronic Computer Technology, pp. 296–300, February 2009

A Genetic Programming-Based Imputation Method for Classification with Missing Data

Cao Truong Tran[✉], Mengjie Zhang, and Peter Andreae

School of Engineering and Computer Science, Victoria University of Wellington,
PO Box 600, Wellington 6140, New Zealand
{cao.truong.tran,mengjie.zhang,peter.andreae}@ecs.vuw.ac.nz

Abstract. Many industrial and real-world datasets suffer from an unavoidable problem of missing values. The ability to deal with missing values is an essential requirement for classification because inadequate treatment of missing values may lead to large errors on classification. The problem of missing data has been addressed extensively in the statistics literature, and also, but to a lesser extent in the classification literature. One of the most popular approaches to deal with missing data is to use imputation methods to fill missing values with plausible values. Some powerful imputation methods such as regression-based imputations in MICE [36] are often suitable for batch imputation tasks. However, they are often expensive to impute missing values for every single incomplete instance in the unseen set for classification. This paper proposes a genetic programming-based imputation (GPI) method for classification with missing data that uses genetic programming as a regression method to impute missing values. The experiments on six benchmark datasets and five popular classifiers compare GPI with five other popular and advanced regression-based imputation methods in MICE on two measures: classification accuracy and computation time. The results showed that, in most cases, GPI achieves classification accuracy at least as good as the other imputation methods, and sometimes significantly better. However, using GPI to impute missing values for every single incomplete instance is dramatically faster than the other imputation methods.

Keywords: Missing data · Imputation methods · Genentic programming · Symbolic regression · Classification

1 Introduction

Missing values are a common issue in many datasets [24,29]. For example, about 45 % of datasets in the UCI repository [3], which is one of the most popular data repositories for benchmarking machine learning tasks, contain missing values [15].

There are various reasons why data often contains missing values. For instance, respondents may refuse to answer some questions in a social survey; due to mechanical failures while collecting data, some results may be missing

© Springer International Publishing Switzerland 2016
M. Heywood et al. (Eds.): EuroGP 2016, LNCS 9594, pp. 149–163, 2016.
DOI: 10.1007/978-3-319-30668-1_10

in an industrial experiment; medical databases often suffer from missing values, where almost every patient's record lacks some values because not all possible tests can be run on every patient.

Missing data may cause a number of serious problems [6]. The non-applicability of data analysis methods is one of the most serious problems because the majority of existing data analysis methods require complete data. Therefore, these data analysis methods are not able to work directly with original data containing missing values. Furthermore, missing data may result in biased results because of differences between missing and complete data.

One of the most popular approaches to missing data is to use imputation methods that fill missing values with plausible values. For instance, mean imputation fills in missing values in each feature with the average of the complete values in the same feature. Imputation methods provide complete data that can be used by any data analysis methods. Therefore, imputation methods are a popular approach to handling missing data and a number of imputation methods have been developed in literature [24, 29].

Classification is one of the most important tasks in machine learning and data mining [18]. The ability to deal with missing values is an essential requirement for classification because the majority of classifiers are not able to classify missing data. For example, there are numerous decision tree-based classifiers, but only C4.5 [28] and CART [8] can handle missing data. Furthermore, inadequate treatment of missing values may result in large errors on classification [13].

Regression-based imputation is an approach to imputing missing values, where a regression method is used to generate a regression function that is then used to estimate missing values. For example, log-linear models are often used to impute discrete missing features, and linear or polynomial models are often used to impute continuous missing features [14]. One of the most powerful regression-based imputation methods is MICE [36]. MICE uses a number of regression methods to estimate missing values such as Bayesian linear regression [26], classification and regression tree (CART) [8] and random forest regression [23]. Regression-based imputation methods in MICE are often suitable for batch imputation. However, they are very expensive to impute missing values for every single incomplete instance that is required in classification problems. The main reason is that they take a long time to rebuild regression functions when they need to impute missing values for new single incomplete instances.

Genetic programming (GP) is an evolutionary technique to generate solutions in the form of computer programs for a problem [22]. GP has successfully been used for symbolic regression. In [33], GP is used as a non-parametric regression method to build a multiple imputation. The empirical results show that, in most cases, GP-based multiple imputation achieves better prediction accuracy and better classification accuracy than other popular and advanced imputation methods. However, just as regression-based imputations in MICE, the GP-based multiple imputation is expensive for imputing missing values for every single incomplete instance in the unseen set for classification. Therefore, developing an efficient GP-based imputation method to estimate missing values in single incomplete instances for classification tasks should be investigated.

1.1   Research Goals

The goal of this paper is to propose an efficient GP-based imputation method to estimate missing values for classification tasks (GPI). GPI will be compared with other popular and advanced regression-based imputation methods in MICE [36] on different classifiers to address the following objectives:

1. Whether GPI can achieve better classification accuracy than the other popular and advanced imputation methods; and
2. Whether GPI is more efficient than the other popular and advanced imputation methods.
3. What classifiers should combine with GPI to achieve good performance.

1.2   Organisation

The rest of the paper is organised as follows. Section 2 discusses related work. Section 3 introduces GPI, a GP-based imputation method for classification with missing data. Section 4 presents experiment design. Section 5 shows results and analysis. Section 6 draws conclusions and discusses potential future work.

2   Related Work

This section discusses the background to our work including classification with missing data, imputation methods, GP-based symbolic regression.

2.1   Classification with Missing Data

Most approaches to classification with missing data can be divided into four groups [15]:

Deletion approach eliminates all incomplete instances before applying classifiers. This approach provides complete data for classifiers, but instances containing missing values are not included in the classification process.

Imputation approach fills missing values with plausible values before using classifiers. As a result, missing data is transferred to complete data that can be classified by any classifiers. Furthermore, most imputation methods help to improve classification accuracy when compared to classification without imputation [13]. Therefore, using imputation methods is a popular approach to classification with missing data.

Model-based approach models data distribution of input data that is then used to classify both complete and incomplete instances by using the Bayesian decision theory [7]. Although this approach can classify incomplete instances, it requires making assumptions about the joint distribution of all variables in the model [15].

Machine learning approach builds classifiers that can classify incomplete instances directly. For example, C4.5 can classify incomplete instances by using a probabilistic approach to handling missing values in both the training set and test set. In the training stage, each value of each feature is assigned a weight: if a feature value is known, then the weight is assigned one; otherwise, the weight of any other values for that feature is the frequency of that values. In the testing stage, if a test case is unknown, from the current node, it finds all the available branches and decides the class label by using the most probable value [28].

2.2   Imputation Methods

The purpose of imputation methods is to fill missing values with plausible values. By using imputation methods, missing data is transformed into complete data that can be then analysed by any data analysis methods. Therefore, using imputation methods is a popular approach to handling missing data [24, 29, 30].

Many imputation methods have been proposed in literature. For example, one simple imputation method is mean imputation that fills in missing values in each feature with the average of the complete values in the same feature. Mean imputation maintains the mean of each feature, but it under-represents the variability in the data because all missing values in each feature have the same value. Another simple imputation method is hot deck imputation, where missing values are filled with complete values from the most similar instance. An advantage of hot deck imputation is to replace missing values by real values from the data. However, this method ignores all global properties of the data because it uses information of only the most similar instance [2]. One more sophisticated imputation method is expectation maximization-based imputation method [24] that uses the expectation maximization algorithm to estimate a maximum likelihood variance-covariance matrix and vector of means that are then used to impute missing values [16].

One of the most flexible and powerful imputation methods is multivariate imputation by chained functions (MICE) [35, 36]. Multiple imputation by chained functions, which is the first step in MICE, is designed to generate multiple imputed datasets. MICE uses a set of regression models to build regression functions that are then used to estimate missing values. Initially, missing values in each feature are filled randomly with complete values in the feature. After that, each feature containing missing values is regressed on other features to compute a better estimate for the feature. The process is repeated several times for all features containing missing values to generate one imputed dataset. The whole procedure is repeated N times to generate N imputed datasets. After that, the N imputed datasets are combined to provide the final imputed data [38]. MICE software [9] makes it easy to use this method.

This paper compared our proposed imputation method with five multiple imputation methods in MICE: multiple Bayesian linear regression imputation, multiple linear regression non-Bayesian imputation, multiple linear regression using bootstrap imputation, multiple classification and regression trees (CART) imputation and multiple random forest regression imputation. Bayesian linear regression is a linear regression that applies Bayesian inference to do statistical analysis [26]. CART [8] is a decision tree-based learning technique that has ability to produce either classification or regression trees. CART uses the sum of squared errors as a splitting criterion to build a tree. The same as CART, random forest [23] is also a learning method for both classification and regression. Random forest builds a set of decision trees, and outputs are then the mode of the classes (classification) or mean prediction (regression) of the individual trees.

2.3 Genetic Programming-Based Symbolic Regression

Regression analysis is a process of inferring the relationship between a dependent variable and one or more independent variables in the form of a function of the independent variables. Many techniques for doing regression analysis have been developed [12]. Regression analysis methods can be subdivided into parameter regression methods and non-parametric regression methods. In parametric regression methods, a regression function has a form with a finite number of unknown parameters. In contrast, in non-parametric regression methods, a regression function does not take a predetermined form, but it is constructed according to information from the data. Regression analysis has been widely applied to prediction and forecasting [21].

Symbolic regression is a kind of regression analysis that search on mathematical expressions space to find a regression function that best fits a given dataset. Symbolic regression is non-parametric regression because instead of requiring a predetermined model structure, it attempts to discover both model structures and corresponding parameters. The original purpose of GP is to evolve computer programs; hence, GP is an ideal choice for symbol regression [4]. GP has been successfully been used for symbolic regression [4,5,20,34].

To apply GP to solve symbol regression, regression functions are initially formed by randomly combining mathematical building blocks such as mathematical operators, functions, and constants. After that, new functions are formed by recombining previous functions using GP operators such as crossover and mutation [4].

Traditional parameter regression techniques require a pre-defined model structure. After that, they try to optimize parameters for the pre-defined model structure. Hence, conventional parameter regression techniques not only require domain knowledge to identify a pre-defined model structure, but also may be affected by human bias. In contrast, instead of requiring priori assumptions, GP-based symbolic regression attempts to discover both model structures and model parameters. Although GP-based symbolic regression often takes much longer to discover an appropriate model structure and parameter than traditional regression techniques, the end result is likely to be a selection of high-scoring models [37].

3 Genetic Programming-Based Imputation for Classification with Missing Data

The proposed algorithm has two phrases: the training process and the imputation process. The training process builds regression functions for each feature. After that, the imputation process uses the built regression functions to estimate missing values for each incomplete instance.

Algorithm 1 shows the main body of the training process. The input of the training process is the training data and a number indicating the required number of evolved regression functions for each feature. Firstly, the complete instances are extracted from the training data. Next, for each feature in the data, each instance in the complete data is divided into 2 parts: first part containing only the feature and the second part containing all of other features. After that, GP uses the complete data to build a regression function for the feature in the first part in terms of the features in the second part. The last step is repeated several times to generate a set of regression functions for each feature.

Algorithm 1. Training Process

Input:
X: training data
N: number of regression functions required for each feature
Output: N regression functions with fitness values for each feature
1 Extract complete data X_c from data X.
2 **for** $d=1$ **to** *NumberOfFeatures* **do**
3 $\quad$ Partition each instance in X_c into two parts: the first part contains only the dth feature; the second part contains all of other features.
4 $\quad$ **for** $i=1$ **to** N **do**
5 $\quad\quad$ - Use GP, the first part of X_c as a expected results and the second part of X_c as an input to build a regression function of the dth feature on the other features.
6 $\quad\quad$ - Store the regression function and its fitness.
7 $\quad$ **end**
8 **end**
9 **return** The N regresion functions with fitness values for each feature

---

Algorithm 2 shows the main body of the imputation process. The input of the imputation process is one instance containing missing values that need to be imputed, a set of regression functions for each features and a vector containing the mean values of all features. For each feature having missing value in the instance, it searches for the best regression function. It first counts the number of features in each regression function that have missing values in the instance, and choose the regression function with the fewest missing features and the highest fitness. After that, if the best regression function does not contain features having missing values in the instance, then the function is used to estimate missing value in the feature. Otherwise, missing values of features in the function are

---

Algorithm 2. Imputation Process

   Input:
   I: one instance containing training missing values
   Regression functions with fitness values of all features
   M: a vector containing mean values of all features
   Output: Instance I with missing values replaced by imputed values

1 **for** $d=1$ **to** *NumberOfFeatures* **do**
2   **if** *the dth feature in I is missing* **then**
3   **for** *each regression function of the dth feature* **do**
4   Count the number of features in the regression function that have missing values in I.
5   **end**
6   Identify the highest fitness regression function for the dth feature that contains the fewest number of features having missing values in I.
7   **if** *The number of missing features in the best regression function > 0* **then**
8   Fill temporarily missing features in the best regression function by their mean values in M.
9   **end**
10   Use the best regression function to impute missing value in the dth feature
11   **end**
12 **end**
13 **return** Instance I with imputed values

---

temporarily filled by their mean values, before the function is used to estimate missing value in the feature.

The key point of the proposed algorithm is that the training process produces a set of regression functions for each feature rather than a single regression function. The purpose of evolving a set of regression functions is that when an incomplete instance has multiple missing values, a regression function may depend on features having missing values. Therefore, for a particular incomplete instance, a set of regression functions provides the imputation process more chances to choose the most suitable regression function for each feature to impute missing values. The ability to evolve a set of different regression functions for each feature is an advantage of GP over traditional regression methods which would build the same regression function on each repetition.

4 Experiment Design

4.1 Method

The main objective of the experiments was to compare GPI with other imputation methods on classification. To achieve this, an experimental setup was designed as shown in Fig. 1. In case of complete data, firstly, missing values are introduced into compete data to generate missing data. Next, missing data is divided into training missing data and testing missing data. Training missing data and testing missing data are then put into an imputation method to generate imputed training data and imputed testing data. After that, the imputed training data is put into a classifier to build a classification model that is then used to classify the imputed testing data. The detailed configuration for each step of the experiment is shown below.

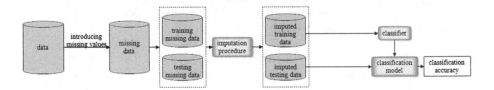

Fig. 1. Classification with missing data by using an imputation method before applying a classifier

4.2   Datasets

The experiments used six benchmark datasets selected from the UCI machine learning repository [3]. Table 1 summarises the main characteristics of each dataset including the number of instances, the number of features and the number of classes.

Table 1. Datasets used in the experiments

Dataset	#Instances	#Features	#Classes
Hepatitis	115	19	2
Ecoli	336	7	8
Leaf	340	15	30
Parkinsons	267	44	2
Seeds	210	7	3
Vertebral	310	6	3

The first dataset has 48.39 % incomplete instances which contain at least one missing value. For each of the last five complete datasets, perform 30 times: choose randomly 50 % features of the dataset, and then put randomly 10 % missing values into the chosen features. Therefore, for each dataset, 30 artificial missing datasets were generated, and a total of 150 (30×5) artificial missing datasets were used in the experiments.

Since none of the datasets in the experiments comes with a specific test set and the number of examples in some datasets is relatively small, a ten-fold cross-validation approach was used to evaluate the performance of induced classification models. With the first dataset, which contains natural missing values, a ten-fold cross-validation approach was performed 30 times on the dataset. For each of the last five datasets, ten-fold cross-validation was performed one time on the 30 missing datasets. As a result, for each dataset, 300 couples of training set and testing set were generated.

4.3   Benchmark Imputation Methods for Comparison

The experiments compared GPI to five regression-based imputation methods in MICE [36]: norm, nob, boot, cart and rf. Table 2 summaries the main characteristics of the imputation methods. The experiments used MICE's implementation in [9] for the five imputation methods by setting their parameters as the default values. In the five imputation methods, each feature was repeatedly regressed on other features 20 times as suggested in [35]. With each missing dataset, each of the imputation methods was performed 10 times to procedure 10 imputed datasets before combining by calculating average of them to make a final imputed dataset.

Table 2. Benchmark imputation methods.

Methods	Description	Scale type
norm	Bayesian linear regression-based multiple imputation	numeric
nob	Linear regression, non-Bayesian-based multiple imputation	numeric
boot	Linear regression using bootstrap-based multiple imputation	numeric
cart	Classification and regression trees-based multiple imputation	any
rf	Random forest -based multiple imputation	any

4.4   Classification Algorithms

The experiments compared GPI with the other imputation methods on five popular classifiers in the data mining community: C4.5 [28], k-Nearest Neighbors (kNN) [11], Naive Bayes (NB) [27], Support Vector Machine (SVM) [10] and Multilayer Perceptron (MLP) [19]. For the classifiers, WEKA's implementation [17] was used and all parameters were set to WEKA's defaults.

4.5   GP Settings

The experiments used the ECJ package [25] to implement GP. Table 3 shows the parameters of GP in GPI. For each dataset, 300 couples of training set and test set were generated. For each couple of training set and test set, GP repeated 10 times to generate 10 regression functions for each feature. Therefore, GP run 3000×NumberFeatures times on each dataset.

Table 3. GP parameters in the experiments.

Parameter	Value
Function set	$+, -, \times, /$ (protected division)
Variable terminals	all features except one feature being regressed
Constant terminals	Random float values
Population size	1000
Initialization	Ramped half-and-half
Selection type	Tournament(size=7)
Generations	100
Crossover probability	60 %
Mutation probability	30 %
Reproduction rate	10 %
Elitism	Yes

5 Results and Analysis

5.1 Classification Accuracy

Table 4 presents the average classification accuracy and corresponding standard deviation of the different imputation methods combined with different classifiers on six datasets. With the first dataset, which contains natural missing values, the average classification accuracy was computed on 30 times performing ten-fold cross-validation on the dataset. For each of the other five datasets, the average classification accuracy was computed on 30 versions of the dataset each with 10 % missing values on 50 % of the features. In order to compare the classification accuracy of GPI with the other methods, t-tests at 95 % confidence level were conducted. "T" columns in Table 4 shows the result of the significance tests of the colums before them against GPI, where "+" means GPI was significantly more accurate, "=" means not significantly different, and "-" means significantly less accurate.

Figure 2 summarises the results from Table 4 by comparing GPI with each of the other imputation methods. For each of the other imputation methods (norm, nob, boot, cart and rf), it shows the fraction of entries in the column for that method where GPI is significantly better, similar or significantly worse than the other imputation method. It is clear from Fig. 2 that GPI is at least as good as all the other methods. GPI performs significantly better than boot in almost

Table 4. Classification accuracy comparison of GPI with other imputation methods on different classifiers. The T columns indicate significant tests of the columns before them against GPI.

	Data	GPI	norm	T	nob	T	boot	T	cart	T	rf	T
C4.5	Hepatitis	80.05±1.69	78.43±1.93	+	79.14±1.74	=	78.17±2.36	+	78.68±1.68	+	78.19±1.62	+
	Ecoli	82.00±1.36	81.55±1.47	=	81.36±1.64	+	81.41±1.46	+	82.09±1.42	=	82.12±1.47	=
	Leaf	59.90±2.10	59.56±2.34	=	58.77±2.50	+	58.99±2.52	+	59.78±2.04	=	59.18±1.77	=
	Parkinsons	85.52±2.14	85.19±2.37	=	84.70±2.08	=	85.23±1.67	=	84.75±2.19	=	84.74±1.70	=
	Seeds	91.90±1.14	91.84±1.12	=	91.76±1.25	=	91.52±1.51	=	91.34±1.38	+	91.15±1.66	+
	Vertebral	80.37±1.59	79.74±2.00	=	79.93±1.99	=	79.98±1.99	=	80.45±1.84	=	80.69±2.03	=
kNN	Hepatitis	84.93±0.78	85.10±1.15	=	85.25±0.81	=	84.77±0.91	=	84.48±0.83	+	84.34±0.77	+
	Ecoli	85.29±1.14	85.24±1.00	=	85.16±1.22	=	85.19±1.02	=	84.76±1.20	+	85.09±1.08	=
	Leaf	55.35±1.15	55.21±1.71	=	55.14±1.73	=	55.23±1.20	=	55.19±1.10	=	55.31±1.50	=
	Parkinsons	89.61±1.34	88.10±1.80	+	88.17±1.64	+	8.13±1.62	+	88.16±1.72	+	88.27±1.44	+
	Seeds	92.25±0.79	91.95±0.66	+	92.19±0.89	=	92.01±0.79	=	91.85±0.84	+	91.95±0.80	+
	Vertebral	77.33±1.19	76.65±1.19	+	76.82±1.30	+	76.84±1.26	+	76.98±1.28	=	76.78±1.32	=
NB	Hepatitis	83.40±1.83	84.35±0.66	−	84.41±0.81	−	84.53±0.83	−	84.00±0.84	=	84.16±0.71	=
	Ecoli	84.03±1.36	83.10±1.62	+	83.27±1.52	+	83.23±1.74	+	84.37±1.16	=	84.75±1.21	=
	Leaf	70.42±1.48	68.38±1.76	+	68.49±1.74	+	68.37±1.81	+	70.35±1.33	=	0.06±1.44	=
	Parkinsons	69.29±0.92	69.56±0.94	−	69.51±0.94	=	69.56±1.03	=	69.56±0.94	=	69.52±0.78	=
	Seeds	90.06±0.55	90.00±0.52	=	90.12±0.60	=	89.90±0.59	=	89.92±0.63	=	89.93±0.74	=
	Vertebral	81.26±1.64	80.44±2.74	+	80.17±2.56	+	80.47±2.56	+	81.20±1.79	=	81.05±1.95	=
MNP	Hepatitis	84.02±1.80	81.23±1.97	+	81.89±1.25	+	81.81±1.59	+	79.07±1.36	+	82.17±1.30	+
	Ecoli	84.55±1.27	84.28±1.37	=	84.38±1.35	=	84.33±1.23	=	84.19±1.20	+	84.05±1.47	=
	Leaf	74.57±1.59	74.35±2.17	=	73.79±1.94	+	73.81±1.95	+	74.37±2.05	=	74.10±1.84	=
	Parkinsons	90.44±1.75	90.55±1.71	=	90.51±1.71	=	90.80±1.67	=	90.92±1.82	=	90.96±1.63	=
	Seeds	94.36±1.33	94.25±1.55	=	94.25±1.14	=	94.30±0.99	=	94.00±1.48	=	94.04±1.28	=
	Vertebral	84.38±1.37	83.90±1.63	+	83.65±2.04	+	83.78±1.60	+	84.39±1.38	=	84.34±1.54	=
SVM	Hepatitis	85.82±0.81	85.33±1.07	+	85.73±1.01	=	85.49±0.66	=	85.80±0.94	=	86.04±1.33	=
	Ecoli	82.51±1.11	82.52±1.06	=	82.47±1.15	=	82.64±1.01	=	82.51±1.05	=	82.55±1.29	=
	Leaf	50.43±1.24	50.14±1.78	=	49.87±1.65	+	49.76±1.49	+	51.07±1.57	−	50.63±1.68	=
	Parkinsons	87.53±1.02	87 58±0.86	=	87.52±0.94	=	87.56±1.06	=	87.47±0.83	=	87.54±0.93	=
	Seeds	93.25±0.68	93.30±0.44	=	93.34±0.56	=	93.26±0.54	=	93.25±0.76	=	93.19±0.73	=
	Vertebral	75.20±1.26	74.96±1.37	=	75.03±1.33	=	75.19±1.32	=	75.39±1.13	=	75.20±0.98	=

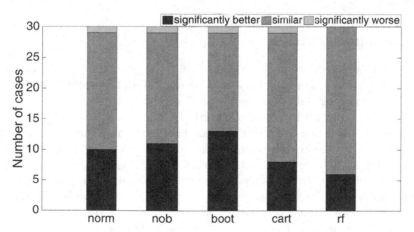

Fig. 2. Fraction of cases that GPI is significantly better or worse than norm, nob, boot, cart and rf.

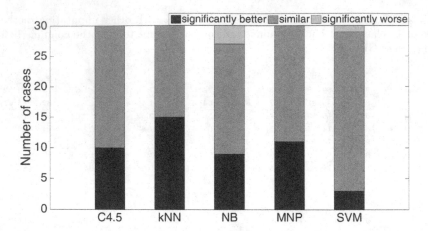

Fig. 3. Fraction of cases that GPI is significantly better or worse than the other imputation method for each classifier.

half of the cases, but is better than rf in only a quarter of the cases. GPI is significantly worse in very few cases- only one case for norm, nob, boot and cart.

Figure 3 summarises the results from Table 4 by comparing GPI with all of the other imputation methods on each classifier. It is clear from Fig. 3 that the choice of classifier makes only a little difference. With kNN, C4.5 and MNP, GPI is never significantly worse than the other methods and better in one third to half of the cases. With NB, GPI is more frequently worse than any the other classifier. With SVM, GPI is more consistently similar to the other imputation methods.

In summary, GPI is almost always at least as good as than the other imputation methods, and sometimes significantly better, especially with kNN as the classifiers.

5.2   Computation Time

Regression-based imputation methods in MICE [36] are expensive for classification. The main reason is that they require rebuilding the regression functions using all the training data and the new instance each time when they need to estimate missing values in a new instance. This may not be significant in the training stage, but when using the learned classifier in an application, each new instance to be classified requires imputation methods to impute any missing values. It is not feasible in many application for the classification to take many seconds per instance. Table 5 shows the average computation time of GPI and the other imputation methods to impute the missing values for one instance.

It is clear from Table 5 that GPI is at least thousand times faster than the other imputation methods in almost all of the datasets (it is only about 600 times faster than three other methods on the Vertebral dataset). The reason is that GPI builds symbolic regression functions to impute incomplete instances in

Table 5. The average computation time (millisecond) of GPI and the other imputation methods to impute the missing values of one instance

Data	GPI	norm	nob	boot	cart	rt
Hepatitis	2.58	3.34×10^3	3.38×10^3	3.45×10^3	1.29×10^4	3.53×10^4
Ecoli	0.61	6.22×10^2	6.23×10^2	6.26×10^2	2.83×10^3	8.51×10^3
Leaf	0.89	2.97×10^3	2.98×10^3	3.00×10^3	1.10×10^4	2.60×10^4
Parkinsons	0.86	3.88×10^3	3.90×10^3	3.95×10^3	1.23×10^4	2.88×10^4
Seeds	0.61	5.95×10^2	5.85×10^2	5.89×10^2	2.45×10^3	7.12×10^3
Vertebral	1.10	6.25×10^2	6.17×10^2	6.19×10^2	2.70×10^3	7.72×10^3

a training process. After that, GPI uses the built symbolic regression function directly to impute missing values in new incomplete instances. In other words, unlike regression-based imputation methods in MICE, GPI does not require rebuilding the regressions to impute missing values in new incomplete instances. Therefore, GPI spends no time rebuilding the regressions and only spends time to estimate missing values based on the built symbolic regressions.

In summary, GPI is dramatically more efficient for estimating missing values in new incomplete instances than the regression-based imputation methods in MICE, and achieves this with no loss of accuracy.

6   Conclusion and Future Work

This paper proposed GPI algorithm, which uses GP as a symbolic regression method to build regression functions for estimating missing values. GPI has two processes: the train process and the imputation process. The training process builds a set of regression functions for each feature that are then used to estimate missing values in the imputation process. The experiments on six benchmark datasets and five popular classifiers compared GPI with five other popular and advanced regression-based imputation methods in MICE on two measures: classification accuracy and computation time. The results showed that, in most cases, GPI achieves at least similar classification accuracy to the other imputation methods, and sometimes significantly better. GPI makes the greatest improvement with KNN, but little improvement with SVM and NB. However, using GPI to impute missing values for new incomplete instances is dramatically faster than the other imputation methods, typically by a faster of thousand.

In this paper, GPI used standard GP to build regression functions for features. Therefore, one future work could apply advanced GP-based regression methods such as GP with interval arithmetic and linear scaling [20], gradient descent GP [32], semantic GP [34] to improve GPI. Furthermore, two common issues in GP are bloat and overfitting. Hence, another possible future work is to apply bloat control methods [31] and overfitting control methods [1] to enhance GPI.

References

1. Agapitos, A., Brabazon, A., O'Neill, M.: Controlling overfitting in symbolic regression based on a bias/variance error decomposition. In: Coello, C.A.C., Cutello, V., Deb, K., Forrest, S., Nicosia, G., Pavone, M. (eds.) PPSN 2012, Part I. LNCS, vol. 7491, pp. 438–447. Springer, Heidelberg (2012)
2. Andridge, R.R., Little, R.J.: A review of hot deck imputation for survey nonresponse. Int. Stat. Rev. **78**, 40–64 (2010)
3. Asuncion, A., Newman, D.: UCI machine learning repository (2007). http://www.ics.uci.edu/~mlearn/MLRepository.html
4. Augusto, D.A., Barbosa, H.J.: Symbolic regression via genetic programming. In: Sixth Brazilian Symposium on Neural Networks, 2000, Proceedings, pp. 173–178 (2000)
5. Barmpalexis, P., Kachrimanis, K., Tsakonas, A., Georgarakis, E.: Symbolic regression via genetic programming in the optimization of a controlled release pharmaceutical formulation. Chemometr. Intell. Lab. Syst. **107**, 75–82 (2011)
6. Barnard, J., Meng, X.L.: Applications of multiple imputation in medical studies: from AIDS to NHANES. Stat. Methods Med. Res. **8**, 7–36 (1999)
7. Berger, J.O.: Statistical Decision Theory and Bayesian Analysis. Springer Science & Business Media, New York (2013)
8. Breiman, L., Friedman, J., Stone, C.J., Olshen, R.A.: Classification and Regression Trees. CRC Press, Boca Raton (1984)
9. Buuren, S., Groothuis-Oudshoorn, K.: MICE: multivariate imputation by chained equations in R. J. Stat. Soft. **45**, 1–67 (2011)
10. Cortes, C., Vapnik, V.: Support-vector networks. Mach. Learn. **20**, 273–297 (1995)
11. Cunningham, P., Delany, S.J.: k-Nearest Neighbour classifiers. In: Multiple Classifier Systems, pp. 1–17 (2007)
12. Draper, N.R., Smith, H., Pownell, E.: Applied Regression Analysis, vol. 3. Wiley, New York (1966)
13. Farhangfar, A., Kurgan, L., Dy, J.: Impact of imputation of missing values on classification error for discrete data. Pattern Recogn. **41**, 3692–3705 (2008)
14. Farhangfar, A., Kurgan, L.A., Pedrycz, W.: A novel framework for imputation of missing values in databases. IEEE Trans. Syst. Man Cybern. Part A: Syst. Hum. **37**, 692–709 (2007)
15. García-Laencina, P.J., Sancho-Gómez, J.L., Figueiras-Vidal, A.R.: Pattern classification with missing data: a review. Neural Comput. Appl. **19**, 263–282 (2010)
16. Graham, J.W.: Missing data analysis: making it work in the real world. Ann. Rev. Psychol. **60**, 549–576 (2009)
17. Hall, M., Frank, E., Holmes, G., Pfahringer, B., Reutemann, P., Witten, I.H.: The WEKA data mining software: an update. ACM SIGKDD Explor. Newslett. **11**, 10–18 (2009)
18. Han, J., Kamber, M., Pei, J.: Data Mining, Southeast Asia Edition: Concepts and Techniques. Morgan Kaufmann, San Francisco (2006)
19. Hornik, K., Stinchcombe, M., White, H.: Multilayer feedforward networks are universal approximators. Neural Netw. **2**, 359–366 (1989)
20. Keijzer, M.: Improving symbolic regression with interval arithmetic and linear scaling. In: Genetic programming, pp. 70–82 (2003)
21. Kleinbaum, D., Kupper, L., Nizam, A., Rosenberg, E.: Applied regression analysis and other multivariable methods. Cengage Learning (2013)

22. Koza, J.R.: Genetic Programming: On the Programming of Computers by Means of Natural Selection. MIT Press, Cambridge (1992)

23. Liaw, A., Wiener, M.: Classification and regression by randomforest. R News **2**, 18–22 (2002)

24. Little, R.J.A., Rubin, D.B.: Statistical Analysis with Missing Data. Wiley-Interscience, New York (2002)

25. Luke, S., Panait, L., Balan, G., Paus, S., Skolicki, Z., Bassett, J., Hubley, R., Chircop, A.: ECJ: A java-based evolutionary computation research system (2006) Downloadable versions and documentation can be found at the following http:// cs.gmu.edu/eclab/projects/ecj

26. Minka, T.: Bayesian linear regression. Technical report, 3594 Security Ticket Control (1999)

27. Murphy, K.P.: Naive Bayes classifiers. University of British Columbia (2006)

28. Quinlan, J.R.: C4.5: Programs for Machine Learning. Morgan Kaufmann Publishers Inc., San Francisco (1993)

29. Schafer, J.L.: Analysis of Incomplete Multivariate Data. Monographs on Statistics & Applied Probability. Chapman & Hall/CRC, New York (1997)

30. Schafer, J.L.: Analysis of Incomplete Multivariate Data. CRC Press, New York (1997)

31. Silva, S., Dignum, S., Vanneschi, L.: Operator equalisation for bloat free genetic programming and a survey of bloat control methods. Genet. Program. Evolvable Mach. **13**, 197–238 (2012)

32. Topchy, A., Punch, W.F.: Faster genetic programming based on local gradient search of numeric leaf values. In: Proceedings of the Genetic and Evolutionary Computation Conference (GECCO-2001), vol. 155162 (2001)

33. Tran, C.T., Zhang, M., Andreae, P.: Multiple imputation for missing data using genetic programming. In: Proceedings of the 2015 on Genetic and Evolutionary Computation Conference, pp. 583–590 (2015)

34. Uy, N.Q., Hoai, N.X., O'Neill, M., Mckay, R.I., Galván-López, E.: Semantically-based crossover in genetic programming: application to real-valued symbolic regression. Genet. Program. Evolvable Mach. **12**, 91–119 (2011)

35. Van Buuren, S., Oudshoorn, C.: Multivariate imputation by chained equations. MICE V1. 0 user's manual. Leiden: TNO Preventie en Gezondheid (2000)

36. Van Buuren, S., Oudshoorn, K.: Flexible multivariate imputation by MICE. Technical report, PG/VGZ/99.054: TNO Prevention and Health, Leiden (1999)

37. Vladislavleva, E.J., Smits, G.F., Den Hertog, D.: Order of nonlinearity as a complexity measure for models generated by symbolic regression via pareto genetic programming. IEEE Trans. Evol. Comput. **13**, 333–349 (2009)

38. White, I.R., Royston, P., Wood, A.M.: Multiple imputation using chained equations: issues and guidance for practice. Stat. Med. **30**, 377–399 (2011)

Plastic Fitness Predictors Coevolved with Cartesian Programs

Michal Wiglasz[(✉)] and Michaela Drahosova

Faculty of Information Technology, Brno University of Technology,
Božetěchova 2, 612 66 Brno, Czech Republic
{iwiglasz,idrahosova}@fit.vutbr.cz

Abstract. Coevolution of fitness predictors, which are a small sample of all training data for a particular task, was successfully used to reduce the computational cost of the design performed by cartesian genetic programming. However, it is necessary to specify the most advantageous number of fitness cases in predictors, which differs from task to task. This paper introduces a new type of directly encoded fitness predictors inspired by the principles of phenotypic plasticity. The size of the coevolved fitness predictor is adapted in response to the learning phase that the program evolution goes through. It is shown in 5 symbolic regression tasks that the proposed algorithm is able to adapt the number of fitness cases in predictors in response to the solved task and the program evolution flow.

Keywords: Fitness predictors · Cartesian genetic programming · Coevolution · Phenotypic plasticity

1 Introduction

Cartesian genetic programming (CGP) is a specific form of genetic programming (GP) and has been successfully applied to a number of challenging real-world problem domains [7]. In CGP, as well as in GP, every evolved program must be executed to find out what it does. Each program in the population is assigned a fitness value, representing the degree to which it solves the problem of interest. Often, but not always, the fitness is calculated over a set of *fitness cases*. A fitness case consists of potential program inputs and target values expected from a perfect solution as a response to these program inputs. The outputs of the evolved program are then compared with the desired outputs for given inputs. The choice of how many fitness cases (and which ones) to use is often a crucial decision since whether or not the evolved program will generalize over the entire domain depends on this choice.

In the case of digital circuit evolution, which is a typical task for CGP, it is necessary to verify whether a candidate n-input circuit generates correct responses for all possible input combinations (i.e., 2^n assignments). It was shown that testing just a subset of 2^n fitness cases does not lead to correctly working circuits [5].

In the *symbolic regression* tasks, the goal of GP system design and GP parameters' tuning is to obtain a solution with predefined accuracy and robustness.

© Springer International Publishing Switzerland 2016
M. Heywood et al. (Eds.): EuroGP 2016, LNCS 9594, pp. 164–179, 2016.
DOI: 10.1007/978-3-319-30668-1_11

In this case, k fitness cases are evaluated during one fitness function call, where k typically goes from hundreds to ten thousands. The time needed for evaluating a single fitness case depends on a particular application. Usually, in order to find a robust and acceptable solution a large number of fitness evaluations has to be performed. In order to reduce the evaluation time, *fitness approximation* techniques have been employed, e.g. fitness modeling [6].

Closely related concept to the fitness modeling is a *fitness prediction*, which is a low cost adaptive procedure utilized to replace the fitness evaluation. A framework for reducing the computation requirements of symbolic regression using fitness predictors has been introduced for standard genetic programming by Schmidt and Lipson [9]. The method utilizes a coevolutionary algorithm which exploits the fact that one individual can influence the relative fitness ranking between two other individuals in the same or a separate population [4]. The state of the art of coevolutionary principles has recently been summarized in the chapter of Handbook of Natural Computing [8].

Inspired by [9], we have introduced coevolving fitness predictors to CGP and have shown that by using them, the execution time of symbolic regression can significantly be reduced [12]. Fitness predictors have been represented as a constant-size array of pointers to elements in the fitness case set and operated using a simple genetic algorithm. The same coevolutionary CGP and Hillis' *competitive coevolution* approach [4] adapted for CGP have been used in the evolutionary image filter design [11]. Although the time of evolution has also been reduced, a large number of experiments had to be accomplished in order to find the most advantageous size of the fitness predictor (the number of fitness cases in predictor) for this particular task.

To solve this problem, we have introduced a new type of indirectly encoded fitness predictors which can automatically adapt the number of fitness cases used to evaluate the candidate programs [10]. However, during the evolution of fitness predictors, also large fitness predictors have to be evaluated (and then refused for a larger size), and thus plenty of fitness case evaluations have been wasted.

In this paper, we integrate *phenotypic plasticity* principles into coevolution. The phenotypic plasticity is the ability of an individual to learn how to utilize its genotype in order to adapt to the environment [1]. It was shown that a proper rate of environmental change may reduce the learning cost while evolving the solution [2,3]. Inspired by these principles, we introduce a new type of fitness predictors, operated using a simple genetic algorithm (GA), using the phenotypic plasticity in order to adapt the number of fitness cases for candidate solution evaluations and thus regulate the rate of environmental change. In the case of fitness prediction, a stable environment contains a complete fitness cases set, a highly changing environment only a few of them.

The paper is organized as follows. Section 2 introduces cartesian genetic programming and coevolution of fitness predictors. In Sect. 3, a new approach to fitness predictor encoding is presented. The proposed approach is evaluated using 5 symbolic regression benchmarks. Experimental results are discussed in Sect. 4. Finally, conclusions are given in Sect. 5.

2   Fitness Prediction in CGP

In standard CGP, candidate programs are represented in the form of directed acyclic graph, which is modeled as a matrix of $n_c \times n_r$ programmable elements (nodes). Each node is programmed to perform one of n_a-input functions defined in the set Γ. The number of primary inputs, n_i, and outputs, n_o, of the program is defined for a particular task. Each node input can be connected either to the output of a node placed in previous l columns or to one of the program inputs. Feedback is not allowed. The search is usually performed using a simple $(1 + \lambda)$ evolutionary algorithm, where usually $\lambda = 4$. Every new population consists of the best individual of the previous population and its λ offspring created using a mutation operator which modifies up to h genes of the chromosome. The state of the art of CGP has recently been summarized in a monograph [7].

In the case of *symbolic regression*, the set of fitness cases is usually constructed from experimentally obtained data. Then each of k fitness cases from the set is used to evaluate each candidate program (see Fig. 1). The fitness function of candidate program is often defined as the relative number of hits. Formally,

$$f(s) = \frac{1}{k} \sum_{j=1}^{k} g(y(j)), \text{ where} \tag{1}$$

$$g(y(j)) = \begin{cases} 0 \text{ if } |y(j) - t(j)| \geq \varepsilon \\ 1 \text{ if } |y(j) - t(j)| < \varepsilon \end{cases} \tag{2}$$

and y is a candidate program response, t is a target response and ε is a user-defined acceptable error. The fitness evaluation is the most time consuming part in standard CGP (as well as tree-based GP).

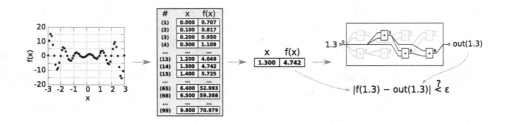

Fig. 1. Fitness evaluation of a candidate cartesian program.

2.1   Fitness Predictor

In order to reduce the total number of evaluations during each one fitness function call, fitness predictor in the form of small subset of the fitness case set have been introduced to CGP [12]. An optimal fitness predictor is sought using a simple genetic algorithm (GA) which operates with a population of fitness

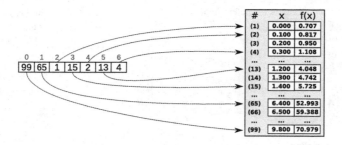

Fig. 2. Fitness predictor representation.

predictors. Every predictor is encoded as a constant-size array of pointers to elements in the training data (see Fig. 2). In addition to one-point crossover and mutation, a randomly selected predictor replacing the worst-scored predictor in each generation has been introduced as a new genetic operator of GA. The goal of the evolution of predictors is to minimize the relative error of fitness prediction and the expensive exact fitness evaluation.

2.2 Coevolution of Cartesian Programs and Fitness Predictors

The aim of coevolving fitness predictors and programs is to allow both solutions (programs) and fitness predictors to enhance each other automatically until a satisfactory problem solution is found. There are two concurrently working populations: (1) candidate programs (syntactic expressions) evolving using CGP and (2) fitness predictors evolving using GA. The overall scheme of the coevolutionary algorithm is shown in Fig. 3.

Evolution of candidate programs is based on principles of CGP. The fitness function for CGP is defined as the relative number of hits. There are, in fact, two fitness functions for candidate program s. While the exact fitness function $f_{exact}(s)$ utilizes the complete set of fitness cases, the predicted fitness function $f_{predicted}(s)$ employs only selected fitness cases. Formally,

$$f_{exact}(s) = \frac{1}{k} \sum_{j=1}^{k} g(y(j)) \tag{3}$$

$$f_{predicted}(s) = \frac{1}{m} \sum_{j=1}^{m} g(y(j)) \tag{4}$$

where k is the number of fitness cases in the set of fitness cases and m is the number of fitness cases in the fitness predictor. The $f_{predicted}$ is used to evaluate the candidate programs in the population. The f_{exact} is used during the predictor training.

The predictor training is accomplished as follows. The *archive of trainers* is generated and updated in response to the candidate program evolution. It consists of candidate programs with evaluated f_{exact} and is divided into two

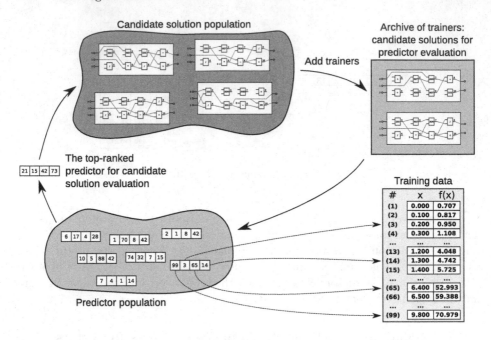

Fig. 3. Coevolution of candidate solutions and fitness predictors.

parts: The first part contains copies of top-ranked programs (with different fitness) obtained during the program evolution and the second part is periodically updated with randomly generated programs to ensure genetic diversity of the archive. The size of the archive is kept constant during the coevolution and each new trainer replaces the oldest one in the corresponding part of the archive.

The fitness value of predictor p is calculated using the *mean absolute error* of the exact and predicted fitness values of programs in the archive of trainers:

$$f(p) = \frac{1}{u} \sum_{i=1}^{u} |f_{exact}(i) - f_{predicted}(i)| \tag{5}$$

where u is the number of candidate programs in the archive of trainers. The predictor with the best fitness value is used to predict the fitness of candidate programs in the population of candidate programs [9].

3   Proposed Method

In this paper, we propose a new approach to fitness predictor encoding. The number of fitness cases required to obtain a satisfactory solution varies from benchmark to benchmark. In order to apply coevolutionary CGP to different tasks, it is required to perform numerous experiments to find the most advantageous number of fitness cases in fitness predictors.

It can be observed that the population of solutions goes through various phases as the population's ability to adapt to the problem changes over the time [2]. A *lower fitness* phase needs less stimuli to improve solutions, but the same amount of stimuli does not lead to converge during the *higher fitness* phase. This property is discussed by Ellefsen [2], in order to reduce the learning cost. In this paper, the number of fitness cases in predictors is changed according to the latest development in the population of the candidate programs.

3.1   Plastic Directly Encoded Predictor

We propose directly encoded fitness predictors with an adaptive number of fitness cases for candidate solution evaluations. To be able to modify their size, we employ the principles of phenotypic plasticity. This allows the individual to produce different phenotypes from the same genotype, depending on the environmental conditions [2]. In the plastic fitness predictors, the phenotype is constructed by including only selected subset of genes.

The predictor genotype is a constant-size circular array of pointers to elements in the training data. Its size is equal to the total number of fitness cases. In order to produce the phenotype, the genes are read sequentially from specified position (*offset*). The genotype may contain duplicate gene values. Therefore, the gene with a value, which is already included in the phenotype, is skipped in order to prevent duplicate fitness case in predictor. The reading stops after it has processed the number of genes specified by the *readLength* variable. The *readLength* value is determined by the flow of the candidate program evolution.

The *offset* is determined by an extra gene included in the genotype, evolved by a special mutation operator, which adds a small Gaussian random number to the current value. Figure 4 shows an example of phenotype construction when 6 out of 10 available genes are used.

The evolution of predictors is directed by the genetic algorithm (GA). The crossover operator is modified so the split point is always selected within the active part of the genotype, which increases phenotype diversity.

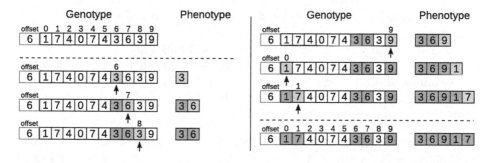

Fig. 4. Predictor phenotype construction with *offset* = 6 and *readLength* = 6.

3.2 Predictor Size Adaptation

The predictor size is adapted through the *readLength* variable. Its value is changed according to the latest development in the population of the candidate programs. It can be observed that the population goes through various phases as the population's ability to adapt to the problem changes over the time. If the ability is higher, the overall fitness increases towards better solutions, if it is lower, the fitness remains almost constant. In this case the evolution probably reached some local optimum.

The phase of evolution can be described in terms of the evolution speed which we express as follows:

$$v = \frac{\Delta f_{exact}}{\Delta G}, \tag{6}$$

where ΔG is the number of generations between two last fitness changes of CGP population parent (top-ranked programs) and Δf_{exact} is the difference of exact fitness values of these parents. Although the evolution of programs is guided by the predicted fitness, the speed can be negative, because it is calculated from the exact fitness.

It is necessary to set the lower boundary of the predictor size. If the prediction is based on only a few fitness cases (in extreme cases on only one fitness case), over-fitting of predictors occurs. The prediction inaccuracy can be expressed as the absolute difference between predicted and exact fitness:

$$I = |f_{predicted} - f_{exact}|, \tag{7}$$

In the case the prediction inaccuracy exceeds given threshold I_{thr}, the number of fitness cases should be increased.

The *readLength* value is updated each time a new solution with better predicted fitness than parent individual is found. It can be also updated after a user-specified number of generations during which a new solution is not found. The evolution speed and prediction inaccuracy is updated and a corresponding rule is selected. The rules are based on the following assumptions:

1. If the inaccuracy exceeds the threshold ($I > I_{thr}$), the size is increased.
2. If the fitness remains unchanged ($v \approx 0$), the predictor size is decreased, which should help the evolution to leave a local optimum.
3. If the fitness decreases ($v < 0$), the evolution is probably leaving a local optimum and decreasing the size can accelerate this process.
4. If the fitness increases ($v > 0$), the predictor size is increased to make the prediction more accurate.

The purpose of these rules is to find the lowest possible predictor size while the evolution still converges. The new *readLength* value is obtained by multiplication of the previous value and a coefficient, which is selected using described rules. Experimentally obtained values of the coefficients are specified in Sect. 4.2.

4   Results

In this section, 5 symbolic regression benchmarks are introduced. Next, we present experimental results, in particular the proposed predictor behaviour and the comparisons of the proposed approach with the previously presented approaches to coevolutionary and standard CGP.

4.1   Benchmark Problems

Five symbolic regression benchmark functions (F1 – F5, see Fig. 5) were selected as training data sources for evaluation of the proposed method:

$$F1 : f(x) = x^2 - x^3, \qquad\qquad x = [-10 : 0.1 : 10]$$

$$F2 : f(x) = e^{|x|} \sin(x), \qquad\qquad x = [-10 : 0.1 : 10]$$

$$F3 : f(x) = x^2 e^{\sin(x)} + x + \sin\left(\frac{\pi}{x^3}\right), \qquad x = [-10 : 0.1 : 10]$$

$$F4 : f(x) = e^{-x} x^3 \sin(x) \cos(x) \left(\sin^2(x) \cos(x) - 1\right), \qquad x = [0 : 0.05 : 10]$$

$$F5 : f(x) = \frac{10}{(x-3)^2 + 5}, \qquad\qquad x = [-2 : 0.05 : 8].$$

To form the training data, 200 equidistant distributed samples were taken from each function. Functions F1 – F5 are taken from [12] and all functions F1 – F5 were used in order to evaluate coevolution of CGP and both directly and indirectly encoded predictors [10,12].

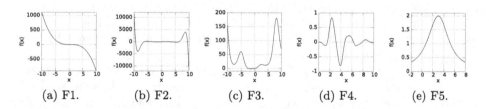

(a) F1.   (b) F2.   (c) F3.   (d) F4.   (e) F5.

Fig. 5. Symbolic regression benchmark functions used for evaluation.

4.2   Experimental Setup

The setup of the program evolution is used according to literature [12], i.e. $\lambda = 12$, $n_i = 1$, $n_o = 1$, $n_c = 32$, $n_r = 1$, $l = 32$, every node has two inputs (i_1, i_2), $\Gamma = \{i_1 + i_2, i_1 - i_2, i_1 \cdot i_2, \frac{i_1}{i_2}, \sin(i_1), \cos(i_1), e^{i_1}, \log(i_1)\}$ and the maximum number of mutations per individual is $h = 8$. The program fitness function is defined as the relative number of hits (see Eqs. 3 and 4). For the benchmarks, the user-defined acceptable errors ε are as follows: F1, F2: 0.5; F3: 1.5; F4, F5: 0.025. The acceptable number of hits is 96 %.

Table 1. Rules used to adapt the *readLength* value.

Priority	Condition	Coefficient
1	$I > I_{thr}$	1.2
2	$|v| \leq 0.001$	0.9
3	$v < 0$	0.96
4	$0 < v' \leq 0.1$	1.07
5	$v > 0.1$	1

The predictor size is adapted as follows: The *readLength* value is initialized with 5 genes (the influence of the initial value is discussed in Sect. 4.3), its minimum is limited to 5 and the maximum is the total number of fitness cases. The value is updated after a new top-ranked program is found, or after 5000 generations since last update. The new *readLength* value is given as *readLength · coefficient*. Experimentally obtained coefficient values are shown in Table 1. The threshold $I_{thr} = 15$ is chosen. Conditions are set according to assumptions in Sect. 3.2. If more conditions are fulfilled at the same time, the value is updated according to the priority (see Table 1).

4.3 Ability to Adapt the Number of Fitness Cases

In order to confirm that the proposed algorithm is able to adapt the predictor size on a given task, we plot the progress of the average number (out of 100 independent runs) of fitness cases in top-ranked predictor during the evolution flow with respect to the initial predictor sizes. It can be seen in Fig. 6 that the size converges to the similar value independently of an initial size and the final predictor size differs for each benchmark.

The success rate is the same for each initial size setting. In the case of benchmarks F1 – F3, a larger initial size leads to more fitness case evaluations required to find an acceptable solution, see Fig. 6. This does not hold for benchmarks F4 and F5, where all settings lead to a comparable number of evaluations. The reason is that the predictor size converges in approximately 10^5 generations, while it takes much more time (approx. $3.7 \cdot 10^6$ generations) to find a satisfactory solution (see Table 2), so the effect of different predictor size in the beginning of the evolution is negligible. Note that a satisfactory solution for the benchmark F1 is found in less generations than it is necessary for the predictor size to converge.

In general, it is advantageous to begin with a lower number of fitness cases in predictor, which in some cases leads to a lower number of evaluations and thus the design process acceleration. On the other hand, if the initial size is too low to find an acceptable solution, it will be automatically increased without a significant impact on the run time.

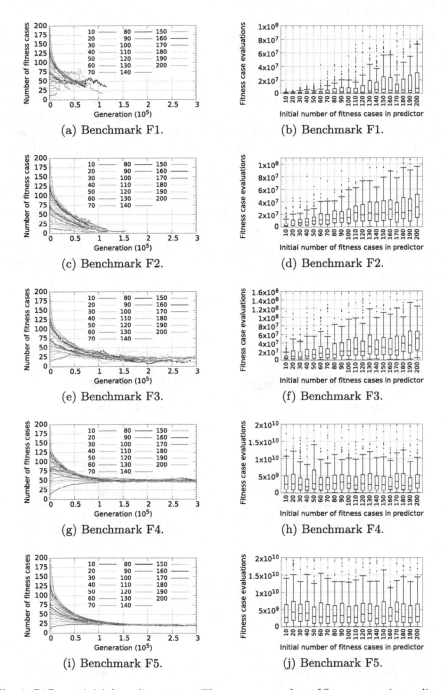

(a) Benchmark F1.

(b) Benchmark F1.

(c) Benchmark F2.

(d) Benchmark F2.

(e) Benchmark F3.

(f) Benchmark F3.

(g) Benchmark F4.

(h) Benchmark F4.

(i) Benchmark F5.

(j) Benchmark F5.

Fig. 6. Different initial predictor sizes: The average number of fitness cases in predictors and the number of fitness case evaluations necessary to find an acceptable solution.

4.4 Predictor Behaviour

In this section, we discuss how a predictor selects a subset of training data
capable of guiding the evolution towards the satisfactory solution. We plot the
distribution of fitness cases selected by predictors during the whole coevolutionary process out of 100 independent runs. Figure 7 show the frequency of fitness
cases addressed by the top-ranked predictors during the coevolution flow. It can
be seen that for benchmarks F1 and F2 predictors focus more on peaks and
valleys than on flexes. On the other hand, in the case of F3 – F5, the samples
are well distributed over the data set. Considering all fitness cases addressed
by the predictor focused on the interesting regions (peaks and valleys) of the
training data, the predictor would represent the maximum error. Note that this
characteristic is desired in the Hillis' competitive coevolutionary approach [4],
but is improper while requiring the predicted fitness corresponding to the exact
fitness. Furthermore, fitness cases addressed by the fitness predictors are variable in response to the program evolution flow. The program evolution forces
the predictors to contain two types of fitness cases, some of them are easy, others
difficult, for a particular program.

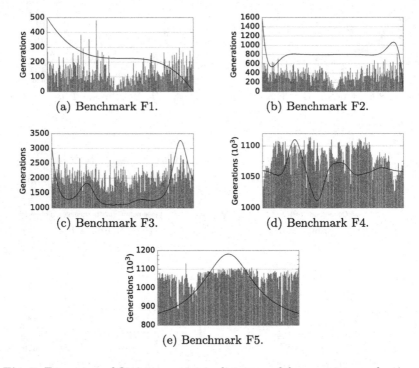

Fig. 7. Frequency of fitness cases in predictors used for programs evaluations.

4.5   Comparison of the Predictor Size

Indirectly encoded fitness predictors based on the principles of CGP (below FP_{indir}) were proposed in order to overcome the problem with selection of the most advantageous number of fitness cases used for fitness evaluation. In FP_{indir}, the predictor size parameter is included in the fitness function. Most of the sizable predictors are then rejected, but evaluated during the predictor training, which results into wasted evaluations. In order to reduce the computational cost of predictor fitness evaluations during the training, a limit of predictor size was introduced. Then, the maximum size of fitness predictor evolved using FP_{indir} was 50 fitness cases.

The size of the proposed adaptive directly encoded predictors (FP_{adapt}) varies only a little in the following generations, depending on coefficients (see Table 1). The number of fitness cases in the active predictor is thus changed only in small steps and no limit of the predictor size is necessary.

Figure 8 shows the number of fitness cases in the top-ranked predictor during the coevolution flow (left part of figures shows FP_{indir}, right part FP_{adapt}). In general, the preferred number of fitness cases differs from benchmark to benchmark. It can be seen that for the benchmarks F1 and F2 (in which only some of the first predictors are used) the preferred size of fitness predictor is the maximum value (50 fitness cases) for FP_{indir} and near to the initializing value for FP_{adapt} approach (6 – 7 fitness cases). For the benchmark F3, the maximum

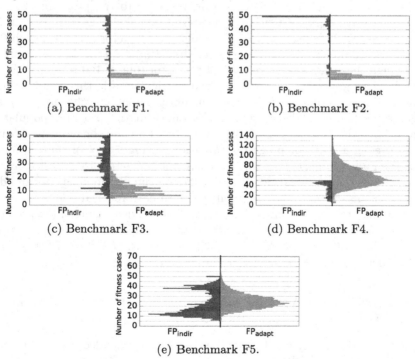

(a) Benchmark F1.

(b) Benchmark F2.

(c) Benchmark F3.

(d) Benchmark F4.

(e) Benchmark F5.

Fig. 8. Number of fitness cases in predictors used for program evaluations.

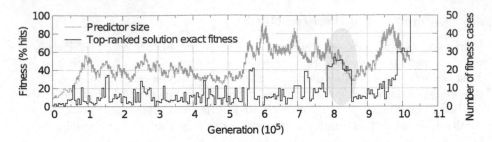

Fig. 9. Relation between the exact fitnesses of top-ranked candidate program and the size of predictor during a typical run for the F5 benchmark.

value is also preferred in FP_{indir} approach (because the evolution of predictors does not have enough time to adapt), but in FP_{adapt} the preferred value is between 7 and 12 fitness cases. Benchmark F4 is an example of how the limit of the predictor size in FP_{indir} could be restrictive. FP_{adapt} predictor size is distributed around 52 fitness cases, whereas FP_{indir} leads to predictors using 45 to 50 fitness cases and must not exceed 50. For benchmark F5, we can observe two peaks (in 12 and 38 fitness cases) for FP_{indir} predictors, but only one peak distributed around 25 fitness cases for FP_{adapt}. Note that FP_{indir} evolution allows fast changes of predictor size in contiguous generations and thus cause skips between distant values of predictor size in response to the program evolution flow. Conversely, FP_{adapt} evolution provides only small changes of predictor size in contiguous generations. The FP_{adapt} preferred predictor size, for benchmark F5, lies between preferred predictor sizes of the FP_{indir} approach, in the middle.

Although the average preferred size of predictor (out of 100 runs) in FP_{adapt} approach converges to the single value for a particular task, this trend is not so obvious while analyzing a single run. During a single run, the predictor size changes in response to the current development in the program population. Figure 9 shows the exact fitness of top-ranked program and the number of fitness cases in predictor used for program evaluation during a typical coevolutionary run for the F5 benchmark. It can be seen that the predictor size is first increased towards the preferred value and then it reacts on the development of candidate program. In this example the evolution seems to have reached a local optimum after approximately $8 \cdot 10^5$ generations which leads to decreasing of the predictor size. Around generation $8.5 \cdot 10^5$ the fitness of top-ranked program drops significantly as the evolution left the local optimum and the number of fitness cases starts to increase again, to increase accuracy of the fitness prediction.

4.6 Comparisons of Various Approaches to Fitness Prediction in CGP

The proposed coevolution employing adaptive directly encoded fitness predictors (FP_{adapt}) is compared with the original fixed-size directly encoded predictors

Table 2. Comparison of standard CGP (CGP_{STD}), coevolutionary CGP with directly encoded constant-size (FP_{const}) and adaptive predictors (FP_{adapt}) and coevolutionary CGP with indirectly encoded CGP-based predictors (FP_{indir}). For each benchmark, the best result is marked in bold font.

	Algorithm	F1	F2	F3	F4	F5
Success rate	CGP_{STD}	100 %	100 %	91 %	5 %	27 %
	FP_{const}	100 %	100 %	100 %	33 %	43 %
	FP_{adapt}	100 %	100 %	100 %	99 %	87 %
	FP_{indir}	100 %	100 %	100 %	100 %	90 %
Generations to converge	CGP_{STD}	$8.66 \cdot 10^3$	$3.09 \cdot 10^4$	$1.17 \cdot 10^5$	$4.13 \cdot 10^6$	$3.25 \cdot 10^6$
(median)	FP_{const}	$2.08 \cdot 10^3$	$1.07 \cdot 10^4$	$\mathbf{2.60 \cdot 10^4}$	$1.13 \cdot 10^7$	$7.32 \cdot 10^6$
	FP_{adapt}	$3.06 \cdot 10^3$	$1.24 \cdot 10^4$	$4.10 \cdot 10^4$	$2.42 \cdot 10^6$	$5.00 \cdot 10^6$
	FP_{indir}	$\mathbf{1.00 \cdot 10^3}$	$\mathbf{2.25 \cdot 10^3}$	$4.11 \cdot 10^4$	$\mathbf{1.47 \cdot 10^6}$	$\mathbf{1.74 \cdot 10^6}$
Fitness case evaluations	CGP_{STD}	$2.09 \cdot 10^7$	$7.45 \cdot 10^7$	$2.82 \cdot 10^8$	$9.96 \cdot 10^9$	$7.84 \cdot 10^9$
to converge (median)	FP_{const}	$\mathbf{4.41 \cdot 10^5}$	$3.09 \cdot 10^6$	$7.40 \cdot 10^6$	$2.31 \cdot 10^9$	$2.18 \cdot 10^9$
	FP_{adapt}	$6.27 \cdot 10^5$	$\mathbf{1.26 \cdot 10^6}$	$\mathbf{4.60 \cdot 10^6}$	$1.47 \cdot 10^9$	$1.53 \cdot 10^9$
	FP_{indir}	$7.43 \cdot 10^5$	$1.60 \cdot 10^6$	$1.90 \cdot 10^7$	$\mathbf{8.05 \cdot 10^8}$	$\mathbf{8.78 \cdot 10^8}$

(FP_{const}), indirectly encoded CGP-based predictors (FP_{indir}) and standard CGP without coevolution (CGP_{STD}).

FP_{const} is used according to literature [12], i.e. 12 fitness cases in chromosome, 32 individuals in predictor population, 2-tournament selection, a single-point crossover and the mutation probability 0.2. The same setup is used for FP_{adapt}, except the number of fitness cases, which is variable.

The algorithms are compared in terms of the success rate (the number of runs, giving a solution with predefined quality), the number of generations and the number of fitness case evaluations to converge (in order to compare the computational cost). Table 2 gives the median values calculated of 100 independent runs for each benchmark F1 – F5.

It can be seen in the Table 2 that both adaptive approaches, FP_{adapt} and FP_{indir}, have the highest success rate in all benchmarks. The difference is in the number of generations and fitness case evaluations required to converge. As described in Sect. 4.4, the FP_{adapt} uses fewer fitness cases than FP_{indir} for benchmarks F1 – F3. For benchmarks F1 and F2, this leads to a larger number of generations to converge using FP_{adapt} compared to FP_{indir}, fewer fitness case evaluations have to be performed using FP_{adapt}. This does not hold for the benchmark F3, where the number of generations is similar for both approaches. Nevertheless, for benchmarks F4 and F5, FP_{indir} needs fewer fitness case evaluations to converge, but still comparable in the order of magnitude. The size of fitness predictors in the FP_{indir} approach is limited to 50 fitness cases, to reduce larger predictor evaluations. However, FP_{adapt} approach prefers, for benchmark F4, more fitness cases in the predictor for this particular task (see Fig. 8), therefore the cost-reducing limit in FP_{indir} approach might be restrictive for more complex tasks.

In comparison with both FP_{adapt} and FP_{indir} approaches, FP_{const} required the lowest number of evaluations for the benchmark F1. In this case the satisfactory solution is found before the predictor size can adapt. However, let's note that many experiments have to be performed to find the most advantageous size of the predictor using FP_{const} approach for these benchmark tasks, while the FP_{adapt} and FP_{indir} adjust the size of the predictor during each single run in response to a particular task.

Finally, all three coevolutionary approaches beats CGP_{STD} in terms of number of fitness case evaluations to converge and thus accelerate the design process performed by CGP.

5   Conclusions

We have introduced the use of coevolution of cartesian programs with a new type of directly encoded predictors with the adaptive number of fitness cases. The proposed fitness predictors employ phenotypic plasticity and are able to modify the number of fitness cases used for program evaluation in dependence on the phase of program evolution.

Applied to the 5 symbolic regression tasks, we have found the proposed approach to outperform the original constant-size predictors, which use only 12 fitness cases for program evaluation, in terms of success rate and computational cost, expressed as the number of fitness case evaluations required to converge. We have shown that the proposed algorithm is able to adapt the predictor size on the solved problem in response to the development in candidate program evolution. As a result, it is possible to use coevolutionary CGP on a new task without the time-consuming experiments aimed at finding the most advantageous predictor size for the particular task.

Compared to coevolutionary CGP with indirectly encoded fitness predictors, the proposed predictor evolution does not produce predictors with larger predictor sizes than necessary. This reduces the number of necessary fitness case evaluations, while maintaining comparable program accuracy and robustness.

While symbolic regression is good to investigate the system behaviour, our future work will be devoted to applying the proposed approach to more complex problems, such as image filter design, and let the proposed approach and the approach employing indirectly encoded fitness predictors compete in the field in which the behaviour of the system is not so obvious.

The CGP has been applied to many different problem domains, predominantly in evolutionary design and optimization of logic networks. Hence the proposed approach will also be useful for evolvable hardware purposes and in real-world applications.

Acknowledgements. This work was supported by the Czech Science Foundation project 14-04197S. The authors thank the IT4Innovations Centre of Excellence for enabling these experiments.

References

1. Baldwin, J.M.: A new factor in evolution. Am. Nat. **30**(354), 441–451 (1896)
2. Ellefsen, K.O.: Balancing the costs and benefits of learning ability. In: Advances in Artificial Life, ECAL 2013, vol. 12, pp. 292–299. MIT Press (2013)
3. Ellefsen, K.O.: Evolved sensitive periods in learning. In: Advances in Artificial Life, ECAL 2013, vol. 12, pp. 409–416. MIT Press (2013)
4. Hillis, W.D.: Co-evolving parasites improve simulated evolution as an optimization procedure. Physica D **42**(1), 228–234 (1990)
5. Imamura, K., Foster, J.A., Krings, A.W.: The test vector problem and limitations to evolving digital circuits. In: Proceedings of the 2nd NASA/DoD Workshop on Evolvable Hardware, pp. 75–79. IEEE Computer Society (2000)
6. Jin, Y.: A comprehensive survey of fitness approximation in evolutionary computation. Soft Comput. J. **9**(1), 3–12 (2005)
7. Miller, J.F.: Cartesian Genetic Programming. Springer, Berlin (2011)
8. Popovici, E., Bucci, A., Wiegand, R.P., de Jong, E.D.: Coevolutionary principles. In: Rozenberg, G., Bäck, T., Kok, J.N. (eds.) Handbook of Natural Computing, pp. 988–1028. Springer, New York (2011)
9. Schmidt, M.D., Lipson, H.: Coevolution of fitness predictors. IEEE Trans. Evol. Comput. **12**(6), 736–749 (2008)
10. Sikulova, M., Hulva, J., Sekanina, L.: Indirectly encoded fitness predictors coevolved with cartesian programs. In: Machado, P., Heywood, M.I., McDermott, J., Castelli, M., García-Sánchez, P., Burelli, P., Risi, S., Sim, K. (eds.) Genetic Programming. LNCS, vol. 9025. Springer, Heidelberg (2015)
11. Sikulova, M., Sekanina, L.: Acceleration of evolutionary image filter design using coevolution in cartesian GP. In: Coello, C.A.C., Cutello, V., Deb, K., Forrest, S., Nicosia, G., Pavone, M. (eds.) PPSN 2012, Part I. LNCS, vol. 7491, pp. 163–172. Springer, Heidelberg (2012)
12. Šikulová, M., Sekanina, L.: Coevolution in cartesian genetic programming. In: Moraglio, A., Silva, S., Krawiec, K., Machado, P., Cotta, C. (eds.) EuroGP 2012. LNCS, vol. 7244, pp. 182–193. Springer, Heidelberg (2012)

Short Presentations

Search-Based SQL Injection Attacks Testing Using Genetic Programming

Benjamin Aziz(✉), Mohamed Bader, and Cerana Hippolyte

School of Computing, University of Portsmouth, Portsmouth PO1 3HE, UK
benjamin.aziz@port.ac.uk

Abstract. Software testing is a key phase of many development methodologies as it provides a natural opportunity for integrating security early in the software development lifecycle. However despite the known importance of software testing, this phase is often overlooked as it is quite difficult and labour-intensive to obtain test datasets to effectively test an application. This lack of adequate automatic software testing renders software applications vulnerable to malicious attacks after they are deployed as detected software vulnerabilities start having an impact during the production phase. Among such attacks are SQL injection attacks. Exploitation of SQL injection vulnerabilities by malicious programs could result in severe consequences such as breaches of confidentiality and false authentication. We present in this paper a search-based software testing technique to detect SQL injection vulnerabilities in software applications. This approach uses genetic programming as a means of generating our test datasets, which are then used to test applications for SQL injection-based vulnerabilities.

Keywords: Genetic programming · Search-based testing · SQL injections

1 Introduction

Over the past few years, many organisations have rapidly adopted Web applications to solve increasingly complex business problems making them widely available on the Internet. However, with this high usability comes the risk of rendering such business applications a natural target to malicious minds. Applications available through Web browsers have become exponentially more vulnerable in recent years [5] due to a wide range of attacks such as buffer overflow attacks, cross-site scripting, SQL Injection attacks and many others.

Among the top ten Web application vulnerabilities published by the Open Web Application Security Project (OWASP) [14] are SQL Injection Attacks (SQLIAs). In 2011, the National Institute of Standards and Technology's National Vulnerability Database [12] reported 289 SQL injection Vulnerabilities (SQLIVs) in websites, including those of IBM, Hewlett-Packard, Cisco, WordPress, and Joomla. In December 2012, security experts in SANS Institute

© Springer International Publishing Switzerland 2016
M. Heywood et al. (Eds.): EuroGP 2016, LNCS 9594, pp. 183–198, 2016.
DOI: 10.1007/978-3-319-30668-1_12

reported a major SQL injection attack that affected approximately 160,000 websites using Microsoft's Internet Information Services (MS-IIS), ASP.NET, and SQL Server frameworks.

An SQL injection is an attack in which malicious code is inserted into strings that are later passed to an instance of the SQL Server for parsing and execution. These types of attacks are particularly harmful as they could give attackers direct access to the database thus enabling them to leak out very sensitive and confidential information with severe consequences for any organisation. More worrying is that SQLIAs have now been documented for well over a decade [8] but still remain an active method of attacks as a result of the lack of effective techniques for detecting and preventing such attacks. Numerous techniques, such as defensive programming and sophisticated input validation, have been proposed and implemented to prevent some types of SQLIAs. However, such techniques have proven incapable of withstanding new forms of SQLIAs as attackers continue to find new exploits that can avoid the checks programmers put in place. Moreover, defensive programming has proven to be very labour-intensive as it requires constant interaction with testers therefore making it an expensive exercise in terms of resources required.

Software testing is a key phase of many development methodologies as it provides a natural opportunity for integrating security early in the software development lifecycle. Inadequate testing of software applications during their development renders these applications vulnerable attacks such as SQLIAs when they are deployed into a live production environment. However despite the known importance of software testing, this stage is often considered lightly as it could be a laborious task. Most software development teams adopt a manual approach in generating test cases thus limiting the number of test cases which could be built and executed within the project's budget.

Automated software testing techniques, such as Search-Based Software Testing (SBST), are essential for the development of complex systems as they aim to lower the cost of writing tests by enabling users to generate tests automatically. We therefore present a new approach using SBST techniques for the detection of SQLIVs in SQL-based database systems. This new approach will help solve two of the biggest issues with existing SQLIA detection techniques; being fully automated in terms of test case generation, it requires very little interaction with testers, and it provides adequate levels of robustness and flexibility to deal with new emerging patterns of SQLIA attacks.

The rest of the paper is organised as follows. In Sect. 2, we give an overview of literature covering work related to the problem of test generation for SQLIVs. In Sect. 3, we introduce the various types of SQLIAs and the technique they each use. In Sect. 4, we introduce the design of the GP grammar corresponding to SQLIAs. In Sect. 5, we define our anti-SQLIA system design and discuss its implementation. Finally, in Sect. 6, we discuss the results obtained from the implementation of the system and in Sect. 7 we conclude the paper.

2   Related Work

A wide range of approaches have been proposed by researchers to address SQL injection-based threats to Web applications. Appelt et al. [1] proposed a black-box automated testing approach targeting SQLIVs, called U4SQLi. The approach rests on a set of mutation operators that manipulate legitimate inputs to create new test inputs to trigger new SQLIAs. With this approach, it is possible to use a combination of different mutation operators to generate a wide range of attacks and generate inputs that contain new attack patterns, thus increasing the likelihood of detecting vulnerabilities. The approach follows a similar pattern with our proposed approach in that it uses SBST to detect SQLIVs. The difference is with the algorithm used, U4SQLi uses genetic algorithms whereas in our case, we use grammar-based genetic programming.

Su et al. [17] propose a grammar-based approach to detect and stop queries containing SQLIAs by implementing their SQLCheck tool. The proposed approach tracks users' input using a special symbol to mark the beginning and end of each input string. This annotated query is called an *augmented query*. The main idea is to forbid the augmented query from modifying the syntactic structure of the rest of the query. This is achieved by constructing an augmented grammar for this augmented query based on the grammar for a standard SQL statement. At runtime, the augmented query is validated based on this grammar, and is therefore rejected if it does not conform to the grammar.

Shahriar and Zulkernine [15], presented a mutation-based testing approach for SQLIV testing. The authors proposed the use of nine injection operators that inject SQLIVs in application source code. The nine mutation operators were divided into two categories. The first category consists of four operators, which inject faults into WHERE conditions of SQL queries, and the second category consists of five operators, which inject faults into database API method calls. These operators generate mutants, which can be killed with test data containing SQLIAs. This approach is similar to the technique presented by Appelt et al. [1], in that they both use mutants to test the presence of SQLIVs in SQL databases. However, it differs from the approach of [1] in that the mutation is applied to the source code of the application rather than the test input. Hence, it would not adequately solve the problem, which we are trying to address in making testing a less expensive process by automatically generating test cases, as it would require high levels of user involvement to conduct testing using this approach.

Chan et al. [3] presented fault-based testing of SQL database applications. This technique used seven mutation operators to represent faults of entity relationships model of a database driven application. These operators were used to modify the cardinality of queries (e.g., replace "SELECT count (column1)" with "SELECT count (column2)"), replace attributes with similar types (e.g., change one column name with another of a similar data type) and replace participation constraints (e.g., replace EXIST with NOT EXIST) and so on. Like the MUSIC tool proposed by Shahriar and Zulkernine [15], this approach mutates the code and not the test cases, and therefore it would not fulfil our desired goal of having an automated way of generating test data to detect SQLIVs.

Shin et al. [16] proposed an approach for SQL injection vulnerability detection, automated by a prototype tool called SQLUnitGen. The tool combines static analysis, runtime detection and automatic testing to identify input manipulation vulnerabilities. Kosuga et al. [9] presented a technique named Sania, for detecting SQLIVs during the development and debugging phases. Sania investigates HTTP requests and SQL queries to try to discover SQLIAs by constructing parse trees of intended SQL queries written by developers. Terminal leafs of parse trees would represent vulnerable spots, which are filled by possible attack strings. The difference between the initial parse tree and the modified parse tree generated from user-supplied attack strings results in warnings of SQLIAs. Both of [9,16] differ to our case as neither injects SQLIAs. There is also a high level of user involvement in both of these approaches and therefore making them inadequate for our objective, which is to automate the generation of test cases.

Ciampa et al. [4] proposed an approach to perform penetration testing of Web applications. This approach differs to ours and many other existing tools in that it does not randomly generate test data but relies on a knowledge base of heuristics to guide the generation of test data. This is achieved by firstly analysing the Web application with the aim of determining its hyperlinks structure and of identifying its input forms. Then it starts seeding a series of standard SQLIAs with the objective of letting the Web application report an error message. Such standard attacks consist of a set of query strings that are not dependent on the Web application. It then matches the output produced by the Web application against an (extensible) library of regular expressions related to error messages that databases can produce. It continues the attack using text mined from the error messages with the objective of identifying likely table of field names, until it is able to retrieve (part of) the database structure. A limitation of this approach is that it depends on known SQL injection patterns to detect errors during testing. This practice have sometimes proven insufficient to test an application as it is incapable of handling unlearned or new attacks. Also there might be a large number of different representations for the same pattern, for example, using different encodings, which may not be captured in the knowledge base.

Tuya et al. [21] proposed a set of mutation operators for SELECT queries and then tested the mutants using a set of queries drawn from the NIST SQL conformance test suite [13]. They then perform further experiments, aimed at reducing the cost of testing using two different approaches: reducing the number of mutants (selective mutation) and reducing the number of test cases (by selecting the order in which mutants are killed). The number of test cases are reduced by ordering mutants from the most difficult to the easiest to be killed.

SQL DOM was proposed by McClure and Kruger [11] as a way of automating the defensive coding testing technique since the manual approach has often proven to be labour-intensive and error-prone. SQL DOM contains a set of classes that enable automated data type validation and escaping. When using this tool, developers provide their own database schema and construct SQL statements using the SQL DOM API. The tool has proven to be quite useful when developers need to use dynamic queries instead of parameterised queries for greater flexibility. However, this solution could only be used with a new software application under development, as it would require considerable amount of refactoring to get it to work with legacy systems. However, our approach could easily be adapted to any project as it entails black-box testing, which does not require knowledge of the application code.

Thomas et al. [20] proposed an automated vulnerability removal approach, which finds potentially vulnerable (dynamic) SQL statements in programs and replaces them with parameterised SQL statements. Similar to the work of [11], this approach is based on white-box testing requiring some knowledge of the internal structure of the application.

Boyd et al. [2] proposed a tool called SQLrand, which prevents injection attacks that contain keywords. Developers construct queries, which use randomised keywords rather than the normal SQL keywords. This is different from our work in that it only protects against attacks that contain SQL keywords, so although an attacker may not be able to inject code containing keywords without the secret key to randomisation, this approach would not prevent an attacker from injecting other codes, which do not contain SQL keywords.

Finally, Halfond et al. [7] implemented the AMNESIA (Analysis for Monitoring and Neutralising SQL Injection Attack) tool to detect and prevent SQLIAs. The tool first scans the application code to identify hotspot points in the application code that issue SQL queries to the underlying database. For each hotspot, a model is built which represents all the possible SQL queries that may be generated at that hotspot. Calls are then added to the runtime monitor for each hotspot in the application. If a generated query is not consumed by the query model, then it is considered an attack.

3   SQL Injection Attacks (SQLIAs)

SQLIAs occur when an attacker exploits an SQLIV by changing the intended logic underlying an SQL query through inserting new SQL characters or keywords into the query. Here, we describe some common SQLIA examples, although our approach is general enough to cover any other SQLIAs.

3.1   Tautologies

Tautology attacks are attacks where the attacker injects code into conditional statements so that these would always evaluate to a logical True value. The consequences of such attacks generally depend on how the result of a query is used by an application. These types of attacks are generally used to bypass authentication and to return all data in a particular table. For example, an attacker could submit,

```
anything' OR 'x=x
```

instead of a password in an input HTML form. The resulting SQL query becomes:

```
SELECT id FROM users WHERE username='Joe' AND password='anything' OR 'x=x'
```

This transforms the entire WHERE clause into a tautology, which could return every record in the users database.

3.2 Union Query

Union queries provide more flexibility in allowing legitimate queries to retrieve additional information from the database. In such attacks, the attacker injects a statement of the form UNION ⟨injected query⟩ to the original query. For example:

```
' Union Select cardNo FROM creditCards WHERE acctNo = 7909
```

when injected into the username field, modifies the original query to become:

```
SELECT id FROM users WHERE username = '' Union Select cardNo FROM
creditCards WHERE acctNo = 7909 -- AND password =
```

When this modified SQL query is executed by the application, the original query would return null and the injected query returns the creditCard number for the given account.

3.3 Piggyback Queries

Piggyback queries are similar to Union queries in that they append additional queries to the original query. However unlike Union queries, the intention is not to modify the original query but to just introduce new queries, which piggyback on the original query. For example, in the case of our authentication query, an attacker may submit the following query:

```
SELECT id FROM users WHERE username='Joe' AND password='';
drop table users;
```

which as expected, may result in the users table being dropped.

3.4 Malformed Queries

Malformed query attacks are used by an attacker to gather information about the database. These attack types are often deployed in conjunction with other SQLIAs such as Union and Piggyback queries, which require the attacker to have some *a priori* knowledge of the database's schema. Malformed queries take advantage of error messages returned by an SQL server, for example the attacker could inject,

```
convert(int,(SELECT top 1 name FROM sysobjects WHERE xtype='u'))
```

into a pin input field for an authentication query. This would result in a new *malformed* SQL query:

```
SELECT id FROM users WHERE username = 'Joe' AND password = 'xxx' AND
pin = convert(int, (SELECT top 1 name FROM sysobjects WHERE xtype='u'))
```

If we assume that the database is indeed an SQL Server, the error message returned may be of the sort "Microsoft OLE DB Provider for SQL Server (0x80040E07) Conversion failed when converting the nvarchar value 'CreditCards' to data type int". This error message provides useful information to aid other attacks, since it informs the attacker that in fact an SQL server is running at the backend. Additionally, the second part of the error message reveals that the first user-defined table in the database is actually 'CreditCards'. The attacker could also use a similar approach to find all the columns in the table.

3.5 Inference Queries

The final type of SQLIAs we discuss here is called *Inference queries*, and these are similar to malformed queries in that they allow the attacker to discover information about the SQL database. With this type of attacks, code is injected to allow the application to behave differently based on the results of a particular query. For example, if the attacker inputs the following two queries at two different times,

```
'legalUser' AND 1=0 - -'
'legalUser' AND 1=1 - -'
```

then an original SQL query would be modified into two different versions,

```
SELECT id FROM users WHERE username='legalUser' AND
1=0 --' AND password=" AND pin=0
SELECT id FROM users WHERE username= 'legalUser' AND
1=1 -- ' AND password="AND pin=0
```

Let us suppose the application is an insecure application. When the first query is run, and since $1 = 0$ is always False, the application will return a login error. However, at this stage, it is impossible for the attacker to determine whether the error message was actually the result of the application validating the input correctly and blocking the attack attempt. If the attacker submits the second query, which always evaluates to True, and there is no login error, the attacker can confirm that the username parameter is indeed vulnerable.

4 Design of the GP Grammar

In this section, we propose our automated technique for detecting SQLIVs. The technique applies the concept of Genetic Programming (GP) to evolve legitimate inputs to create new inputs, which could trigger SQLIAs in an application under test. The proposed technique is driven by an evolutionary computation system called ECJ [10], a framework that supports a variety of evolutionary computation techniques such as genetic programming and genetic algorithms. The ECJ framework was chosen as it is based on a well-engineered structure, which makes heavy use of Java inheritance, abstraction and pattern-oriented design, has great flexibility, with nearly all classes (and all of their settings) dynamically determined at runtime by a user-provided parameter file. This means it is possible to support common functions of GP such as population initialisation, fitness, selection and variation operators without requiring additional user-written code.

We now discuss the preparatory steps for defining our ECJ-based GP.

4.1   Terminal Sets

Terminal (and function) sets, specify the language used to evolve programs used in GP. A terminal set represents the leaves of the parse tree, which corresponds to the program inputs and it typically consists of variables, constants and functions with no arguments. Our implementation included only one terminal node called X, which holds an alphanumeric string. X is the program external input, a string which will be evolved to generate SQL injection test cases. This node could be used as a child for each of the seven internal nodes defined in the function set.

4.2   Functions Sets

Function sets are the interior nodes of the parse tree of the GP. They are usually all the functions allowed in the program and are driven based on the nature of the problem domain. Our function set contains a total of seven functions. These functions are classified into two broad categories based on the purpose of the functions; i.e. whether they are *behaviour-changing* or *syntax-repairing*. The definition of these functions was inspired by previous mutation testing frameworks by Appelt et al. [1], matching six of the mutation operators defined in [1].

Behaviour-Changing Functions. An SQLIA occurs when an attacker changes the intended effect of an SQL query by inserting new SQL keywords or operators into the query. This class of functions from our function set are intended to evolve our legitimate inputs with SQLIAs to take advantage of vulnerabilities in an application. Although there are a wide variety of behaviour-changing attacks, we have restricted this function set to three examples, which represent the most forms of simple SQLIAs. These examples include the AND, OR and SEMI funcitons. The OR function accepts one input and appends "OR x=x" to the end of the input where x could be a random number or any string enclosed in quotes. For the AND function, it accepts one input and appends "AND x=y" to the end of the input, where x and y could be different random numbers or strings enclosed in quotes. When this resulting output is added to the WHERE clause of an SQL query, it changes the behaviour of the query so that it always evaluate to False therefore no records are returned. Finally, SEMI accepts one input and appends a semicolon ";" followed by an additional SQL statement. When this resulting string is added to the WHERE clause of the SQL query, it changes the behaviour of the query by including a new and distinct query that *piggybacks* itself to the original one. We show the relationship between SQLIAs and these functions in Table 1.

Table 1. SQL injection attacks and proposed functions

Attack	Function
Tautologies	OR
Union queries	SEMI
Piggybacked queries	SEMI
Inference attacks	AND
Malformed queries	OR, SEMI, AND

Syntax-Repairing Functions. An SQLIA attempt will only be successful if the resulting query is syntactically correct. Often, malicious inputs may cause SQL syntax errors to appear when these are combined with the original SQL statements, which hinder attempts to undermine the system. Therefore, this next class of functions, PARA, CMT, QUOTE and DoubleQUOTE, evolve inputs with the goal of *repairing* SQL syntax errors when these are encountered. Such functions are not used on their own, but in combination with the behaviour-changing functions. First, we define PARA as a function that accepts a valid input and appends a closing parenthesis to the end of the input. This is often needed as sometimes the input provided is inserted within parenthesis used as a parameter in an SQL function call or within a nested SELECT statement. In such cases, a vulnerability can only be exploited if the opening parenthesis is matched with a closing one. Next, the CMT function adds an SQL comment (i.e. double dashes −− or a hash character #), ensuring that anything which follows the comment is not executed. The QUOTE function accepts one input and adds a single quote (') to the end of the input. This is usually necessary for string inputs, which are enclosed in quotes in the predefined SQL statement. Finally, DoubleQUOTE accepts a single input and adds a double quote (") to the end of the input. This is usually necessary for string inputs, which are usually enclosed in quotes in the SQL statement.

4.3 Fitness Function

Whilst the first two preparatory steps define the primitive set for the GP, and therefore indirectly define the search space that the GP will explore, these two steps are unable to instruct the GP system about which elements or regions of the search space are good. This is the task of the *fitness measure*, the most difficult and most important concept of genetic programming. The fitness function determines how well a program is able to solve the problem. It varies greatly from one type of program to the next. In our GP implementation, the fitness of an individual is assessed by determining the number of possible injection types that this individual could generate when inserted into the WHERE clause of an SQL query and calculating the sum of all those injection types.

$$Fitness = \sum(1/(1+S)) \tag{1}$$

where S is the number of different injection types found in one individual. The more injections generated the smaller the fitness value and the fitter the individual is. If an individual reaches a fitness of 1, it means the generated string is likely to possess a syntactical error and would therefore not result in an SQLIA, whereas a fitness of 0 represents a definite SQLIA.

4.4 Parameters

The fourth preparatory step specifies the control parameters for the run. The most important control parameter is the *population size*. Most of the parameters defined in the *koza.params* seem adequate for our implementation. The important parameters inherited by *koza.params* are as follows.

Initialisation. Our implementation used Koza's halfBuilder technique to create the initial population. HalfBuilder is essentially a mixture of grow and full technique, with a ramp from 2 to 6 inclusive. The ramp is essentially an initial random number between 2 and 6 inclusive, which is the *maximum* tree size.

Selection. In our implementation, *tournament selection* was used for selecting individuals with a tournament size of 7. Tournament selection is a technique whereby a number of individuals are chosen at random from the population, then they are compared with one other to determine which will be the parent.

Operators. Our implementation used a mixture of mutation, reproduction and crossover to evolve our inputs.

4.5 Termination and Solution Designation

The fifth preparatory step consists of specifying the termination criterion and the method of designating the result of the run. In our implementation, the criterion was the *maximum number of generations* to be run. The single individual was harvested and designated as the result of the run, which was returned as the solution to our problem domain.

5 System Design and Implementation

Our testing system consists of three main components: An AntiSQLInjection tool, an ECJ-based test generator and a vulnerable Web service. Our AntiSQLInjection tool uses heavily an open source tool called General SQL Parser [6] that functions as a database proxy, which serves to intercept communications between the target system and its database in order to identify if an input is potentially harmful or not. On the other hand, the Web service in this case was the subject application, i.e. SuiteCRM [19]. We next outline how various components were implemented to map our problem to the ECJ framework.

Figure 1 shows the key components of our testing system.

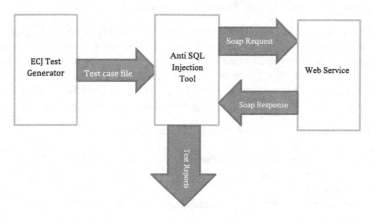

Fig. 1. Components of our system showing data flows

5.1   Representation of Individuals

When setting-up a GP problem in ECJ, we begin by defining the individuals and the species that they belong to. Species available in ECJ vary from integer vectors, to object lists and trees. An evolutionary run in ECJ would allow one population an unlimited number of sub-populations containing an array of the defined individuals, the species and fitness functions. Here, individuals are defined as SQL injected input strings. The implementation used Abstract Syntax Trees (ASTs) to represent our individuals, as this is able to hold an arbitrary number of inputs. Our AST was defined using the built-in ECJ facility for defining and generating ASTs.

The tree structure in ECJ is typically defined by its types, tree constraints, node constraints and a function set, which were all specified in the parameters file. The tree constraints contain data elements shared by the tree. This includes the definition of the tree, its return types and function set. The type essentially defines all the possible node types, which could be found in a tree. ECJ's type objects are of two kinds: atomic types and set types. An atomic type is just a single object (in fact, it is theoretically just a symbol) [10]. A set type is a set of atomic types. Node constraints define what nodes can be children of other nodes and how many children each node could have. The function set is used to link the node constraints and their implementation in the Java class files.

We identified four atomic types and one set type:

- *strValue*: this type holds a string value, used as a terminal node
- *intValue*: this type holds an integer value, used as a terminal node

- *syntax*: this type is the output of a string where a syntax-repairing function is applied, hence, it is the return value of a syntax-repairing function
- *behaviour*: this type is a string where a behaviour-changing function is applied, hence, it is the return value of a behaviour-changing function
- *before-or-after*: this is a set type with two members, *strValue* and *syntax*. This would serve as a child to any behaviour changing function

The node constraints for our tree structure when applied to a GPNode would describe three properties of the node: Number of children, the type of children and the type of the parent of the children. Our tree consists of tree nodes representing the function sets ()behaviour-changing and syntax-repairing), which were implemented as java class files in the ec.app.sqli package. Their entries in our parameter files are as follows, along with their respective constraints:

```
# We have one function set, of class GPFunctionSet
       gp.fs.size = 1
       gp.fs.0 = ec.gp.GPFunctionSet
# We call the function set "f0"
       gp.fs.0.name = f0
# We have seven functions in the function set. They are:
       gp.fs.0.size = 7
       gp.fs.0.func.0 = ec.app.sqli.X
       gp.fs.0.func.0.nc = nc0
       gp.fs.0.func.1 = ec.app.sqli.Or
       gp.fs.0.func.1.nc = nc1
       gp.fs.0.func.2 = ec.app.sqli.And
       gp.fs.0.func.2.nc = nc1
       gp.fs.0.func.3 = ec.app.sqli.Cmt
       gp.fs.0.func.3.nc = nc3
       gp.fs.0.func.4 = ec.app.sqli.Quote
       gp.fs.0.func.4.nc = nc2
       gp.fs.0.func.5 = ec.app.sqli.Para
       gp.fs.0.func.5.nc = nc2
       gp.fs.0.func.6 = ec.app.sqli.Semi
       gp.fs.0.func.6.nc = nc1
```

5.2   Evaluation and Fitness

After the representation of an individual is defined, the ECJ framework requires the user to define a problem, which evaluates individuals and assigns fitness values to them. As was mentioned earlier, the behaviour of each node type is defined in a separate Java class with a single crucial method overridden:

```
public void eval(final EvolutionState state,
        final int thread, final GPData input, final ADFStack stack,
        final GPIndividual individual, final Problem problem)
          {DoubleData rd = ((DoubleData)(input));
           children[0].eval(state,thread,input,stack,individual,problem);
                    rd.x = rd.x +" or 1=1";}
```

The method is called when the GPNode is being executed in the course of executing the tree. The execution proceeds depth-first like the evaluation of a standard parse tree. The *eval*() method has several arguments fairly straightforward to understand. The GPData argument is a simple data object passed around amongst the GPNodes when they execute one another. This is how data is passed from one node to another. The GPData object is defined in the parameter file as *eval.problem.data = ec.app.sqli.StringData*. During evaluation, our GPData object passes through the tree. The object contains a simple string "*x*". Therefore, when this object arrives at the "OR" node, it becomes the case of simply retrieving x and concatenating it with "OR 1=1", which is then returned, so that the value could then be used by some other syntax-repairing function. The result of the syntax-repairing function is the return value of the tree, which is stored in a text file and serves as the generated test case for our GP run. This process will be done based on the maximum number of runs we have specified in our parameters file.

Like most other grammar-based evolutionary GP systems implemented using the ECJ framework, our implementation uses KozaFitness. This is a fitness function that stores an individual's fitness. In KozaFitness, standardised fitness and raw fitness are considered the same (there are different methods for these, but they return the same thing). Standardised fitness f ranges from 0.0 inclusive (the best) to infinity exclusive (the worst). Adjusted fitness converts standardised fitness, using the formula $adj_f = 1/(1+f)$, into a scale from 0.0 exclusive (worst) to 1.0 inclusive (best). Our standardised fitness, which is passed to the ECJ system was derived from the definition of 1 earlier in Sect. 4.3.

6   Results and Analysis

There is no standard benchmark application for testing the existence of SQLIVs. However we evaluated the effectiveness of our approach on one open source system called SuiteCRM [19]. SuiteCRM provides a software suite for the management of popular customer relationships. The application was implemented using PHP, with a MySQL backend database, and provides a SOAP-based Web Service API. The use of this application was motivated by related work [1], in which a version of SugarCRM [18] was used to evaluate the mutation testing approach proposed in [1]. Our evaluation approach is based on the following:

1. First, we identify the distinct list of injectable parameters for the collection of Web services from our subject application. We define an injectable parameter as an input parameter to a Web application whose value is used to build part of a query that is sent to the database. We identified a total of 4 injectable parameters that are being used across the 26 Web services.
2. We identify legitimate inputs for each of the injectable parameters so that these are passed to our GP tool to generate SQLIAs, which will used to test for SQLIVs.
3. We create a repository of SQL statements for each injectable parameter to mimic a similar structure to what exists in the subject application.

4. We generate test cases from our GP tool from the legitimate inputs identified for each parameter in 2.
5. And finally, we use test cases obtained from our GP tool and run these through our AntiSQLInjection tool for each injectable parameter to conduct our tests.

Though most of the run parameters were already predefined by inheriting the Koza parameter file, it is worth mentioning here the values used for each parameter during a GP run, as shown in Table 2.

Table 2. GP experiment parameters

Parameter	Value
Population size	1024
Generation size	50
Crossover rate	90 %
Mutation rate	10 %
Selection method	Tournament selection
Tournament size	7
Number of runs	20

During the experiment evaluation, we investigated the first research question, the effectiveness of our technique in detecting and preventing SQLIAs. The evaluation did not include all parameters of the subject application as it would have taken a substantial amount of time to execute every parameter. The focus therefore was only on the vulnerable parameters which were already identified in a related work [1]. Starting with two legitimate inputs for each parameter, our GP system generated 2048 malicious inputs for each such parameter. These malicious inputs were then applied using the AntiSQLInjection tool. For each parameter, we tracked the number of inputs, which were able to exploit the vulnerability in our subject application. The test results are shown in Table 3.

Table 3. Results of our SQLIA tests

Operation	Parameter	Successful (Flagged)	Unsuccessful (Unflagged)	Syntax
get_entry_list	query	136	261	1651
get_entry_list	order_by	202	690	1257
search_by_module	assigned_user_id	56	1233	778
get_relationships	related_module_query	478	115	1455
get_entries_count	query	145	343	1560
set_relationships	value	1203	0	845

The table shows for each parameter of our chosen subject the number of unsuccessful attacks (Unsuccessful), the number of successful attacks (Successful), and the number of our generated inputs, which resulted in a syntax error and would therefore not be successful in launching an attack. As the table shows, our tool was able, for all vulnerable parameters of our subject application, to successfully create test cases, which would enable us to exploit vulnerabilities of the application.

Whilst our tool was able to launch successful attacks for all parameters, we did note an exceptionally high number of inputs resulting in syntax errors. This could perhaps be corrected in the future by having stricter type constraints on our GP implementation. We also noted a significant number of syntactically correct, yet un-flagged inputs. With further investigation we discovered that this was an issue related to the AntiSQLInjection tool and not our GP system as when these same inputs were applied directly to our subject application, they were able to cause successful attacks. This implies a weakness in the implementation of the AntiSQLInjection tool, where it is sometimes unable to identify attack patterns.

7 Conclusion and Future Work

We presented a GP-based approach for automatically generating test cases for the detection of SQLIAs. Our approach is general in the sense that it is capable of generating any malicious SQL query possible to construct based on the syntax of SQL commands. We demonstrated how our solution can be implemented using open source toolkits and we discussed the results against a subject application, namely SuiteCRM [19]. For future work, we would like fine-tune our implementation to obtain better rates of syntactic errors in the generated test cases, which would improve the performance of our tool. We would also like to widen the set of functions used in generating SQLIAs, including functions related to other forms of SQL-based attacks such as cross-site scripting.

References

1. Appelt, D., Alshahwan, N., Nguyen, C.D., Briand, L.C.: Black-box SQL injection testing. Technical report, University of Luxembourg and University College London (2014)
2. Boyd, S.W., Keromytis, A.D.: SQLrand: preventing SQL injection attacks. In: Jakobsson, M., Yung, M., Zhou, J. (eds.) ACNS 2004. LNCS, vol. 3089, pp. 292–302. Springer, Heidelberg (2004)
3. Chan, W., Cheung, S., Tse, T.: Fault-based testing of database application programs with conceptual data model. In: Fifth International Conference on Quality Software, (QSIC), pp. 187–196, September 2005
4. Ciampa, A., Visaggio, C.A., Di Penta, M.: A heuristic-based approach for detecting sql-injection vulnerabilities in web applications. In: Proceedings of the 2010 ICSE Workshop on Software Engineering for Secure Systems, SESS 2010, pp. 43–49. ACM, New York (2010)

5. Fossi, M., Turner, D., Mack, E.J.T., Adams, T., Blackbird, J., Entwisle, S., Graveland, B., McKinney, D., Mulcahy, J., Wueest, C.: Symantec global internet security threat report: trends for 2009. Technical report XV, Symantec, April 2010

6. Gudu Software: GSP: general SQL parser. http://www.sqlparser.com (Accessed 14 October 2015)

7. Halfond, W.G.J., Orso, A.: AMNESIA: analysis and monitoring for neutralizing sql-injection attacks. In: Proceedings of the 20th IEEE/ACM International Conference on Automated Software Engineering, ASE 2005, pp. 174–183. ACM, New York (2005)

8. Forristal, J.: NT web technology vulnerabilities. Phrack Mag. **8**(54), December 1998

9. Kosuga, Y., Kernel, K., Hanaoka, M., Hishiyama, M., Takahama, Y.: Sania: syntactic and semantic analysis for automated testing against SQL injection. In: Twenty-Third Annual Computer Security Applications Conference, ACSAC 2007, pp. 107–117, December 2007

10. Luke, S., Panait, L., Balan, G., Paus, S., Skolicki, Z., Bassett, J., Hubley, R., Chircop, A.: ECJ: a java-based evolutionary computation research system. https://cs.gmu.edu/eclab/projects/ecj/ (Accessed 14 October 2015)

11. McClure, R., Kruger, I.: SQL DOM: compile time checking of dynamic SQL statements. In: Proceedings of 27th International Conference on Software Engineering, ICSE 2005, pp. 88–96, May 2005

12. NIST: National vulnerability database: automating vulnerability management, security measurement and compliance checking. https://nvd.nist.gov (Accessed 14 October 2015)

13. NIST: Structured Query Language (SQL) test suite. http://www.itl.nist.gov/div897/ctg/sql_form.htm (Accessed 14 October 2015)

14. OWASP: OWASP Top 10–2010-the ten most critical web application security risks. the open web application security project (2010). https://www.owasp.org/images/0/0f/OWASP_T10_-_2010_rc1.pdf

15. Shahriar, H., Zulkernine, M.: MUSIC: mutation-based SQL injection vulnerability checking. In: The Eighth International Conference on Quality Software, QSIC 2008, pp. 77–86, August 2008

16. Shin, Y., Williams, L., Xie, T.: Sqlunitgen: test case generation for SQL injection detection. Technical report NCSU CSC TR, 21:2006, North Carolina State University (2016)

17. Su, Z., Wassermann, G.: The essence of command injection attacks in web applications. In: Conference Record of the 33rd ACM SIGPLAN-SIGACT Symposium on Principles of Programming Languages, POPL 2006, pp. 372–382. ACM, New York (2006)

18. SugarCRM: SugarCRM. https://www.sugarcrm.com (Accessed 14 October 2015)

19. SuiteCRM: SuiteCRM – CRM for the world. https://suitecrm.com (Accessed 14 October 2015)

20. Thomas, S., Williams, L., Xie, T.: On automated prepared statement generation to remove SQL injection vulnerabilities. Inf. Softw. Technol. **51**(3), 589–598 (2009)

21. Tuya, J., Suárez-Cabal, M.J., De La Riva, C.: Mutating database queries. Inf. Softw. Technol. **49**(4), 398–417 (2007)

Grammar Design for Derivation Tree Based Genetic Programming Systems

Stefan Forstenlechner[(✉)], Miguel Nicolau,
David Fagan, and Michael O'Neill

Natural Computing Research and Applications Group, School of Business,
University College Dublin, Dublin, Ireland
stefan.forstenlechner@ucdconnect.ie,
{miguel.nicolau,david.fagan,m.oneill}@ucd.ie

Abstract. Grammar-based genetic programming systems have gained interest in recent decades and are widely used nowadays. Although researchers normally present the grammar used to solve a certain problem, they seldom write about processes used to construct the grammar. This paper sheds some light on how to design a grammar that not only covers the search space, but also supports the search process in finding good solutions. The focus lies on context free grammar guided systems using derivation tree crossover and mutation, in contrast to linearised grammar based systems. Several grammars are presented encompassing the search space of sorting networks and show concepts which apply to general grammar design. An analysis of the search operators on different grammar is undertaken and performance examined on the sorting network problem. The results show that the overall structure for derivation trees created by the grammar has little effect on the performance, but still affects the genetic material changed by search operators.

Keywords: Grammar design · Derivation trees · Genetic programming

1 Introduction

Grammars are an important representation in computer science, especially for programming languages and compilers. Grammars have gained popularity in genetic programming (GP) over time as they overcame some of traditional GP's limitations. Grammar based genetic programming is widely used nowadays [14]. A search space for a problem in GP can easily be defined with a grammar and even problem specific information can be added to bias the search in a certain region [18,27]. Much research has been conducted on different grammars that can be used in GP to solve certain problems. Nevertheless, how to design a grammar for GP remains an under explored research area. McKay et al. wrote in a survey about grammar-based genetic programming [14]:

> "While experienced practitioners of each representation form have some
> tacit understanding of how to choose grammars, there is little explicit

© Springer International Publishing Switzerland 2016
M. Heywood et al. (Eds.): EuroGP 2016, LNCS 9594, pp. 199–214, 2016.
DOI: 10.1007/978-3-319-30668-1_13

knowledge. More explicit knowledge may lead to more structured methodologies (and interactive software support) to incrementally find good representations for new problem domains, and even to partial or complete
automation of the process."

However some limited studies have been undertaken, including Whigham
[27], Hemberg [6], Murphy [17] and Nicolau [19].

This paper makes a step towards analysing grammar design and how it influences the search process. Section 2 gives an overview of how grammars have been
used in evolutionary systems. Section 3.3 explains the experimental setup. The
results are presented and discussed in Sect. 4. Finally, conclusion is presented in
Sect. 5 as well as possible future work.

2   Grammars in Evolutionary Systems

2.1   Grammar Guided Genetic Programming

Many different grammar based genetic programming system have been introduced. The broad term Grammar Guided Genetic Programming (GGGP) will
be used in this paper to address these systems.

Some important GGGP systems are mentioned in this section. Whigham's
CFG-GP system [27] uses a context-free grammar to specify the syntax of solutions. He also defined specialized crossover and mutation operators for the search.
The search operators manipulate the derivation trees that are created when a
sentence is derived from a grammar.

Another well-known grammar-based system is grammatical evolution (GE),
which also uses CFGs normally in Backus-Naur Form (BNF) [5,22], but has also
been adapted to employ other grammars mentioned below.

In contrast to CFG-GP, GE uses an integer string as representation instead of
the derivation tree. A mapping process is used to generate a derivation tree from
the linear representation, which describes the phenotype. The search operators
used in GE primarily operate on the integer string. Other systems integrated
context-sensitive grammars into GP [25], logic grammars [8,28], tree adjoining
grammars [7,16], attribute grammars [1]. Also shape grammars have been used
for design by O'Neill et al. [21].

Depending on the grammar used, it might provide special features; context
free grammars, for example, can be used to interpret non-terminals differently
in different contexts.

Major benefits of GGGP systems are that the closure property is implicitly
given by the grammar similar to strongly typed GP [15] and bias can easily be
incorporated into the grammar. Bias is often an unwanted property of representations or operators. In grammars bias can be used as an advantage [27]. A
grammar can be adapted to influence the search based on a priori knowledge
about a problem. Bias can be subtle, by increasing the frequency of a symbol
in a grammar or more definite by forcing a certain structure on all problems.
For example, the first few bits of a binary string can be defined to be all zeros.

The risk of bias is that the grammar might not cover the whole search space or worse that the global optimum is not even in the search space any more [14].

GGGP systems can be classified by their representation into tree based grammar guided and linearised grammar guided systems. This paper mainly focuses on tree based genetic programming systems, but differences to linearised representations about grammar design are mentioned.

2.2   Grammar Design

GGGP systems can be used similar to fixed and variable length GAs [23] as well as a generalization of GP, as shown by Whigham in his thesis [27]. A variable length GA representation for binary strings in BNF can be achieved by a grammar, as shown in Fig. 1.

```
<string> ::= <string><bit> | <bit>
<bit> ::= 0 | 1
```

Fig. 1. Variable length GA like representation for binary strings in BNF.

A fixed length representation in BNF is not more difficult than a variable length representation, but more rules are required, as for every position in the representation a separate rule has to be created, as shown in Fig. 2. A single rule with one production with a fixed number of <bit> non-terminal symbols could be used, but then crossover in a derivation tree based system could only change single bits. A linearised grammar guided system like GE can represent this grammar with exactly the same number of integers as bits are in the representation, because the mapping process only uses an integer when deciding which production to choose. As all rules except <bit> have only a single rule, exactly n (number of *stringparts*) integers are needed. In a derivation tree based system, crossover can only exchange subtrees with the same symbol, therefore the structure of the derivation trees non-terminal symbols will never change.

```
<stringpart0> ::= <stringpart1><bit>
<stringpart1> ::= <stringpart2><bit>
...
<stringpartN> ::= <bit>
<bit> ::= 0 | 1
```

Fig. 2. Fixed length GA representation for binary strings in BNF.

Standard GP consists of a function and terminal set, with the important property of closure, which means that every function has to be capable to evaluate any possible input it gets. Any GP representation with a function set

```
<tree> ::= f₁ tree ... tree | f₂ tree ... tree | ... | fₓ tree ... tree
         | t₁ | t₂ | ... | t_y
```

Fig. 3. General standard GP grammar in BNF.

$f_1, f_2, ..., f_x$ and terminal set $t_1, t_2, ..., t_y$ can be represented with the general grammar in Fig. 3. Note that any function can have an arbitrary number of inputs and that only a single rule is required to represent the function and terminal set of GP due to the closure property.

For a detailed discussion about grammars for GA and GP representations as well as the schema theorem for such grammars, see Whigham's thesis [27].

2.3   Structure in Grammars

Creating variable length derivation trees requires direct or indirect recursion in a grammar as shown in Sect. 2.2. A direct left recursion, e.g. <rule> ::= <rule><part> | <part> does that trick. <rule> and <part> can be replaced with any rule like <string> and <bit> see Fig. 1, <code> and <line> such as Santa Fe Ant Trail problem [22], <for> and <code> for program synthesis [20], <design> and <component> for creating designs [13], <int_constant> and <number> for integer constant creation [4], etc.

If a derivation tree is drawn that has been created by this rule, it is more similar to a "list" than a tree, as depicted in Sect. 3.2 for a similar grammar in Fig. 6. It will be a list of <part> non-terminals. The reason is that it only expands in one direction (unless there is an indirect recursion from <part> to <rule>). The question about this fairly commonly used rule is, if it should be expressed in another way. As the operators applied to the derivation tree are tree based, the grammar might improve the search if it would express more tree-like structures as in standard GP. For this purpose, we choose a problem, described in Sect. 3.1, which can be expressed with a short grammar. Multiple grammars, discussed in Sect. 3.2 are presented. All of them cover the search space for the problem, but present different properties.

3   Experimental Setup

3.1   Sorting Network

For the grammars and the experiments in this study we use the sorting network problem [10, 24]. The reason why we choose this problem is that it is a real-world problem and simple grammars can be written which cover all possible solutions. At the same time a grammar for sorting networks has properties that apply to other grammars and can be generalized from.

A sorting network can be seen as a sorting algorithm in hardware with a fixed number of inputs and outputs. The output of a sorting network is the input in sorted order. The sorting network consists only of wires and comparator modules.

Comparators take two wires as input and swap the values of these wires if they are not in the correct order, otherwise they return the input.

When drawing a sorting network, the wires are represented as straight lines and the comparator modules are connections between the wires. An optimal sorting network with four inputs is shown in Fig. 4. It is important to notice that the order of the connections makes a difference. For example, if you would put the last connection (on the right side) before the first connection, the sorting network would not return correct values for all inputs.

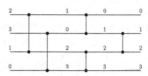

Fig. 4. Sorting network with four inputs and five comparators.

A sorting network has three properties. The number of inputs, the size or number of connections and its depth. The depth is the number of steps it takes to complete the network. One step can execute multiple comparisons at once, if the comparisons are independent of each other, which means they do not depend on the output of the other comparisons in the same step. For example, the network in Fig. 4 has 4 inputs, 5 connections and is of depth 3.

Testing the correctness of sorting networks is a computational expensive task. Fortunately not all possible combinations of inputs have to be tested, which would result in a runtime complexity of $n!$. The zero-one principle [9] says that if all combinations of 0, 1 as inputs are sorted correctly by a network, all other arbitrary values will be sorted correctly. This reduces runtime complexity to 2^n.

3.2 Experimental Grammar Design

In this section we present multiple grammars which can be used in GGGP systems to evolve sorting networks. The phenotype the grammars create are lists with an even amount of numbers and at least one pair of numbers, e.g. 0 2 1 3 0 1 2 3 1 2 represents the sorting network in Fig. 4. Therefore, a grammar needs a recursive rule to create arbitrary amount of pairs of numbers and always has to append two numbers to the phenotype at once.

Section 2.3 showed a common grammar used to address such a grammar design problem, but this grammar may hinder the search operators by forcing them to exchange large parts of individuals, due to the list like structure of the derivation trees. To address this issue, we assume that grammars that describe more tree like structures, may be beneficial to find optimal solutions, as derivation tree based operators can exchange variable amounts of genetic material anywhere in an individual.

The grammars are context free grammars and written in (BNF). Note that all grammars have a rule <node>, which can be expressed as a number from $0 - 3$ as the sorting network presented in Sect. 3.1 and for reasons of simplicity. Depending on the number of inputs of the network this rule has to be adapted.

Grammar 1 (G1). A simple grammar is depicted in Fig. 5. The first production of <snet> is a left recursion and the second production stops the recursion, as shown in Sect. 2.3. In both productions two non-terminals <node> are used so that two numbers are added in every recursion step. This grammar consist of only two rules and one production which creates the structure of the derivation tree. An example of a derivation tree created with G1 is shown in Fig. 6. The derivation tree has more in common with a list of pairs of numbers than a tree. The rule <snet> is only responsible for creating the structure or more specifically the length of the list.

```
<snet> ::= <snet> <node> <node> | <node> <node>
<node> ::= 0 | 1 | 2 | 3
```

Fig. 5. Simple grammar, which works similar to a list (G1).

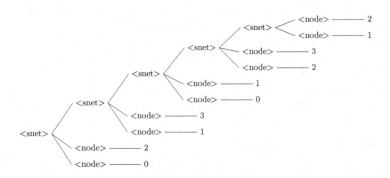

Fig. 6. Grammar 1 derivation tree for the optimal sorting network with four inputs, see also Fig. 4.

Crossover can only exchange single numbers, if applied to <node>. If applied to <snet> all pairs of numbers from the first parent will be in the child up to the crossover point, where all the pairs of numbers from the second parent will be inserted. In case of the phenotype, subtree crossover on G1 is similar to single-point crossover in genetic algorithms, limited to crossover after an even amount of numbers. It is not possible to exchange sorting network comparisons somewhere in the middle of a list of numbers with subtree crossover in G1.

Grammar 2 (G2). The next grammar is very similar to G1, see Fig. 7. Again the structure the derivation tree creates, looks more like a list than a tree. The difference to G1 is that every pair of numbers is encapsulated in a separate rule `<nodes>`, which represents a single compare operation.

```
<snet>   ::= <snet> <nodes> | <nodes>
<nodes>  ::= <node> <node>
<node>   ::= 0 | 1 | 2 | 3
```

Fig. 7. Simple grammar, which works similar to a list with the benefit that pairs of nodes can be exchanged (G2).

The benefit of G2 is that crossover can exchange a single compare operation between parents, even if located in the middle of the list of comparisons. Mutation can also replace a comparison, whereas in G1 it either changes a single number or every number after the mutation point in the tree, which can be very destructive.

Grammar 3 (G3). A grammar that can generate more tree-like derivation structures is shown in Fig. 8. Additionally to the left recursion, its complement a right recursion has been added, as well as production that can have two child `<snet>` nodes. Therefore, binary trees can be created with G3. Note that every production of `<snet>` also creates a pair of numbers. An example of a derivation tree created with G3 is shown in Fig. 9.

The benefit of G3 is that the tree structure provides subtree crossover with more possibilities for crossover points. Instead of operating on a "list", where

```
<snet> ::= <snet> <node> <node> | <node> <node> <snet>
         | <snet> <node> <node> <snet> | <node> <node>
<node> ::= 0 | 1 | 2 | 3
```

Fig. 8. Tree-like grammar, where every production contains a pair of nodes (G3).

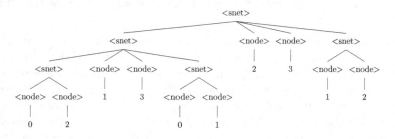

Fig. 9. Grammar 3 derivation tree for the optimal sorting network with four inputs, see also Fig. 4.

crossover takes the first part of the first parent and the second part from the second parent, crossover on G3 can exchange any number of compare-exchange operations anywhere in the tree. Furthermore, mutation might not be as destructive as in G1 and G2 any more, because smaller subtrees might be mutated instead of the whole tail of a "list".

Grammar 4 (G4). The fourth grammar provides the same tree structure as G3, but with the same modification that has been added in G2. A separate rule `<nodes>` has been added which encapsulates a pair of numbers, to easily exchange a single compare operation anywhere in the tree (Fig. 10).

```
<snet>   ::= <snet> <nodes> | <nodes> <snet>
           | <snet> <nodes> <snet> | <nodes>
<nodes> ::= <node> <node>
<node>  ::= 0 | 1 | 2 | 3
```

Fig. 10. Binary tree-like grammar, where every node contains a pair of nodes and a pair of nodes can be exchanged individually (G4).

Grammar 5 (G5). The last grammar is short and simple, but it can also create binary trees. It does not need a separate rule `<nodes>` to exchange single comparisons, because every pair of nodes can already be exchanged on its own when their parent node is exchanged (Fig. 11).

```
<snet> ::= <snet> <snet> | <node> <node>
<node> ::= 0 | 1 | 2 | 3
```

Fig. 11. Binary tree-like grammar. Nodes are not in the structure of the tree and pairs of nodes can still be exchanged individually (G5).

G5 probably provides the easiest and most readable way to create binary trees with a grammar. Additionally, it can easily be adapted to any n-ary trees by adding any number of non-terminal `<snet>` as production to the rule `<snet>`.

Derivation Tree Sizes. As grammars define the structure of the derivation trees, they also define the number of nodes and the depth of derivations trees needed to form a solution. In the grammars above, a pair of numbers represents a comparison operation in a sorting network. The minimum number of nodes and the minimum depth required for representing a certain number of comparisons is given in Table 1, which will be used in Sect. 3.3. Note that every production will be treated as a single node with one child node for every non-terminal in the production. This does not change the behaviour of the derivation trees or search operators, but decreases the number of nodes in a tree.

Table 1. Minimum number of nodes and minimum depth for each grammar given a certain number of comparisons (c)

Grammar	Number of nodes	Depth
G1	$3 * c$	$c + 1$
G2	$4 * c$	$c + 2$
G3	$3 * c$	$\lceil \log_2(c + 1) \rceil + 1$
G4	$4 * c$	$\lceil \log_2(c + 1) \rceil + 2$
G5	$4 * c - 1$	$\lceil \log_2(c) \rceil + 1$

Grammar Design Details. The grammars presented in this section have been written concerning the derivation tree they will create and how search operations which use the derivation tree might behave on them. These grammars can also be used by other GGGP systems which have a linear representation like GE, but keep in mind that systems with a linear representation do behave differently. For example, G1 and G2 should yield the same results in GE, because the rule nodes that has been added, has only one production rule. The mapping process in GE automatically replaces non-terminals with its production rule, if only one is available. The same applies for G3 and G4.

One additional change that may improve the grammars would be to change the rule <nodes> to all possible comparisons for a given number of inputs of a sorting network as depicted in Fig. 12. All given grammars can represent all combinations of pairs of numbers which are n^2. If only all possible compare-exchange operations are used, then it reduces the number of pairs to $\frac{n^2 - n}{2}$, because duplicates can be removed (for example, 0 1 and 1 0 represent the same operation). And pairs with the same number twice are ignored as comparisons with the same input would not do anything. It might still be beneficial to allow one pair with the same number or an empty production of <nodes>, so that a compare-exchange operation can be deleted.

```
<nodes> ::= 0 1 | 0 2 | 0 3
          | 1 2 | 1 3
          | 2 3
```

Fig. 12. All possible comparisons in a sorting network with four inputs.

This and further optimizations of the rule <nodes> have not been investigated, because they would be problem specific, whereas changing the structure of the derivation tree through the grammar and encapsulating non-terminals into new rules can be used in any grammar.

3.3   Experiments

The experiments performed for this paper have been executed with HeuristicLab [26] and a plugin which we added that can be found on GitHub[1].

Two experiments are performed on the grammars presented in Sect. 3.2. As the grammars define different structures, the derivation tree operators, crossover and mutation, may behave differently. They are going to add and remove different amounts of genetic material and choose other nodes to exchange material.

Experiment 1. The first experiment is to analyse the grammars and the difference of genetic material that gets exchanged between individuals depending on the grammars. One time we only use crossover, the second time we only use mutation and the last time we use both operators. No evaluation is performed and selection is done randomly as we are only interested how the search operators behave. In all three cases we use 100 % probability for crossover and mutation. The derivation trees are limited to 50 compare-exchange operations for this experiment, see Table 1 for the number of nodes.

Experiment 2. In the second experiment, we want to know if any of the grammars has a performance advantage over the others. Therefore, fitness is measured and tournament selection is used. We choose a sorting network with twelve inputs as problem to compare the grammars. The optimal number of comparators is not yet known, but it has been proven that it lies between 37 to 39 comparators [2]. Due to the zero-one principle the 12 input sorting network has 4096 training cases. As fitness function we minimize the number of incorrect sorted inputs plus the number of used comparators divided by 100, as in Koza et al. [11]. Therefore, the main objective will be to sort the inputs correctly and the subsidiary goal is to minimize the size of the sorting network. We limit the maximum number of comparators to 59, which is 1.5 times the upper bound rounded up.

General Settings. The settings of the experiments are summarized in Table 2. The differences between the experiments are marked with superscripts.

The initialisation of the individuals is done with the Probabilistic Tree-Creation 2 (PTC2) [12], because PTC2 gives us the possibility to limit the number of nodes of the initial trees and not only the depth. Setting a max depth for the initialisation, like it is done for ramped half-and-half initialisation and also grow or full method, would not give a fair comparison between the grammars. The grammars produce different structures and therefore derivation trees can have completely different amount of nodes for a certain depth, which would make it impossible to compare the results. Additionally, Daida et al. [3] showed that standard GP with binary trees searches rather sparse than dense trees. Therefore, we decided to define the number of compare-exchange operations that are allowed and calculated the required number of nodes in a derivation tree for every grammar. So we set the number of nodes for every experiment individually depending on the grammar, according to Table 1. Therefore, the

---

[1] https://github.com/t-h-e/HeuristicLab.CFGGP.

Table 2. Experimental parameter settings. [1]Experiment 1. [2]Experiment 2.

Parameter	Setting
Runs	100^1, 50^2
Generations	100
Population size	1000
Population initialisation	PTC2 [12]
Tournament size	7
Internal crossover probability	0.9
Mutation probability	$100\%^1$, $5\%^2$
Elite size	0^1, 1^2
Maximum compare-exchange operations	50^1, 59^2

structure of the derivation tree is not limited by depth and arbitrary trees only limited by the number of nodes can be created.

4   Results

This section presents the results of the experiments. Note that no fitness evaluation was used in experiment one, because only the behaviour of the search operators was observed.

4.1   Experiment 1

Changing grammars to more tree-like structures has an obvious effect on crossover and mutation, which is that smaller amounts of genetic material can be exchanged as Fig. 13 shows. The plots for the experiments where only crossover or mutation was used are omitted, as they are quite similar. The only difference is that the trees are shrinking over time in the experiments where only crossover is used. The reason is that the trees are limited by a maximum number of nodes. When crossover selects a subtree from the second parent, it has to select a subtree which does not violate this limit. Therefore, the chance of creating an overall smaller tree is more likely, when the tree is already rather big. But the size of the trees stabilizes after 30 generations. When using mutation only, the same amount of genetic material is removed and added again. In the case of using crossover and mutation, this phenomena is only observed up to the tenth generations, but crossover continues to remove more genetic material than it is adding. Mutation counteracts crossover by adding more material.

In the case of G1, crossover and mutation take place on `<snet>` most of the time, as this is the node which is most frequent in the trees. The most frequent exchanged symbols by crossover with G2 are `<snet>` and `<nodes>`. The frequency of `<nodes>` is slightly higher as there is always one more `<nodes>` than `<snet>` non-terminal. For G3, G4 and G5, it is obviously `<snet>` and the child node

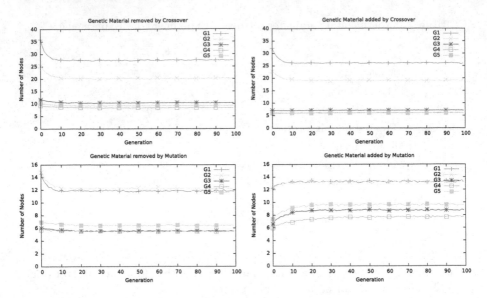

Fig. 13. Genetic material added and removed when using crossover and mutation.

(from the second parent) is mostly `<node> <node>`. The reason is that crossover chooses uniformly from all nodes in the tree and therefore smaller trees are more likely to be selected, as there is a higher amount of smaller subtrees. Because we use Koza's crossover where a probability is used to decide whether an internal node or a leaf node should be chosen, with a 90 % probability for internal nodes. As crossover still favours smaller subtrees, `<node> <node>` gets exchanged most frequently. For mutation there is no such probability, which explains why leaf nodes are changed more frequently.

The first experiment showed that crossover on grammars which create tree-like structures can exchange smaller amounts of genetic material between individuals. Mutation changes smaller amounts as well. The use of the extra rule `<nodes>` had also on effect on the change of genetic material, see e.g. G1 and G2 in Fig. 13, but not as much as the tree-like structure.

4.2   Experiment 2

When we look at the results in Table 3, we can see that G2, G4 and G5 are doing better than G1 and G3. G2 and G4 have the extra rule `<nodes>` to be able to exchange comparators individually, which G5 can also do without that extra rule. The tree structure of the grammar seems to give G3 a slight advantage over G1, but the difference is not statistically significant, similar to the difference between G4 and G2. Although G5 also has a tree structure, it is not doing better than G2.

A more difficult sorting network with 14 inputs is used to repeat the same experiment with G2, G4 and G5, to check if it might create a statistically

Table 3. Results for sorting networks with 12 inputs with the average best fitness, standard deviation, median, best individual and success ratio over 50 runs.

	Average best fitness ± Std dev	Median	Best	Success ratio
G1	82.604 ± 42.794	77.58	18.56	0 %
G2	31.851 ± 21.662	26.58	0.58	4 %
G3	65.600 ± 47.892	51.57	14.58	0 %
G4	23.528 ± 15.472	19.57	0.53	8 %
G5	34.130 ± 28.255	28.59	0.55	2 %

significant difference. The results, shown in Table 4, are very similar. On the one hand, G4 was again doing slightly better. On the other hand, G5 is doing worse than G2. So there is no way to say that the tree structure improves the results.

Table 4. Results for sorting networks with 14 inputs with the average best fitness, standard deviation, median and best individual over 50 runs. No correct sorting network was found.

	Average best fitness ± Std dev	Median	Best
G1	769.833 ± 303.508	724.74	256.74
G2	245.515 ± 143.939	214.75	26.77
G3	575.554 ± 300.543	539.76	102.75
G4	230.196 ± 119.649	201.76	8.73
G5	297.675 ± 197.856	224.75	40.75

After examining the results from the experiments, we noticed that the main reason G2, G4 and G5 are doing better is that they can exchange comparators individually. The Koza style crossover favours internal nodes over leaf nodes, but that does not change the fact that smaller trees are exchanged more often. Therefore single comparators are exchanged quite frequently when using these grammars, whereas leaf nodes rarely get exchanged. So we performed an additional experiment where crossover was changed to select nodes in the tree with an equal probability to see the effect of exchanging even smaller bits of information. The results are shown in Table 5. Now that single nodes can be exchanged more frequently, the results have completely changed and improved for all grammars.

Using crossover which selects from all nodes with equal probability is useful for this specific problem as single numbers are exchanged frequently. If more rules are used which created bigger subtrees in the non-recursive part, this crossover might have a negative effect.

Experiment 2 indicates that the structure of the grammar has only little influence in performance, as G2 shows similar results as G4 and G5. Adding the

Table 5. Results for sorting network with 12 inputs with subtree crossover that chooses from all nodes in the tree with equal probability.

	Average best fitness ± Std dev	Median	Best	Success ratio
G1	9.719 ± 8.621	8.58	0.50	20 %
G2	11.809 ± 11.209	8.58	0.53	22 %
G3	11.767 ± 12.827	8.58	0.53	18 %
G4	12.725 ± 11.410	11.56	0.53	16 %
G5	13.286 ± 11.907	11.55	0.52	16 %

extra rule `<nodes>` improved the performance, because it encapsulated a small piece of information for the problem and was exchanged more often than leaf nodes. Changing the crossover improved the performance on the sorting network problem, but for grammars where the non-recursive part might express a deeper derivation tree, it might not have any effect.

5   Conclusion and Future Work

This paper presented some general concepts on how to design a grammar, especially possibilities on how to write grammars that produce variable length phenotypes, so that the derivation tree does not become "list-like". The grammars were analysed in terms of the behaviour of the applied search operators. Crossover and mutation were able to exchange arbitrary amounts of genetic material within these trees in grammars that created tree-like structures. The second set of experiments analysed the impact of grammar design for tree based GGGP on performance on sorting networks, particularly in the definition of recursive rules for derivation tree based operators. Although the Koza style crossover helps exchange bigger amounts of genetic material, it interferes with the search in this problem instance. If Koza's crossover should be used for tree based grammar guided GP systems, cannot be inferred by these experiments alone. It might be beneficial for problems which create bigger subtrees in the non-recursive part of a grammar.

Nevertheless, the results of the experiments showed that the structure of the underlying derivation tree created by a grammar seems to have no or only little effect on the search given the search operators employed in this study, if the grammar is not biased towards certain solutions and the language of the grammars is equivalent. This conclusion can be seen positive, as this means that no particular attention has to be paid to this aspect, when designing grammars.

Further investigation is needed, if grammars with non-recursive parts that create bigger subtrees than the rule `<nodes>`, show the similar results.

Acknowledgments. This research is based upon works supported by the Science Foundation Ireland, under Grant No. 13/IA/1850.

References

1. Cleary, R., O'Neill, M.: An attribute grammar decoder for the 01 multiconstrained knapsack problem. In: Raidl, G.R., Gottlieb, J. (eds.) EvoCOP 2005. LNCS, vol. 3448, pp. 34–45. Springer, Heidelberg (2005)

2. Codish, M., Cruz-Filipe, L., Frank, M., Schneider-Kamp, P.: Twenty-five comparators is optimal when sorting nine inputs (and twenty-nine for ten). CoRR (2014)

3. Daida, J., Hilss, A.: Identifying structural mechanisms in standard genetic programming. In: Cantú-Paz, E., et al. (eds.) Genetic and Evolutionary Computation — GECCO 2003. LNCS, vol. 2724, pp. 1639–1651. Springer, Heidelberg (2003)

4. Dempsey, I., O'Neill, M., Brabazon, A.: Constant creation in grammatical evolution. Int. J. Innovative Comput. Appl. **1**, 23–38 (2007)

5. Dempsey, I., O'Neill, M., Brabazon, A.: Grammatical evolution. In: Dempsey, I., O'Neill, M., Brabazon, A. (eds.) Foundations in Grammatical Evolution for Dynamic Environments. SCI, vol. 194, pp. 9–24. Springer, Heidelberg (2009)

6. Hemberg, E.: University College, D.S.o.C.S.I. An exploration of grammars in grammatical evolution. Ph.D. thesis, University College Dublin, Ireland (2010)

7. Hoai, N.X., McKay, R., Essam, D.: Representation and structural difficulty in genetic programming. IEEE Trans. Evol. Comput. **10**(2), 157–166 (2006)

8. Keijzer, M., Babovic, V., Ryan, C., O'Neill, M., Cattolico, M.: Adaptive logic programming. In: Proceedings of the Genetic and Evolutionary Computation Conference (GECCO-2001), California, USA, pp. 42–49, 7–11 July 2001

9. Knuth, D.E.: The Art of Computer Programming. Sorting and Searching, vol. 3, 2nd edn. Addison Wesley Longman Publishing Co. Inc, Redwood City (1998)

10. Koza, J.R., Andre, D., Bennett, F.H., Keane, M.A.: Genetic Programming III: Darwinian Invention & Problem Solving, 1st edn. Morgan Kaufmann Publishers Inc., San Francisco (1999)

11. Koza, J.R., Bennett I, F.H., Hutchings, J., Bade, S., Keane, M.A., Andre, D.: Evolving sorting networks using genetic programming and the rapidly reconfigurable xilinx 6216 field-programmable gate array. In: Conference Record of the Thirty-First Asilomar Conference on Signals, Systems amp; Computers, vol. 1, pp. 404–410, November 1997

12. Luke, S.: Two fast tree-creation algorithms for genetic programming. IEEE Trans. Evol. Comput. **4**(3), 274–283 (2000)

13. McDermott, J., Swafford, J.M., Hemberg, M., Byrne, J., Hemberg, E., Fenton, M., McNally, C., Shotton, E., O'Neill, M.: String-rewriting grammars for evolutionary architectural design. Environ. Plann. B Plann. Des. **39**(4), 713–731 (2012)

14. McKay, R., Hoai, N., Whigham, P., Shan, Y., ONeill, M.: Grammar-based genetic programming: a survey. Genet. Program. Evolvable Mach. **11**(3–4), 365–396 (2010)

15. Montana, D.J.: Strongly typed genetic programming. Evol. Comput. **3**(2), 199–230 (1995)

16. Murphy, E., O'Neill, M., Galvapez, E., Brabazon, A. : Tree-adjunct grammatical evolution. In: 2010 IEEE Congress on Evolutionary Computation (CEC), pp. 1–8 (2010)

17. Murphy, E.: An exploration of tree-adjoining grammars for grammatical evolution. Ph.D. thesis, University College Dublin, Ireland, 6 December 2014

18. Murphy, E., Hemberg, E., Nicolau, M., O'Neill, M., Brabazon, A.: Grammar bias and initialisation in grammar based genetic programming. In: Moraglio, A., Silva, S., Krawiec, K., Machado, P., Cotta, C. (eds.) EuroGP 2012. LNCS, vol. 7244, pp. 85–96. Springer, Heidelberg (2012)

19. Nicolau, M.: Automatic grammar complexity reduction in grammatical evolution. In: GECCO 2004 Workshop Proceedings, Seattle, Washington, USA (2004)
20. O'Neill, M., Nicolau, M., Agapitos, A.: Experiments in program synthesis with grammatical evolution: A focus on integer sorting. In: 2014 IEEE Congress on Evolutionary Computation (CEC), pp. 1504–1511, July 2014
21. O'Neill, M., McDermott, J., Swafford, J.M., Byrne, J., Hemberg, E., Brabazon, A., Shotton, E., McNally, C., Hemberg, M.: Evolutionary design using grammatical evolution and shape grammars: designing a shelter. Int. J. Des. Eng. **3**(1), 4–24 (2010)
22. O'Neill, M., Ryan, C.: Grammatical Evolution: Evolutionary Automatic Programming in an Arbitrary Language. Kluwer Academic Publishers, Berlin (2003)
23. Ryan, C., Nicolau, M., O'Neill, M.: Genetic algorithms using grammatical evolution. In: Foster, J.A., Lutton, E., Miller, J., Ryan, C., Tettamanzi, A.G.B. (eds.) EuroGP 2002. LNCS, vol. 2278, pp. 278–287. Springer, Heidelberg (2002)
24. Sekanina, L., Bidlo, M.: Evolutionary design of arbitrarily large sorting networks using development. Genet. Program. Evolvable Mach. **6**(3), 319–347 (2005)
25. Tanev, I.: Incorporating learning probabilistic context-sensitive grammar in genetic programming for efficient evolution and adaptation of snakebot. In: Keijzer, M., Tettamanzi, A.G.B., Collet, P., van Hemert, J., Tomassini, M. (eds.) EuroGP 2005. LNCS, vol. 3447, pp. 155–166. Springer, Heidelberg (2005)
26. Wagner, S., et al.: Architecture and design of the heuristiclab optimization environment. In: Klempous, R., Nikodem, J., Jacak, W., Chaczko, Z. (eds.) Advanced Methods and Applications in Computational Intelligence. TIEI, vol. 6, pp. 193–258. Springer, Heidelberg (2013)
27. Whigham, P.A.: Grammatical bias for evolutionary learning. Ph.D. thesis, New South Wales, Australia, Australia (1996)
28. Wong, M.L., Leung, K.S.: Evolutionary program induction directed by logic grammars. Evol. Comput. **5**(2), 143–180 (1997)

Modelling Evolvability in Genetic Programming

Benjamin Fowler$^{(\boxtimes)}$ and Wolfgang Banzhaf

Memorial University of Newfoundland, St. John's, Canada
{b.fowler,banzhaf}@mun.ca

Abstract. We develop a tree-based genetic programming system capable of modelling evolvability during evolution through machine learning algorithms, and exploiting those models to increase the efficiency and final fitness. Existing methods of determining evolvability require too much computational time to be effective in any practical sense. By being able to model evolvability instead, computational time may be reduced. This will be done first by demonstrating the effectiveness of modelling these properties *a priori*, before expanding the system to show its effectiveness as evolution occurs.

Keywords: Genetic programming · Evolvability · Meta-learning · Artificial neural networks

1 Introduction

Genetic Programming (GP) [17] would be more effective and efficient if we could select based on how individuals may contribute to evolutionary processes, not solely based on their fitness. In other words, it would be useful to select individuals that may contribute more to the fitness of future generations, individuals that are more evolvable. Evolvability indicates the capacity of an individual to improve its fitness [1]. We opt to define evolvability as the probability of a mutation operation resulting in a strictly positive fitness change, the reasoning for which is detailed in Sect. 3. However, it is expensive to measure; it is computationally impractical to measure evolvability for individuals and then use evolvability to aid selection processes.

Biologically, evolvability has been defined as the ability of a population to respond to selection [5]. In his review of other works, Pigliucci [23] comes the conclusions that evolvability, however it may be defined, itself evolves, but there is a lack of evidence to see if this is caused by natural selection or other evolutionary mechanisms. Wilder & Stanley [32] show adaptive processes in gene regulatory networks produce evolvability individuals, but divergent processes produce evolvable populations. Altenberg [2] notes that evolutionary computation brought about more biological-based evolutionary interest in evolvability; evolvability in organisms was simply presumed to exist. Altenberg further notes that there were 170 papers published in 2013 alone that mention the evolution of evolvability. Evolvability in genetic programming refers to the ability of an

© Springer International Publishing Switzerland 2016
M. Heywood et al. (Eds.): EuroGP 2016, LNCS 9594, pp. 215–229, 2016.
DOI: 10.1007/978-3-319-30668-1_14

individual or population of programs to produce higher fitness individuals [1]. To encourage more evolvable programs, it would be beneficial to quantify evolvability, and exploit these quantities when judging fitness. Kattan & Ong [16] use Bayesian inference to adjust fitness functions in order to encourage evolvability. Using genotype-phenotype or genotype-fitness mappings could also prove beneficial to the study of these properties [19]. Properties related to evolvability and robustness, such as self-repair, may emerge in artificial systems without modifying the underlying systems to encourage their emergence [22].

We model evolvability using GP properties that are computationally inexpensive to generate, and, once such models are developed, evolvability may be calculated and utilized in the GP selection process to improve evolution. This is accomplished by generating properties related to evolvability, as well as evolvability itself, *a priori* for a specific problem, then developing a machine learning model for evolvability. Evolvability may then be calculated during evolution. GP may be utilized to solve the problem while predicting evolvability values for individuals, which may then be used to influence selection. Section 2 reviews related literature, Sect. 3 describes the problem domain, the system to be used, and the risks in regard to the applicability of the method. Section 4 describes the specific problem parameters that are examined, the design of the experiments that are conducted, presents the results, and discusses the results and future work.

2   Related Work

In genetic programming, to maximize fitness, we favour programs that currently have greater fitness. There are various selection methods [17,24] that apply varying amounts of selection pressure. However, selection is inherently driven by differences in fitness. This process does not directly consider structural properties of programs, such as bloat [25]. Instead, selection is meant to allow more desirable structural properties, which allow greater fitness, to emerge [3]. The process further ignores how changes in genotype changes the phenotype [7], how this affects fitness, and how this might skip optima in the fitness search space [27]. Evolvability is related to these structural properties; by analyzing their interrelatedness, we should gain insight to improve genetic programming by accommodating them, in lieu of ignoring them, by measuring and selecting for evolvability. Basset *et al.* [4] postulate bloat occurs because offspring are not effectively inheriting the phenotype traits from their parents. The notion is that ideally, we want to perform a cross-over on the phenotype, not the genotype.

Altenberg [1] describes a method of measuring evolvability through a transmission function, deriving a formula that describes the probability that a population (not an individual) will produce an individual that has greater fitness than any in the existing population. Essentially, one considers all possible results of genetic operations on all individuals in a population, and computes the probability that an individual will be produced whose fitness surpasses that of the existing population. This is an intuitive method of measuring evolvability; an individual is highly evolvable if its potential offspring are more likely to be more fit.

It also requires extensive computations; instead of conduction one genetic operation on an individual, we need to conduct all possible operations, on each individual. Then, each fitness case needs to be evaluated for each such operation. Thus, measuring evolvability in this way would require many orders of magnitude more computational effort than standard GP.

Pragmatically, exhaustive searches to measure evolvability can be improved upon by using sampling or estimation [29]. This is much more computationally feasible, but the same question is posed when using exhaustive search; why not simply keep the resulting most fit individual? Sampling still adds a significant computational burden, as sufficient samples are required to estimate evolvability, but even adding a single sample doubles the computational time required in standard GP. As such, evolvability is too computationally expensive to measure directly. Instead, current literature efforts to exploit evolvability do so indirectly, without having to measure it, such as defining new evolvability metrics [28] and characterizing evolvability's relatedness to other properties [12]. There has been some success in determining how much to select for evolvability, but only under limited circumstances [30]. Li *et al.* [18] have had success balancing fitness selection with diversity metrics, using multi-objective optimization. Multi-objective approaches using Pareto dominance or hypervolume indicators, with various objective criteria, are well-studied in the literature, generally targeting concepts related to evolvability, such as diversity, rather than evolvability itself [9,26].

3   Approach

An extendible synthetic domain will be most useful for this work. White *et al.* [31] propose a set of benchmark problems to replace ageing, simple problems. Among the list of new synthetic, extendible problems is the order tree problem [13]. A synthetic, extendible problem such as the order tree problem allows for tunable problem difficulty, thus the conditions under which the use of evolvability is most beneficial may be more easily examined.

An order tree domain may be defined as having a size of n. Function nodes and terminal nodes take on values of whole numbers on a range of $[0, n - 1]$. Function nodes all take two arguments. The fitness of a solution is calculated in a top-down fashion. A node will add 1 to the total fitness of the solution if its numeric value is strictly greater than its parent's numeric value, and, in the restricted version of the order tree problem, only if the parent is also adding to the total fitness of the solution. Thus, the optimal solution is an ordered tree, where the root is the functional node valued at 0, its children are valued at 1, and so on. The order tree problem is useful because the difficulty is tunable to n, where difficulty may be increased by increasing n, thus increasing our functional and terminal set. Furthermore, node dormancy is easily determined as a by-product of fitness evaluation. Problem difficulty may be further tuned by adjusting how much fitness is contributed by each node; by weighing higher-valued nodes more greatly (i.e., by increasing fitness greater than 1 for any given node) the fitness structure may be changed. This alters the fitness landscape, and encourages higher-valued nodes to be selected, even though this interferes with finding the optimal solution. A more evolvable solution would still favour lower-valued nodes. This allows for tuning the desirability of evolvability. Tuning the order tree problem in these two ways will demonstrate the problem conditions for the effectiveness of the proposed system.

This work will focus on one representation of GP, tree-based representation, and the modelling of evolvability *a priori*. This will indicate if modelling evolvability is viable, and may be extended to dynamically built models, and data and model sharing between related problems. Eventually, the goal is apply the system to real-world problems where GP is known to excel relative to other algorithms or human efforts, in order to obtain better solutions more quickly. There are many ways to expand after initial efforts in controllable problems are shown to be functional; it may be certain classes of problems are more receptive to the methodology, alternate GP representations may be preferred, or certain structural properties are much more significant than others, and each of these may not be independent with another.

As we are concerned with modelling evolvability, structural properties of individuals in GP which may be easily measured (that is, without a significant increase in computational resources) are of interest. There can be significant sections of individuals in GP which, in addition to not affecting fitness, provide no change in output, regardless of input. These sections are referred to as introns in GP literature. Introns may be categorized by their behaviour; Nordin *et al.* [21] propose several categories. They are categorized based on whether their lack of contribution of fitness is due to the fitness cases themselves, or apply to the entire problem domain, and whether cross-over operations can introduce a change in fitness. Identifying all introns is computationally expensive. However, it is computationally inexpensive to identify a certain type of intron, that occurs when a code section is never executed for any fitness case; these are dormant sections [14]. In tree-based or cartesian GP, these nodes are referred to as dormant nodes, and can account for the majority of the nodes, around 90 %-95 % [14, 20]. Despite the apparent uselessness of dormant sections of code, dormancy is helpful; if dormant nodes are detected and removed, performance actually suffers, and more generations are required to reach comparable solutions [14]. Locality is another structural property in GP, relating to evolvability, robustness, and genotype-phenotype mappings. A problem has high locality if neighbouring genotypes correspond to neighbouring phenotypes [7]. High locality problems are generally easier to solve. Low locality indicates a more rugged search space, which indicates a more difficult search. Furthermore, the ruggedness describes how robust and evolvable the search space is [8, 15, 27, 29]. Neutral genetic operations represent plateaus on the search space. Evolvability and robustness act as counterparts; steep inclines indicate great fitness gains moving toward optima, but also great fitness losses moving away from optima. There is motivation to organize all the structural properties together, to analyze their interactions, for they all affect problem difficulty, the efficiency of the search, and the efficacy of the search.

We propose a narrower scope of tree-based GP. We further limit to algorithms available in the Waikato Environment for Knowledge Analysis (WEKA), an accessible machine learning software suite [10]. This would allow for rapid experimentation on a number of different algorithms, as well as some convenient visualization and analytical tools that may yield insight into the nature and relatedness of the structural properties of the individuals. The predicted evolvability of an individual shall be used to guide the selection process in various ways, to find the most beneficial usage of evolvability. Such a system would only indicate that, with enough data generated *a priori* for a specific problem, models for structural properties could be built which can benefit evolution.

4   Experimental Design and Results

We design and implement a tree-based GP system that records measurements of evolvability by sampling, along with records of other structural properties, such as dormancy, per every generation that occurs during evolution. Further, we model evolvability using these records, then exploit their predicted values during evolution, in order to develop a faster and more efficient GP system. This shall be accomplished by modifying an existing tree-based GP system to track and record additional structural elements, for specific problems. Once generated, evolvability will be modelled using machine learning algorithms. These models will be incorporated into the existing tree-based GP system, to predict evolvability without the need to sample them. Then, the predicted values will be used to guide selection beyond the standard fitness measurements. Initially, we consider the various structural properties of solutions generated by genetic programming for a small parity problem, and a contrived regression problem consisting of a single input variable. Once this system is verified to yield improvements in solution accuracy and efficiency for these simpler problems, the system will be modified further to develop models for evolvability as evolution actually occurs.

We modify a minimalist version of Open BEAGLE, referred to as BEAGLE Puppy [6]. Open BEAGLE is an evolutionary computation framework, developed in C++. BEAGLE Puppy utilizes the core GP algorithms of Open BEAGLE, but is simpler to modify for our purposes, since it is minimalist. It contains a tree-based GP implementation of simple parity and regression problems. Our methodology requires editing the selection process, additional tracking of various statistics (such as dormancy), sampling for evolvability, and eventually, dynamic modelling of evolvability. These are easier to implement by editing core GP algorithms. Furthermore, we are afforded more flexibility by working with a lesser amount of code. Another advantage is working with an efficient object-oriented language. To sample for evolvability, more fitness cases need to be evaluated, which is already the most computationally intensive part of GP. Object-oriented code allows for more easily reusable code; as we expand into more difficult problem domains, we can reuse our efforts building simpler ones.

Modelling evolvability, however, requires faster machine learning algorithms than evolutionary computation provides. These are provided by a machine learning suite, the Waikato Environment for Knowledge Analysis (WEKA) [10].

This implementation can be expanded to allow models to be generated by WEKA as evolution occurs. The accuracy of these models can be monitored until they are sufficiently accurate, in order to stop sampling evolvability, and instead, predict it. Sampling may still be interleaved to ensure the models remain accurate.

4.1 Sampling Accuracy

This subsection describes the effectiveness of altering the fitness mechanism of standard GP to consider evolvability in various ways. This will demonstrate the effectiveness of using sampled evolvability to improve GP. The significance of evolvability on selection will be monitored, so the optimal amount of selection can be used. Once the necessary conditions for the effectiveness of using evolvability in selection has been determined, it can be used to gauge the effectiveness of modelling evolvability.

Calculating precise evolvability is computationally infeasible for practical genetic programming. Instead of calculating all possible results of all possible genetic operations for any given individual genetic program, we elect to instead conduct sampling, where a random subset of all possible genetic operations are applied. Sampling can approximate the precise calculation of evolvability for a fraction of the computational cost. How many samples are necessary to produce a reasonable approximation of the correct evolvability, such that selection errors will occur less than 5 % of the time? How accurate must the approximation be to achieve an improvement when using evolvability to guide selection?

In order to answer these questions, we must first define more experimental parameters. Several evolvability metrics exist. We opt to define evolvability as the probability of a mutation operation resulting in a strictly positive fitness change. This may differ from other metrics in two ways: probability of change instead of magnitudes of change, and excluding neutral changes. Preliminary experiments indicated that selecting for the probability of a positive fitness change were more productive than when neutral changes were included. Similarly, they indicated that using probabilities instead of average magnitude of fitness change were more productive. Mutation operations are considered, in order to evaluate evolvability of individuals without considering how the gene pool of the population would affect measurements, as it would measuring evolvability using cross-over operations. More samples are required to achieve a good approximation if we consider the average magnitude of change of fitness. Furthermore, selecting for greater positive magnitude of fitness change will heavily bias evolution toward lower fitness individuals, as they have the greatest capacity for fitness improvement. We discount neutral changes, as this encourages a bias toward large trees in the order tree problem, as they have many possible neutral mutations. Considering neutral changes to be equivalent to positive ones encourages robustness, but not evolvability.

To determine how many samples are necessary to achieve a reasonable approximation of evolvability, we conduct the following experiment. We vary the number of samples while keeping other experimental conditions consistent, and compare the sampled evolvability to the strongest approximation (using the

Table 1. Evolutionary parameters for varying the number of samples.

Population Size	50	Crossover Probability	0.9
Tournament Size	3	Probability of Non-Terminal Crossover	0.9
Min Initial Depth	3	Standard Mutation Probability	0.05
Max Initial Depth	6	Mutation Max Regen Depth	2
Max Depth	6	Swap Mutation Probability	0.05
Initial Grow Probability	0.5	Probability to Mutate a Function Node	0.5

largest sample size). Even for a smaller order tree problem, it is still computationally infeasible to calculate the correct evolvability. By selecting for fitness, higher fitness individuals are more likely to occur. If we select for evolvability, more evolvable individuals are likely to occur.

The experimental parameters are shown in Table 1. 10000 runs with different random seeds are completed for standard GP, and 1000 runs for everything else. The max depth was raised for the 7th and 8th order tree problems. These parameters are consistent throughout the experiments in this work. We define the mean absolute error of evolvability as follows:

$$MAE = \frac{1}{n} * \sum_{k=1}^{n} |e'_k - e_k| \tag{1}$$

where n is the number of runs, e'_k is the measured evolvability for 1000 samples, and e_k is the measured evolvability of the indicated number of samples.

Figure 1 shows that a reasonable approximation for evolvability occurs when the number of samples is about 100. Similar experiments for higher order tree problems show that holds true. Since the purpose of evolvability for this system is to be used with an altered fitness function in order to guide selection, the required accuracy of sampling and modelling evolvability is proportional to the actual influence evolvability has on selection. Therefore, it is necessary to choose precisely how evolvability will guide selection in order to determine how many samples are sufficient to ensure accurate selection. This can be evaluated by

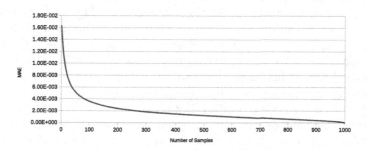

Fig. 1. Mean Absolute Error of Evolvability for number of samples compared to 1000 samples.

using the modified fitness function, and compare which individuals are selected when using a reduced number of samples (or a model) for evolvability with individuals selected using a large number of samples. Discrepancies indicate that an individual was incorrectly selected.

4.2   Selection of Evolvability

We need to determine how to select for evolvability. To determine the optimal selection amount, we conduct the following experiment. We vary the standard GP selection mechanism by using the sampled evolvability in various ways, while keeping other experimental conditions consistent. There are several methods to guide selection with evolvability. One is a threshold for fitness; if fitness of two individuals falls within a specific threshold, then we select the one with greater evolvability. Another is a weighted sum; we sum the fitness and evolvability, each weighted by a specified amount, and select individuals according to their weighted sum. We can allow a generational modifier for using a weighted sum; as the number of generations increase, we select less strongly for evolvability. Using a weighted sum and a generational modifier, we have, formally:

$$F' = \left[\begin{array}{ll} (f + \frac{e*p(g_{max}-g)}{g_{max}}) \text{ if } & g < g_{max} \\ f & \text{otherwise} \end{array} \right] \tag{2}$$

where F' is the adjusted fitness function, f is the standard fitness function, e is evolvability, p is the weight parameter, g_{max} is the maximum generation parameter, and g is the current generation. This translates to the fitness function being modified by the probability of a change being positive multiplied by the weight parameter for the initial population, and where this modifier linearly approaches zero as the generation increases. Upon reaching zero, the modifier becomes zero for the remaining generations, rendering evolvability uninfluential. This is desirable because evolvability should become less significant as the number of generations increase, as standard fitness approaches optimal values. Maximizing standard fitness becomes the only goal when evolution completes. Eventually, we would just want to select for standard fitness. We conduct experiments for different Order Tree problems under varying selection pressures (varying the weight and maximum generation parameters). The other experimental parameters are identical to the previous experiment, as shown in Table 1.

Figures 2, 3, and 4 shows the average maximum fitness as the generation increases, for a subset of the tested problems, for clarity. This indicates that using modified fitness functions that use evolvability in addition to standard fitness outperform using standard fitness functions alone. They indicate the general appropriate proportion of evolvability to use for selection, indicated by the better performing selection pressures. Furthermore, using a greater weight parameter is still useful, provided that a maximum generation parameter is specified, so fitness becomes more dominant as individuals approach higher fitness values. Using extreme values for a weight parameter, even tempered by small maximum

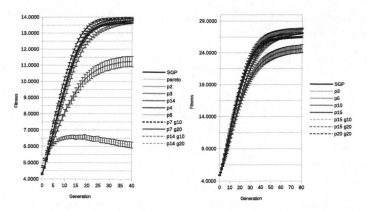

Fig. 2. Fitness over generation for varying selection pressures, for the Order Tree 4 (left) and 5 (right) problem. 'p' indicates the evolvability weight, and 'g' indicates the g_{max} value. Error bars indicate the 95 % confidence interval of standard error of the mean. SGP refers to standard genetic programming.

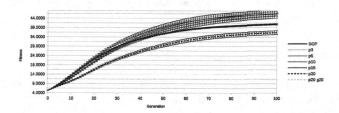

Fig. 3. Fitness over generation for varying selection pressures, for the Order Tree 6 problem.

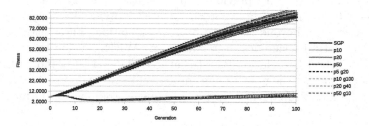

Fig. 4. Fitness over generation for varying selection pressures, for the Order Tree 8 problem.

generation parameter, did not produce fit results. For the higher difficulty problems, a greater emphasis on evolvability improves the results. We also note that selection based on pareto-dominance, where fitness and evolvability are the two

Table 2. Probability of an incorrect selection comparing 100 samples with 1000 samples under various modified fitness functions over 100 runs.

Order	p	g	Mean selection error
4	7	10	0.34495 %
5	10	N/A	2.6304 %
6	5	N/A	3.6288 %
7	10	N/A	4.338 %
8	20	40	0.34230 %

objectives, produces worse results than standard GP. A subset of all the tested values for varying selection pressures are shown, for clarity.

We see in Table 2 that under the varying selection pressures, that using 100 samples for evolvability differs from using 1000 samples less than 5 % of the time. The tested selection pressures were some of the top performing selection methods for their order of problem, as shown in the previous experiments. Establishing a performance baseline for evolvability selection pressure allows us to proceed to modelling evolvability.

4.3   Modelling of Evolvability

Once the effectiveness of using sampled evolvability has been demonstrated and the evolvability selection methods have been evaluated, we must now build a model for evolvability and demonstrate its effectiveness. Firstly, we must describe the attributes we use to build the machine learning models for evolvability. We record a number of attributes associated with individuals. These include generation, tree height, tree size, functional & terminal frequency, number of dormant nodes, dormancy ratio, previous standard fitness, fitness change, and standard fitness. These may all be recorded for each individual without significant computational costs beyond standard fitness calculation. These attributes were subjected to attribute significance testing using WEKA, using the correlation-based filter method Correlation-Based Feature Selection [11], and further tests on WEKA classifiers. The most significant attributes were determined to be generation, size, function frequency, terminal frequency, number of dormant nodes, previous fitness, and fitness.

WEKA offers rapid use of many machine learning classifiers. In order to build a model, we must provide training data and choose a classifier. We can generate training data by running standard GP with the addition of evolvability sampling and selection; this will produce individuals which will be similar to those that will occur when using the model system, ensuring the models will be more accurate in practice. We can evaluate the effectiveness of the different models for evolvability by comparing the mean absolute error between them, also comparing this with the mean absolute error of the evolvability by varying number of samples. Various experiments indicated that a number of machine learning

Table 3. Probability of an incorrect selection comparing a multilayer perceptron model constructed with a varied number of training instances (themselves constructed under a varied number of evolvability samples) with 1000 evolability samples under the 4th Order Tree problem using the p7 g10 fitness function over 1000 runs.

Samples	Training instances	Mean selection error
1000	2000	0.487524 %
1000	4000	0.488446 %
1000	8000	0.483173 %
1000	40000	0.460605 %

models were appropriate for this task, having similar mean absolute error rates. We select the multilayer perceptron (an artificial neural network) for verifying the effect of the number of training instances and number of evolvability samples are required for acceptable mean absolute error rates. Acceptable mean absolute error rates are those which indicate that erroneous selection will occur less than 5 % of the time. We do this by varying the number of training instances, and the amount of evolvability samples used to generate those instances, and measuring the frequency of selection error compared with 1000 samples of evolvability.

Once the conditions required for acceptable selection error rates have been determined, we test the system by comparing the top performing selective conditions in each order tree problem, compared with standard GP and the improvements made by using sampled evolvability, to indicate that modelling evolvability and modifying the standard fitness function, we can improve GP. This will indicate that modelling evolvability is viable.

We see in Table 3 that relatively few training instances are required to build an accurate model of evolvability. Very few selection errors are made when the evolvability used to train the model is accurate; that is, when a large number of samples of evolvability are taken to generate the model. We see in Fig. 5 that the models perform sufficiently well in practice. They are a statistical improvement over standard GP, and fare about as well as sampled evolvability. Even as few as 1000 training instances can build a successful model. Since a training instance is generated for each individual in the population for each generation, a single run with these settings generates 5000 training instances.

In Figs. 6, 7, 8 and 9 we see that this trend holds in higher Order Tree problems; modelling evolvability offers a statistically significant improvement over standard GP, and performs about as well as using samples to calculate evolvability. Using models built 1000 samples of evolvability even performs better than continually sampling evolvability 100 times for each individual.

In conclusion, we have demonstrated the necessary amount of evolvability sampling to generate a sufficiently accurate calculation of evolvability. We have further demonstrated how evolvability may be used to modify the standard fitness function, in order to encourage the selection of evolvable individuals, and how this may generate an overall increase in standard fitness. We have

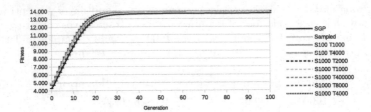

Fig. 5. Fitness over generation for ANN models built from various amounts of training instances and various amounts of evolvability samples for the Order Tree 4 problem. 's' indicates the number of evolvability samples, and 'I' indicates the number of training instances.

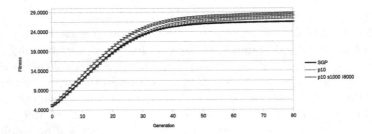

Fig. 6. Fitness over generation comparing standard GP, using sampled evolvability & modelled evolvability with a modified fitness function for the Order Tree 5 problem.

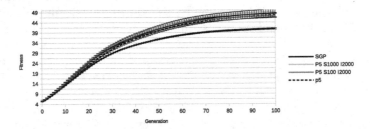

Fig. 7. Fitness over generation comparing standard GP, using sampled evolvability & modelled evolvability with a modified fitness function for the Order Tree 6 problem.

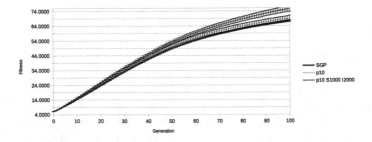

Fig. 8. Fitness over generation comparing standard GP, using sampled evolvability & modelled evolvability with a modified fitness function for the Order Tree 7 problem.

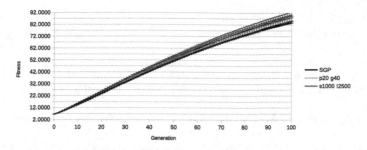

Fig. 9. Fitness over generation comparing standard GP, using sampled evolvability & modelled evolvability with a modified fitness function for the Order Tree 8 problem.

demonstrated how many instances and samples are required to build a sufficiently accurate model of evolvability, in order to predict it to guide selection. Finally, we have shown that modelling evolvability and using it in selection allows for a similar improvement in overall fitness than simply sampling for evolvability. The additional number of evaluations required to sample evolvability to guide selection is prohibitive. The extra computational time required to predict evolvability using an external program is prohibitively computationally expensive, as well. Furthermore, its use is limited in this experiment by gathering training instances *a priori*. However, the results demonstrate that predicting evolvability from relatively few training instances with a relatively few number of samples still leads to improved fitness. This indicates the viability of modelling evolvability in order to improve genetic programming. In future work, the system will be modified to generate training instances for evolvability periodically while evolution occurs, in order to build and update models for evolvability periodically, so that more performance gains can be achieved. This work lays a foundation for the success of such a system.

References

1. Altenberg, L.: The evolution of evolvability in genetic programming. In: Advances in Genetic Programming, pp. 47–74 (1994)
2. Altenberg, L.: Evolvability and robustness in artificial evolving systems: three perturbations. Genet. Program. Evolvable Mach. **15**(3), 275–280 (2014)
3. Banzhaf, W.: Genetic Programming and Emergence. Genet. Program. Evolvable Mach. **15**(1), 63–73 (2013)
4. Bassett, J.K., Coletti, M., De Jong, K.A.: The relationship between evolvability and bloat. In: Proceedings of the 11th Annual Conference on Genetic and Evolutionary Computation. GECCO 2009, NY, USA, pp. 1899–1900. ACM, New York (2009)
5. Flatt, T.: The evolutionary genetics of canalization. Q. Rev. Biol. **80**(3), 287–316 (2005)
6. Gagné, C., Parizeau, M.: Genericity in evolutionary computation software tools: principles and case study. Int. J. Artif. Intell. tools **15**(2), 173–194 (2006)
7. Galván-López, E., McDermott, J.: Defining locality as a problem difficulty measure in genetic programming. Genet. Program. Evolvable Mach. **12**(4), 365–401 (2011)

8. Galván-López, E., Poli, R., Kattan, A., ONeill, M., Brabazon, A.: Neutrality in evolutionary algorithms. What do we know? Evolving Syst. **2**(3), 145–163 (2011)
9. Hadka, D., Reed, P.: Borg: an auto-adaptive many-objective evolutionary computing framework. Evolutionary Comput. **21**(2), 231–259 (2013)
10. Hall, M., Frank, E., Holmes, G., Pfahringer, B., Reutemann, P., Witten, I.H.: The weka data mining software: an update. SIGKDD Explor. Newsl. **11**(1), 10–18 (2009)
11. Hall, M.A., Smith, L.A.: Practical feature subset selection for machine learning (1998)
12. Heywood, M.I.: Evolutionary model building under streaming data for classification tasks: opportunities and challenges. Genet. Program. Evolvable Mach. **16**(3), 283–326 (2015)
13. Hoang, T.H., Hoai, N.X., Hien, N.T., McKay, R.I., Essam, D.: ORDERTREE: a new test problem for genetic programming. In: Proceedings of the 8th Annual Conference on Genetic and Evolutionary Computation. GECCO 2006, vol. 1, pp. 807–814 (2006)
14. Jackson, D.: The identification and exploitation of dormancy in genetic programming. Genet. Program. Evolvable Mach. **11**(1), 89–121 (2009)
15. Jones, T.: Evolutionary algorithms, fitness landscapes and search. Ph.D. thesis, The University of New Mexico (1995)
16. Kattan, A., Ong, Y.S.: Bayesian inference to sustain evolvability in genetic programming. In: Handa, H., Ishibuchi, H., Ong, Y.S., Tan, K.C. (eds.) Proceedings of the 18th Asia Pacific Symposium on Intelligent and Evolutionary Systems. Proceedings in Adaptation, Learning and Optimization, vol. 1, pp. 75–87. Springer, Heidelberg (2015)
17. Koza, J.: Genetic Programming: On the Programming of Computers by Means of Natural Selection. MIT Press, Cambridge (1992)
18. Li, K., Kwong, S., Cao, J., Li, M., Zheng, J., Shen, R.: Achieving balance between proximity and diversity in multi-objective evolutionary algorithm. Inf. Sci. **182**(1), 220–242 (2012)
19. Malan, K.M., Engelbrecht, A.P.: A survey of techniques for characterising fitness landscapes and some possible ways forward. Inf. Sci. **241**, 148–163 (2013)
20. Miller, J.F., Smith, S.L.: Redundancy and computational efficiency in cartesian genetic programming. IEEE Trans. Evol. Comput. **10**(2), 167–174 (2006)
21. Nordin, P., Francone, F., Banzhaf, W.: Explicitly defined introns and destructive crossover in genetic programming. In: Advances in Genetic Programming, pp. 111–134. MIT Press, Cambridge, MA, USA (1996)
22. Öztürkeri, C., Johnson, C.G.: Self-repair ability of evolved self-assembling systems in cellular automata. Genet. Program. Evolvable Mach. **15**(3), 313–341 (2014)
23. Pigliucci, M.: Is evolvability evolvable? Nat. Rev. Genet. **9**(1), 75–82 (2008)
24. Poli, R., Langdon, W., McPhee, N., Koza, J.: A field guide to genetic programming (2008)
25. Silva, S., Dignum, S., Vanneschi, L.: Operator equalisation for bloat free genetic programming and a survey of bloat control methods. Genet. Program. Evolvable Mach. **13**(2), 197–238 (2011)
26. Sindhya, K., Miettinen, K., Deb, K.: A hybrid framework for evolutionary multiobjective optimization. IEEE Trans. Evol. Comput. **17**(4), 495–511 (2013)
27. Smith, T., Husbands, P., Layzell, P., O'Shea, M.: Fitness landscapes and evolvability. Evol. comput. **10**(1), 1–34 (2002)
28. Tarapore, D., Mouret, J.B.: Evolvability signatures of generative encodings: beyond standard performance benchmarks. Inf. Sci. **313**, 43–61 (2015)

29. Wang, Y., Wineberg, M.: Estimation of evolvability genetic algorithm and dynamic environments. Genet. Program. Evolvable Mach. **7**(4), 355–382 (2006)

30. Webb, A.M., Handl, J., Knowles, J.: How much should you select for evolvability?. In: Proceedings of the 2015 European Conference on Artificial Life, pp. 487–494. MIT Press (2015)

31. White, D.R., McDermott, J., Castelli, M., Manzoni, L., Goldman, B.W., Kronberger, G., Jaśkowski, W., O'Reilly, U.M., Luke, S.: Better GP benchmarks: community survey results and proposals. Genet. Program. Evolvable Mach. **14**(1), 3–29 (2013)

32. Wilder, B., Stanley, K.: Reconciling explanations for the evolution of evolvability. Adapt. Behav. **23**(3), 171–179 (2015)

Towards Automated Strategies in Satisfiability Modulo Theory

Nicolás Gálvez Ramírez[1,2](✉), Youssef Hamadi[4],
Eric Monfroy[3], and Frédéric Saubion[2]

[1] LabDII, Universidad Técnica Federico Santa María, Valparaíso, Chile
ngalvez@inf.utfsm.cl
[2] LERIA, Université d'Angers, Angers, France
{ngalvez,saubion}@info.univ-angers.fr
[3] LINA, UMR CNRS 6241, TASC INRIA, Université de Nantes, Nantes, France
eric.monfroy@univ-nantes.fr
[4] LIX, Ecole Polytechnique, Palaiseau, France
youssefh@lix.polytechnique.fr

Abstract. SMT solvers include many heuristic components in order to ease the theorem proving process for different logics and problems. Handling these heuristics is a non-trivial task requiring specific knowledge of many theories that even a SMT solver developer may be unaware of. This is the first barrier to break in order to allow end-users to control heuristics aspects of any SMT solver and to successfully build a strategy for their own purposes. We present a first attempt for generating an automatic selection of heuristics in order to improve SMT solver efficiency and to allow end-users to take better advantage of solvers when unknown problems are faced. Evidence of improvement is shown and the basis for future works with evolutionary and/or learning-based algorithms are raised.

Keywords: SMT · Strategy · Z3 · Learning · Tuning · Evolutionary algorithm · Search-based software engineering

1 Introduction

Optimization tools have played a fundamental role in many fields of Software Engineering during the last fifteen years. The application of various optimization techniques in order to solve specific software engineering problems and improve software performance is now a common practice. The concept of *Search-Based Software Engineering* (SBSE) has been introduced and lead Mark Harman to define a challenge [6] entitled *"Search for strategies rather than instances"*. This challenge aims at avoiding to use specific software engineering optimization algorithms to solve given instances of specific problems, but rather to look for more global strategies. Moreover, another related purpose is to handle efficiently new unknown problems that share properties with already identified problem classes.

© Springer International Publishing Switzerland 2016
M. Heywood et al. (Eds.): EuroGP 2016, LNCS 9594, pp. 230–245, 2016.
DOI: 10.1007/978-3-319-30668-1_15

In theorem proving, SAT modulo theory (SMT) is a generalization of the famous satisfaction problem SAT for logical formulas over one or more theories: Booleans variables may be replaced by formulas expressed over different theories or data structures (e.g. real arithmetic, arrays...) in order to validate a logical formula. The following formula is such an example $\exists a, (a < 10 \vee a = 10 \vee a > 10) \wedge 2a = 20$ where a is an integer. Therefore, SMT solvers need to combine several heuristic algorithms to improve proof efficiency when facing different problems. However, combination of heuristics selected for a given problem (class of problems) may be inefficient for an unexpected problem. Hence, de Moura and Passmore [9] defined the *Strategy challenge in SMT solving*, whose goal is to propose theoretical and practical tools allowing end-users to exert strategic control over heuristic aspects of high-performance SMT solvers. End-users may then generate their own selection of heuristics in order to solve their own problems, without depending on: (1) the built-in heuristics included during the solver building process, (2) the selection of a SMT solver that better suits a specific problem, and (3) the performance of solvers for problems that solvers designers did not have in mind. Note that most of the time, end-users do not have the required knowledge in order to use properly all the heuristics features in SMT solvers. This lack of knowledge is the starting point of our work, whose aim is to automatically generate solver strategies. It corresponds to the automatic selection and ordering of the heuristics processes in order to check the satisfiability of a SMT formula. Improving solver strategies should improve the solver efficiency but also should allow the end-user to handle identified classes of problems.

Therefore, this paper attempts to automatically generate strategies to improve performance of the SMT solver Z3 [8], one of the current well-known SMT theorem provers. Inspired by previous works on parameter setting [7,10], we use here an evolutionary algorithm (EA) to achieve the strategy tuning/configuration. Given a strategy pattern, the EA aims at adjusting some components as well as some numerical parameters of the strategy.

2   Strategies and SMT Logics

2.1   Strategies

In SMT, the notion of *strategy* is still hard to define. According to [9], we may define a strategy as: *a set of heuristics processes that helps to reduce the search space and/or the way how it is explored in order to find well-known solvable instances in a set of problems.* The aim of a strategy is to guide the prover when searching for proof of satisfiability. Figure 1 presents a simplified grammar that summarises the strategy language of Z3. There are 4 terminal elements:

1. A *Solver* (Tactic) checks the satisfiability of the problem. Any solver can be defined as:

$$S : \Phi \times \Pi \to I \tag{1}$$

   where Φ is the set of all SMT goals, Π is the set of all parameters vectors, and $I = \{\texttt{sat}, \texttt{unsat}, \texttt{unknown}, \texttt{timeout}, \texttt{fail}\}$ is the set of possible return

```
1  strategy  = (tactical <tactics>) | solver
2  <tactics> = (tactical <tactics>) |
3              <tactics> <tactics>  |
4              <tactic>
5  <tactic>  = probe | heuristic | solver
```

Fig. 1. Z3 Strategy Language Grammar.

values for a solver. The application of a solver S using its own parameter vector $\alpha \in \Pi$ to a goal $G \in \Phi$ is defined as:

$$S(G, \alpha) = i \in I \tag{2}$$

2. A *Heuristic* (Tactic) transforms the problem into a sequence of subproblems. Any heuristic can be defined as:

$$H : \Phi \times \Pi \to \Phi^n \times \Omega \tag{3}$$

where Ω is a satisfiability model or an unsatisfiability proof converter from the generated subgoals to the original goal. Let Λ be the set of all satisfiability models or possible unsatisfiability proofs for each $G \in \Phi$. Ω can be defined as:

$$\Omega : \Lambda^n \to \Lambda \tag{4}$$

We define the application of a heuristic H with parameter vector $\alpha \in \Pi$ to a problem or goal G as:

$$H(G, \alpha) = G_1, G_2, \ldots, G_n \wedge \Omega(G_1, G_2, \ldots, G_3) \tag{5}$$

3. A *Probe* (Tactic) checks if in its current state the problem has some property. Let Σ be the set of all possible probes and $J = \{\texttt{true}, \texttt{false}\}$ be the set of Boolean truth values. A probe is formalized as:

$$P : \Sigma \to J \tag{6}$$

A probe P applied to a goal G is defined as:

$$P(G) = j \in J \tag{7}$$

4. *Tacticals* are combinators that define how tactics are applied (e.g., time-out, parameter set, behavior according to problem properties) and/or combined. Using a tactical over a set of tactics generates a new complex tactic. Tacticals are defined as:

$$C : 2^T \to T \tag{8}$$

where T is the set of all tactics. We define a tactical C over a set of tactics τ as:

$$C(\tau) = t \in T \tag{9}$$

A strategy is a set of *tactics* structured by combinators (i.e., tacticals). At least one solver must be used in the strategy in order to always generate a satisfiability result. Semantically, Z3 does not make any difference between the different terminal types. Since they can be considered as *tactics*, they are processed similarly. All tactics are also seen as *a set of constraints*. Moreover, when a set of constraints is applied to a problem or goal, a subproblem set is always returned. Satisfiability states or Boolean values presented above are indeed the translation of the returned subproblem. Figure 2 presents semantics rules for Z3. Note that a *goal* is a set of SMT formulas together with their attributes. Then, a *tactic* is applied to a goal in order to return one of the following subproblem set.

$$
\begin{aligned}
goal \quad &= formula\ sequence \times attribute\ sequence \\
tactic \quad &= goal \rightarrow return \\
return \quad &= empty \rightarrow model \\
&| \quad false \rightarrow proof \\
&| \quad goal \rightarrow goal\ sequence \times modelconv \times proofconv \\
&| \quad fail \\
modelconv &= model\ sequence \rightarrow model \\
proofconv &= proof\ sequence \rightarrow proof
\end{aligned}
$$

Fig. 2. Semantics rules in Z3 theorem prover

1. *Empty Set*: when a tactic determines if a problem or goal is satisfiable, an empty set is returned. Other tactics applied to this set also return the empty set. This set is related to the `sat` value that Z3 prints when it terminates the analysis of a problem. This value is associated to a model that proves the satisfiability of the problem.
2. *False set*: if the tactic computes an unsatisfiable problem/goal, a false set is returned. This set is similar to *empty set* but for the `unsat` value and is also associated to a proof.
3. *Goal set*: when a tactic is applied to a problem, the returned subproblem can be:
   (a) the original problem G, if the applied tactic does not change initial goal;
   (b) or a new subset G'.
4. *Fail set*: if the tactic does not have all requirements to properly work, a fail set is returned and the original goal is not processed. This may have two consequences:
   (a) it leads to a global fail result if tactics are joined by conjunction. Z3 interprets it as an `unknown` result.
   (b) it skips the failing tactic if tactics are joined by disjunction, using the next tactic over the original problem or goal G.

Finally, *modelconv* and *proofconv* are converters which allow the prover to design a proof for (un)satisfiability from the subgoals set to the original goal. At the end of the execution, if a final `sat` or `unsat` result could not be inferred from all the subgoals, Z3 interprets this as `unknown` or as `timeout` (if the global time-out was reached).

Syntactically, there exist two ways to apply a strategy (see Fig. 3). In the first case (Fig. 3a), if the end-user strategy is not successful, the default Z3 strategy is applied. In the second case (Fig. 3b), the Z3 default strategy is completely replaced by the end-user strategy. Since we want to create new alternatives to the Z3 default strategies, we use the second syntax in order to automatically generate and evaluate strategies.

```
1  <Problem header and assertions>
2  (apply strategy)
3  (check-sat)
```

```
1  <Problem header and assertions>
2
3  (check-sat-using strategy)
```

(a) Conjunction between a new *strategy* and Z3 default strategy.

(b) Replacing Z3 default strategy with a new *strategy*.

Fig. 3. Z3 syntax to apply strategies.

Example 1. Figure 4 shows a simple problem with a strategy defined using the SMT-LIB language. Line 1, an integer variable a is declared; the next two lines are assertions which represent the following goal declaration:

$$G = \exists a : \{(a < 10 \lor a = 10 \lor a > 10) \land 2a = 20\}$$

In this problem, one probe, two heuristics tactics, and two solver tactics are applied in a linear conjunctive order strategy (between lines 5 and 11). The solver interprets the strategy as follows:

```
1   (declare-const a Int)
2   (assert (or (< a 10 ) (= a 10) (> a 10)))
3   (assert (= (* 2 a) 20))
4   (check-sat-using
5       (and-then
6           (fail-if not(is-ilp))
7           simplify
8           split-clause
9           (try-for sat 100)
10          (using-params smt :random-seed 100)
11      )
12  )
```

Fig. 4. Simple problem in Integer Arithmetic Modulo Theory with an end-user strategy.

1. At line 6, the probe `is-ilp` checks if the problem is in an Integer Linear Programming (ILP) form. The result of the probe will be processed by the tactical `fail-if` as fail if the probe result is true, or as the original goal if it is false. Therefore, using the negation of the probe, (`not is-ilp`) allows to apply the designed strategy only when the goal is in ILP form.

2. The first applied heuristic (`simplify`, line 7), reduces the problem and gives the following subgoal:

$$G' = \exists a : \{(\neg(a \geq 10) \lor a = 10 \lor \neg(a \leq 10)) \land a = 10\}$$

3. The heuristic `split-clause` (line 8) splits disjunctions into a subgoal set, and returns the following:

$$G'_1 = \exists a : \{\neg(a >= 10) \land a = 10\}$$

$$G'_2 = \exists a : \{a = 10 \land a = 10\}$$

$$G'_3 = \exists a : \{\neg(a <= 10) \land a = 10\}$$

4. The solver `sat` (line 9) cannot modify any subproblem within the timeout of 100 ms specified in the tactical `try-for`, therefore, it returns its input subgoal set.

5. The last tactic `smt` (line 10) modifies and solves the whole set, and returns the following set: $G'_1 = false, G'_2 = empty, G'_3 = false$. The use of the tactical `using-params` allows to change the default value of the random seed generating a new parameter vector for the `smt` solver.

The tree in Fig. 5 sketches the application of the strategy tactics. At the end, Z3 can rebuild the original disjunctive problem and returns an *empty set* as the final result. Finally, the translation of this subset is the expected `sat` output.

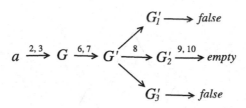

Fig. 5. Strategy application into the linear arithmetic modulo arithmetic example. Numbers over arrows refer to line numbers of tactics of the example of Fig. 4.

The language for end-user strategy design is also found in the source code of the Z3 (default) strategies. Therefore, there exists a common structure for end-users and developers that could be optimized in order to improve software performance and to satisfy both expectations.

2.2   SMT Logics

The SMT-LIB provides an extensive set of benchmarks and defines common standard processes. The **SMT-LIB standard** (version 2.5) defines concepts, formal languages, and a command (script) language [1]. It also introduces the concepts of *Theories* and *Logics* in order to classify *problems*. A *problem* belongs to a *logic*; a *logic* refers to some *theories*; and a *theory* is a specific set of symbols that define a well-known system (i.e. theory of integers, real numbers, arrays, etc.). In [13], a summary of the current recognized logics by SMT-LIB could be found. These logics define the set of problem classes and benchmarks used for research purposes by the SMT community.

In this work, we focus on two basic logics related to Linear Arithmetic: Integer (LIA) and Real (LRA) numbers. The generated strategies are also applied to their quantifier free versions: QF_LIA and QF_LRA.

3   Evolutionary Algorithm

Inspired by previous works on parameter tuning [10,12], our Evolutionary Strategy Generator (ESG) is an evolutionary algorithm which automatically selects tactics within a predefined structure in order to build a strategy for a SMT-LIB logic. ESG goal is to generate a strategy that improves the solving process for a set of instances rather than for a single problem. We designed a static strategy skeleton based on the Z3 default strategy for LIA and LRA logics. The strategy skeleton defines the structure of an *individual* that is used in the *population* later. Figure 6 shows individual strategy skeleton as a tree. Fixing all tacticals values, we define an individual of type X as an array x with $m+n$ genes, where:

1. $\forall i \in \{1, \cdots, m\}$, x_i are *heuristics*; and
2. $\forall j \in \{m+1, \cdots, m+n\}$, x_j are tuples, each with a *solver* and its time-out.

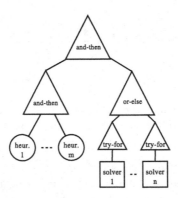

Fig. 6. Individual skeleton of type X: A set of m heuristics is applied, and then the satisfiability is checked up to n solvers.

The sub-tree starting in the *or-else* tactical is used to check the relevance of the (non-present) heuristics tactics in the selected logics. Therefore, we can define a new individual of type Y as an array y with n genes, where:

1. $\forall i \in \{1, \cdots, n\}$, y_i are tuples, each with a *solver* and its time-out.

Two ESG are considered according to the two types of individuals, type X which includes *heuristics*, and type Y that uses only *solvers*. An initial population with s individuals is generated. The size of the population s is defined as the maximum number of possible candidate values that a gene could have.

After creating the initial population, the algorithm iterates the application of two classic variation operators at each cycle: *Wheel-Crossover* and *Gen-Mutation*. The *Wheel-Crossover* is a generalized uniform recombination operator where the value of each gene of the new individual is randomly picked from a randomly selected parent of the population. This new individual replaces the worst individual of the current population. The mutation operator is applied to this new individual and modifies the value of a randomly selected gene by a random value from its domain. Again it replaces the second worst individual of the population if its evaluation is better.

The fitness function is based on the number of solved instances from the benchmark set. We also add the time consumed to solve them as an additional criterion. Each individual fitness evaluation is performed over the whole selected instance set. The fitness function is:

$$h(\Delta, \gamma) = (1 - \gamma)f(\Delta) - \gamma g(\Delta); \gamma \in \{0, 1\}; \forall \Delta \in \{X, Y\} \qquad (10)$$

where:

1. $f(\Delta)$: number of instances solved using the strategy generated for the individual of type Δ.
2. $g(\Delta)$: time elapsed in instances solved using the strategy generated for the individual of type Δ.

Our goal is to maximize the number of solved problems from a set that belongs to a specific logic, or to minimize the time consumption when the amount of solved problems are the same between different individuals. We could define our fitness comparison function between individuals of the same type as:

$$c(\Delta, \Delta') = Max\{h(\Delta, \gamma), h(\Delta', \gamma)\}; \text{ with } \gamma = \begin{cases} 0, & \text{if } f(\Delta) \neq f(\Delta') \\ 1, & \text{if } f(\Delta) = f(\Delta') \end{cases} \qquad (11)$$

Algorithm 1 summarises the ESG procedure. The version based on individuals of type X is the full version **ESG-full**, and the one based on Y is the solver-only version **ESG-solver**.

4   Experimental Setup

We perform several experiments in order to evaluate our strategy generators. All tools are implemented in C and interact with Z3. All experiments were run on

---

Algorithm 1. Evolutionary Strategy Generator, ESG

Input: SMT-LIB logic set of instances, Strategy skeleton
Output: Optimized Strategy
1: Generate Initial Population, *Population*
2: **repeat**
3:    *Child* = Wheel-Crossover(*Population*)
4:    Replace *Worst Population Individual* by *Child*
5:    *Mutated-Child* = Gen-Mutation(*Child*)
6:    **if** *Mutated-Child* is better than *Second Worst Population Individual* **then**
7:       Replace the *Second Worst Population Individual* by *Mutated-Child*
8:    **end if**
9: **until** Ending criterion

---

a workstation with the following specifications: Pentium (R) Dual-Core E5300 at 2,6 Ghz CPU, 8 GB at 800Mhz of RAM, OS Ubuntu 14.04 LTS x64. The theorem prover is Z3 4.4.0 stable, and the compiler is gcc 4.8.4.

Instances. The instances were extracted from 2014 SMT-LIB benchmarks[1]. As mentioned before, we focus on LIA and LRA logics for the strategy generation, and we also tested the generated strategies in the quantifier free versions of both logics (QF_LIA and QF_LRA). Unlike SMT-COMP, we used complete benchmarks of the selected logics. Hence, we include results on the problems whose satisfiability is not known. The best obtained results were compared with the default performance of Z3 on the same instances. Table 1 shows the characteristics of each selected instance set.

Table 1. SMT-LIB Logics: Selected sets of instances characteristics.

SMT-LIB logic	Strategy generation	Strategy testing	Instances with	
			Known results	Unknown results
LIA	✓	✓	46	0
LRA	✓	✓	171	450
QF_LIA	✗	✓	4862	1279
QF_LRA	✗	✓	1474	208

Strategies Generators. As explained before, two kinds of ESG were developed. Both generators build strategies for each selected logic using the following problem subsets: (1) all instances with known final results, and (2) all instances with unknown final results. A cross-validation process will also be performed when a subset is hard to handle, allowing to generate potentially better strategies without consuming resources by checking all instances subset. The subsets

---

[1] Experiments ran before 2015 SMT-LIB benchmarks were released.

used for this purpose are (1) a decile from all instances with known final results, and (2) a decile from all instances with unknown final results. To identify which set is used, we renamed the ESG used for the cross-validation process as **CESG-full** and **CESG-solver**, following the same nomenclature used in Sect. 3.

Parameters. Our ESGs include several values that have to be set. For the static skeleton parameters we select $m = 10$ for the maximum size of heuristics tactics and $n = 5$ for the maximum size of solvers tactics. These values were selected arbitrarily and based on the size of the default Z3 strategies for selected logics: they avoid building too complex strategies (size), but allow enough interesting combinations of heuristics and solvers. Possible values for tuning heuristics and solvers values are :

1. **Heuristics**: skip, simplify, simplify mod., ctx-simplify, ctx-simplify mod., solve-eqs, elim-uncnstr, lia2sat, propagate-ineqs, split-clause.
2. **Solvers**: fail, sat, smt, qe-sat, qe-smt, qflia/qflra.

Despite the use of well-known tactics, it is necessary to explain the following points:

1. In heuristics values, *simplify mod.* and *ctx-simplify mod.* are modified versions of *simplify* and *ctx-simplify* respectively. Both tactics have modified parameter vectors. *lia2sat* is a conjunction of heuristics and not a single one. All this values could be found in the Z3 default strategies.
2. In solvers values, *qflia* and *qflra* are Z3 built-in solvers for quantifier-free linear arithmetic logics (**QF_LIA** and **QF_LRA** respectively), we select them to check if they could be useful in the strategies generated for their quantified version logics.
3. If a *heuristic* or *tuple* (*solver* and its time-out) value is consecutively repeated two or more times, it will be decoded as a single application reducing the strategy size.
4. Finally, time-out for solvers is proportionally tuned w.r.t. the per instance global time-out set.

Table 2 shows the global time-out per instance used in strategy generation and testing execution processes. The first two time-outs are set for generate and test strategies, allowing check how strategies change when available time resource is incremented by an order of magnitude. The last value is set for strategy testing under 2014 SMT-COMP rules. Also, it is not used in the strategy generation process because it would have taken an unexpected time in each run. Finally, because all ESGs include stochastics characteristics, experiments were executed ten times with different random seeds values.

5   Results

In this section, we analyze the results obtained using the automatically generated strategies against Z3 default strategies in the selected SMT-LIB logics. The generated strategies can be found in [3], while Z3 default strategies can be reviewed in Z3 theorem prover source code.

Table 2. Strategies Generation and Testing: Global time-out per instance.

Time-out [s]	Strategy generation	Strategy testing
1	✓	✓
10	✓	✓
1500	✗	✓

5.1   LIA Benchmarks Set

We started running experiments in the easiest instance set. Table 3 shows results of three different strategies applied.

Table 3. LIA Benchmarks Set: Results with different strategies on instances with known results.

	Instance time-out [s]					
Strategy	1		10		1500	
Z3-default	46	0,48	46	0,48	46	0,48
ESG-full	46	0,34	46	0,34	46	0,34
ESG-solver	**46**	**0,33**	**46**	**0,33**	**46**	**0,33**
	solved	time[s]	solved	time[s]	solved	time[s]

We can observe that all strategies help to solve the whole set, but automatically generated strategies outperform Z3 default strategy with a 29,2 % and a 31,3 % time reduction, being the strategy generated by ESG-solver the best. Since there is no performance variation when changing the random seed, we could treat this subset as deterministic. Therefore, there is not a significant difference between both automatically generated strategies. Cross-validated ESGs were not applied because the sample is small and the set is easily solved. Finally, this set does not include instances with unknown results.

5.2   LRA Benchmarks Set

Although the increase of sample size, strategies used to solve LRA subset with known results have a similar behaviour as with the LIA set, including the deterministic performance. As shown in Table 4, all strategies help to solve the entire subset. Again, automatically generated strategies outperforms Z3 default strategy, reducing solving time between 12,3 % and 15,0 %. In this subset, ESG-solver is slightly better than ESG-full, but the difference is not significant. Using the same criteria as in LIA set, it was not necessary to apply cross-validated ESG.

As Table 5 shows, face LRA subset with unknown results was a completely different challenge. No strategy helps to solve the entire subset. ESG-full and

Table 4. LRA Benchmarks Set: Results with different strategies on instances with known results.

Strategy	Instance time-out [s]					
	1		10		1500	
Z3-default	171	1,30	171	1,33	171	1,33
ESG-full	**171**	**1,14**	171	1,14	**171**	**1,13**
ESG-solver	**171**	**1,14**	171	1,13	171	1,13
	solved	time[s]	solved	time[s]	solved	time[s]

Table 5. LRA Benchmarks Set: Results with different strategies on instances with unknown results.

Strategy	Instance time-out [s]					
	1		10		1500	
Z3-default	384	18,00	392	59,20	409	5281,68
ESG-full	386	22,82	409	167,40	428	6953,94
ESG-solver	**391**	**25,52**	**410**	**225,01**	**431**	**8055,69**
CESG-full	384	17,70	403	119,07	414	6083,98
CESG-solver	376	21,60	398	81,81	425	7235,55
	solved	time[s]	solved	time[s]	solved	time[s]

ESG-solver strategies considerably outperform Z3 default strategy performance, helping to solve between 46,3 % and 53,7 % of unsolved instances. Nevertheless, both ESG strategies consume between one and four days in order to provide an optimized strategy. Therefore, cross-validation ESGs were applied in order to find better strategies with a limited time budget. CESG-full and CESG-solver strategies outperform Z3 default strategy in two of three scenarios. Cross-validated strategies help to solve between 10,3 % and 39,0 % of the unsolved instances using Z3 default strategy.

A slight difference is observed between full strategies and solver-only strategies. Figure 7 shows solver performance using strategies generated using different random seeds. In both cases, we could check that all means and most of the solving distributions are allocated around higher values when using full strategies. However, most of the best values correspond to solver-only strategies. Full strategies provide a more stable solving process and reduce the influence of random values, benchmark size or selection process.

Finally, Table 6 shows the statistical results of *two-tailed Wilcoxon signed-ranked test* with significance level of 0.01 applied between full and solver-only versions of each ESG. Used data correspond to all results obtained in all three global time-out limits, matching them in pairs depending on their execution properties. We could observe that there is no significant difference between ESG-full and ESG-solver and CESG-full is significantly better than CESG-solver (despite no reaching the best results).

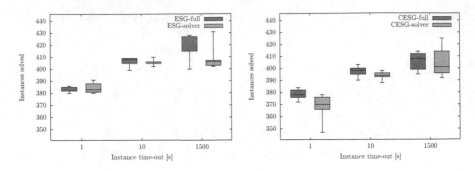

(a) All-instance based strategy generation using different random seeds.

(b) Cross-validation strategy generation using different learning sets.

Fig. 7. Results distribution on LRA instances with unknown results.

Table 6. Wilcoxon Signed-Rank test comparing ESG versions: full and solver-only.

ESG 1	ESG 2	$\sum$ N.Ranks	$\sum$ P.Ranks	Ties	Z-value	p-value
ESG-full	ESG-solver	201,0	150,0	4	-0,648	0,516
CESG-full	CESG-solver	287,5	63,5	4	-2,845	0,004

In conclusion, strategies including heuristics produce a more stable and better average performance than strategies without heuristics. Also, influence of strategies with heuristics in solver are also statistically equal or better than the influence of strategies without heuristics.

5.3   QF_LIA Benchmarks Set

Table 7 shows comparisons on QF_LIA logic using Z3 default strategy and the generated strategy with best result in LIA logic. In both tables we can see that the default strategy is overwhelmingly better than the one generated with LIA and applied to QF_LIA. This huge difference is due to:

1. The Z3 default strategy found in Z3 for QF_LIA is a complex system designed by de Moura and Olney Passmore in order to prove the importance of the strategies in the performance of solver [9]. This strategy is based on the huge knowledge of two of the most recognized researchers in the SMT community. The goal of this work is to try to find an automatic process that could have similar efficiency.
2. The best strategy generated for LIA logics is learned from a little sample in comparison with both QF_LIA subsets. Therefore, is really unlikely to obtain good results using strategies generated from a non-generalizable sample.

Table 7. QF_LIA Benchmarks Set: Comparison between default strategy and best found in LIA logic set.

		Instance time-out [s]					
Result	Strategy	1		10		1500	
known	Z3-default	**2459**	**564,61**	**3510**	**4577,60**	**4668**	**59183,88**
	LIA-ESG-best	1757	353,59	2525	3765,99	4398	288502,85
unknown	Z3-default	**566**	**131,51**	**693**	**612,93**	**1063**	**41305,20**
	LIA-ESG-best	256	21,43	288	179,43	701	197151,57
		solved	time[s]	solved	time[s]	solved	time[s]

5.4   QF_LRA Benchmarks Set

Table 8 shows comparison on QF_LRA using Z3 default strategy and the strategy with best result found in LRA logic. The results in this logic set are completely favorable to the strategy found in LRA. It outperforms Z3 default strategy in all scenarios, helping to solve under 2014 SMT-COMP rules 49,6 % of the unsolvable instances with known results and 65,9 % of unsolvable instances with unknown results. The success of the automatically generated strategy could be explained by:

Table 8. QF_LRA Benchmarks Set: Comparison between default strategy and best found in LRA logic set.

		Instance time-out [s]					
Result	Strategy	1		10		1500	
known	Z3-default	933	70,21	1043	432,89	1343	66401,28
	LRA-ESG-best	**936**	**65,24**	**1095**	**894,41**	**1409**	**41903,69**
unknown	Z3-default	105	3,99	108	20,13	126	13021,12
	LRA-ESG-best	**105**	**3,69**	**111**	**42,35**	**180**	**18977,01**
		solved	time[s]	solved	time[s]	solved	time[s]

1. In this set there is not a complex default strategy in Z3 but only a modified version of the default solver tactical (*smt*).
2. The sample size of the LRA logic set is big enough to check if its properties are related to this logic.

This results raised basis for new ideas which are to design an incremental automatic generation of strategies for a set of related logics, where a strategy from an easier logic could be part of a harder logic, i.e.,. the real arithmetic family (QF_RDL, QF_LRA, LRA, QF_NRA and NRA).

6   Related Work

To our best knowledge, no work directly addresses the automatic generation of strategies in SMT. However, there are several works in different areas that could be revisited in order to generate new interesting research topics in SMT community. Mark Harman analyzes why Evolutionary Algorithms are so much used in SBSE [5]. For instance, let us mention the improvement of the MiniSAT solver by Genetic Programming [11].

Hyperheuristics [2] are also clearly related to our work, as they can be defined as *heuristics to generate heuristics*. The general concept of autonomous search [4] aims at providing tools and methods for helping end-users to use solving and optimization techniques with a minimum amount of expert knowledge. Within this scope, parameter tuning has been widely addressed in the evolutionary computation and optimization communities [7,10,12].

7   Conclusion and Future Work

In this work, we present an automatic strategy generator that allows us to improve performances of the Z3 theorem prover on two selected SMT-LIB logics (LIA and LRA), as well as for a close related logic (QF_LRA). Even if the training time normally takes days as in the LRA with unknown results subset, a cross-validation automatic generator of strategies was also applied to improve default performance, which constitutes an alternative option when a huge set of benchmarks has to be solved. Interesting results have been obtained showing that more improvements could still be expected. In future work, we will turn to genetic programming or grammatical evolution algorithms in order to consider more complex strategies.

Acknowledgment. We want to thanks Christopher Wintersteiger from Microsoft Research for provide us critical information about Z3 theorem prover. Nicolás Gálvez Ramírez is granted by Chilean government: CONICYT-PCHA / Doctorado Nacional / 2013-21130089.

References

1. Barrett, C., Fontaine, P., Tinelli, C.: The SMT-LIB Standard: Version 2.5. Technical report, Department of Computer Science, The University of Iowa (2015). www.SMT-LIB.org
2. Burke, E.K., Gendreau, M., Hyde, M., Kendall, G., Ochoa, G., Ozcan, E., Qu, R.: Hyper-heuristics. J. Oper. Res. Soc. **64**(12), 1695–1724 (2013)
3. Gálvez Ramírez, N., Hamadi, Y., Monfroy, E., Saubion, F.: Towards Automated Strategies in Satisfiability Modulo Theory: Appendix. http://www.inf.utfsm.cl/~ngalvez/strategies.pdf
4. Hamadi, Y., Monfroy, E., Saubion, F.: What is autonomous search? In: van Hentenryck, P., Milano, M. (eds.) Hybrid Optimization. Springer Optimization and Its Applications, vol. 45, pp. 357–391. Springer, New York (2011)

5. Harman, M.: Software engineering meets evolutionary computation. Computer **44**(10), 31–39 (2011)
6. Harman, M.: The role of Artificial Intelligence in Software Engineering. In: 2012 First International Workshop on Realizing Artificial Intelligence Synergies in Software Engineering (RAISE), pp. 1–6 (2012)
7. Hutter, F., Hoos, H.H., Leyton-Brown, K., Stützle, T.: ParamILS: an automatic algorithm configuration framework. J. Artif. Int. Res. **36**(1), 267–306 (2009)
8. de Moura, L., Bjørner, N.S.: Z3: an efficient SMT solver. In: Ramakrishnan, C.R., Rehof, J. (eds.) TACAS 2008. LNCS, vol. 4963, pp. 337–340. Springer, Heidelberg (2008)
9. de Moura, L., Passmore, G.O.: The strategy challenge in SMT solving. In: Bonacina, M.P., Stickel, M.E. (eds.) Automated Reasoning and Mathematics. LNCS, vol. 7788, pp. 15–44. Springer, Heidelberg (2013)
10. Nannen, V., Eiben, A.: Efficient relevance estimation and value calibration of evolutionary algorithm parameters. In: IEEE Congress on Evolutionary Computation, CEC, pp. 103–110 (2007)
11. Petke, J., Langdon, W.B., Harman, M.: Applying genetic improvement to MiniSAT. In: Ruhe, G., Zhang, Y. (eds.) SSBSE 2013. LNCS, vol. 8084, pp. 257–262. Springer, Heidelberg (2013)
12. Riff, M.C., Montero, E.: A new algorithm for reducing metaheuristic design effort. In: 2013 IEEE Congress on Evolutionary Computation (CEC), pp. 3283–3290, June 2013
13. SMT-LIB Community: SMT-LIB Logics, Accessed 10 September 2015. http:// smtlib.cs.uiowa.edu/logics.shtml

Geometric Semantic Genetic Programming Is Overkill

Tomasz P. Pawlak$^{(\boxtimes)}$

Institute of Computing Science, Poznan University of Technology, Poznań, Poland
tpawlak@cs.put.poznan.pl

Abstract. Recently, a new notion of Geometric Semantic Genetic Programming emerged in the field of automatic program induction from examples. Given that the induction problem is stated by means of function learning and a fitness function is a metric, GSGP uses geometry of solution space to search for the optimal program. We demonstrate that a program constructed by GSGP is indeed a linear combination of random parts. We also show that this type of program can be constructed in a predetermined time by much simpler algorithm and with guarantee of solving the induction problem optimally. We experimentally compare the proposed algorithm to GSGP on a set of symbolic regression, Boolean function synthesis and classifier induction problems. The proposed algorithm is superior to GSGP in terms of training-set fitness, size of produced programs and computational cost, and generalizes on test-set similarly to GSGP.

Keywords: Automatic program induction · Geometric semantic genetic programming · Solution space

1 Introduction

Recently in the field of automatic program induction from examples, a new branch of Genetic Programming (GP) called Geometric Semantic Genetic Programming (GSGP) [18] arose. GSGP involves program semantics, meant as a vector of program outcomes produced in the effect of its execution on the given sample program inputs (fitness cases). The key idea behind GSGP is to define fitness function by means of a metric, where the first metric argument is semantics of a program under assessment and the second one is the optimal target semantics. This formulation causes fitness landscape to take shape of a cone, with the target semantics in the apex [24, 25, 27]. GSGP uses specialized search operators that utilize this conic shape to efficiently search the program space. Theoretical analyses have showed that each application of the geometric operators is characterized by beneficial expected improvement of fitness [17] and guaranteed pessimistic bound on change of fitness [25].

Nevertheless, GSGP has crucial design drawback that causes each offspring to be bigger than its parent(s). GSGP crossover leads to exponential growth

© Springer International Publishing Switzerland 2016
M. Heywood et al. (Eds.): EuroGP 2016, LNCS 9594, pp. 246–260, 2016.
DOI: 10.1007/978-3-319-30668-1_16

and GSGP mutation to linear growth of programs over the course of evolution. The size of the produced programs may put the use of GSGP in real world applications under a question. This state of affairs changed, when Castelli *et al.* [2] proposed graph-like encoding of tree programs to efficiently store overgrowth programs in memory. This allowed use of GSGP in many real-world applications [2–4,15,30,31], however the obtained programs are still oversized.

We demonstrate that a final program produced by GSGP implements, indeed, a linear combination of programs from the initial population, i.e., can be considered as an *linear combination of random parts*. We show that qualitatively equivalent results can be achieved by much simpler means than GP. We address all three problem domains, for which GSGP was designed: symbolic regression, Boolean function synthesis and classifier induction [18]. We propose *an exact algorithm*, that guarantees reaching the optimum, in opposition to GSGP, which as a stochastic metaheuristics does not guarantee this. What is more, the exact algorithm produces smaller programs and in shorter time than GSGP.

2   Problem Statement and Solution in Geometric Semantic Genetic Programming

A program $p \in \mathcal{P}$ is a function that maps a set of inputs $\mathcal{I}$ into a set of outputs $\mathcal{O}$, which we denote by $o = p(\mathbf{in})$, where $\mathbf{in} \in \mathcal{I}$ and $o \in \mathcal{O}$. We consider only deterministic programs that feature no side effects, nor memory persistent across executions. Semantics $\mathbf{s} \in \mathcal{S}$ is a vector $\mathbf{s} = [o_1, o_2, ..., o_n]$, where $\forall o_i \in \mathcal{O}$ and $\mathcal{S}$ is a semantic space (a vector space). Semantic mapping is a function $s : \mathcal{P} \to \mathcal{S}$, with property $s(p_1) = s(p_2) \iff \forall_i p_1(\mathbf{in}_i) = p_2(\mathbf{in}_i)$. In other words, semantics $s(p)$ of a program p is a vector of p's outputs when executed on a fixed set of inputs $I \subset \mathcal{I}$, i.e., $s(p) = [p(\mathbf{in}_1), p(\mathbf{in}_2), ..., p(\mathbf{in}_n)]$, $\mathbf{in}_i \in I$.

In GSGP program induction problem $\Pi = (\mathbf{t}, f)$ is an optimization problem, where $\mathbf{t}$ is the target semantics, and f is a metric, such that $f(p) = \|\mathbf{t}, s(p)\|$. The aim of program induction problem is to synthesize program p^*, which semantics equals $\mathbf{t}$, i.e., $\mathbf{t} = s(p^*)$ and $f(p^*) = 0$. Note that Π is stated in terms of learning function $h : \mathcal{I} \to \mathcal{O}$.

GSGP attempts to solve the abovementioned problem by initializing population of random programs, e.g., using ramped half-and-half [9], then iteratively searches the program space using geometric operators defined below (definitions come from [18]).

Definition 1. *Given a parent program p, r-geometric mutation is an operator that produces offspring p' in a ball of radius r centered in p, i.e., $\|s(p), s(p')\| \leq r$.*

Definition 2. *(Algorithms for geometric mutation)* **Symbolic regression:** *Given parent arithmetic program p, an offspring is a program $p' = p + r \cdot (m_1 - m_2)$, where m_1 and m_2 are random arithmetic programs that output values in range $[0, 1]$.* **Boolean:** *Given Boolean parent program p, an offspring is a program $p' = m \vee p$ with probability 0.5, $p' = \overline{m} \wedge p$ otherwise, where m is a* *random minterm.* **Classifier:** *Given parent classifier p, an offspring is a program $p' = IF\, cond\, THEN\, c\, ELSE\, p$, where cond is a random condition that is true only for a single combination of attributes and $c \in \mathcal{O}$ is a random class.*

Definition 3. *Given parent programs p_1, p_2, geometric crossover is an operator that produces offspring p' in a segment between p_1 and p_2, i.e., $\|s(p_1), s(p_2)\| = \|s(p_1), s(p')\| + \|s(p'), s(p_2)\|$.*

Definition 4. *(Algorithms for geometric crossover)* **Symbolic regression:** *Given parent arithmetic programs p_1, p_2, an offspring is a program $p' = mp_1 + (1 - m)p_2$, where m is a function that returns values in range $[0, 1]$.* **Boolean:** *Given Boolean parent programs p_1, p_2, an offspring is a program $p' = (p_1 \wedge m) \vee (\overline{m} \wedge p_2)$, where m is a random Boolean program.* **Classifier:** *Given parent classifiers p_1, p_2, an offspring is a program $p' = IF\, cond\, THEN\, p_1\, ELSE\, p_2$, where cond is a random condition.*

In the essence geometric operators conduct linear transformation of parents, which we formalize below:

Definition 5. *A linear combination over a field is an expression of elements from this field multiplied by constants and added together.* **Symbolic regression:** *Field elements are real numbers; $+$ (arithmetic addition) and $\cdot$ (arithmetic multiplication) meet axioms of addition and multiplication, respectively.* **Boolean:** *($GF(2)$ field) Field elements are 0 and 1; $\oplus$ (xor) and $\wedge$ (and) meet axioms of addition and multiplication, respectively.* **Classifier:** *Field elements are Boolean 0 and 1, classes $c_1, c_2, \ldots$ and null $\varnothing$ (no decision); $\oslash$ and $\oslash$ meet axioms of addition and multiplication, respectively. IF is composition of $\oslash$ and $\oslash$: $IF\, cond\, THEN\, c_1\, ELSE\, c_2 \equiv cond \oslash c_1 \oslash \overline{cond} \oslash c_2$.*[1]

Lemma 1. *An offspring p' obtained from parent p using r-geometric mutation is a linear combination of p and an other random program.*

Proof. The proof comes from Definition 2, each offspring's formula combines p with other parts using only addition and/or multiplication by a constant (e.g., weight). **Symbolic regression:** p' is *addition* of p and program $r \cdot (m_1 - m_2)$. **Boolean:** by expanding $p' = m \vee p$ to $p' = m \oplus p \oplus m \wedge p$, p' is *addition* of p, m and program $m \wedge p$, or by transforming $p' = \overline{m} \wedge p$ to $p' = p \oplus m \wedge p$, p' is *addition* of p and a program $m \wedge p$. **Classifier:** IF term is a weighted *addition* of c and p, where respective weights for c and p are either 0 and 1, or 1 and 0, depending on *cond*'s Boolean value.

Lemma 2. *An offspring p' obtained from parents p_1 and p_2 using geometric crossover is a linear combination of p_1 and p_2.*

---

[1] $\oslash$ for two Booleans is equivalent to $\vee$, for two classes outputs the one with lower id, for a class and $\varnothing$ outputs this class, otherwise $\varnothing$. $\oslash$ for two Booleans is equivalent to $\wedge$, for two classes outputs the one with greater id, for 1 and a class outputs this class, otherwise $\varnothing$. We omitted full tables of these functions for brevity.

Proof. Proof comes from Definition 4, each offspring's formula combines p_1 and p_2 with other parts using only addition and/or multiplication by a constant (e.g., weight).**Symbolic regression:** p' is weighted *addition* of p_1 and p_2, where m and $(1 - m)$ are weights. **Boolean:** by transforming $p' = (p_1 \wedge m) \vee (\overline{m} \wedge p_2)$ to $p' = (p_1 \wedge m) \oplus (\overline{m} \wedge p_2)$, p' is weighted *addition* of p_1 and p_2, where m and $\overline{m}$ are weights. **Classifier:** IF term is a weighted *addition* of p_1 and p_2, where respective weights for c and p are either 0 and 1, or 1 and 0, depending on *cond*'s Boolean value.

Theorem 1. *A program produced by GSGP in any generation after the initial one is a linear combination of programs from the initial population and other random programs.*

Proof. Let P_g denote population and M_g set of random functions added by geometric operators in generation $g \in \mathbb{N}_{\geq 1}$. Proof by induction: each program $p_2 \in P_2$ is linear combination of programs $p_1 \in P_1$ and $m_1 \in M_1$ by Lemmas 1 and 2. A program $p_i \in P_i$ is a linear combination of programs $p_{i-1} \in P_{i-1}$ and random programs $m_{i-1} \in M_{i-1}$. By commutativity of linear combinations each program in P_i is a linear combination of programs in P_1 and $M_1, M_2, .., M_{i-1}$.

This leads to the straightforward conclusion that if the initial population consists of randomly generated programs, e.g., using ramped half-and-half [9], in the essence *a program produced by GSGP is a linear combination of random parts*. However, linear combination of, or even interpolation[2] using random parts can be done analytically with guarantee of constructing the optimal program p^*.

3   Function Learning Using Linear Combination

In this section, we show how to interpolate points from different semantic spaces: $\mathbb{R}^n$, $\{0, 1\}^n$ and $\{c_1, c_2, ..., c_k\}^n$ to obtain the optimal program p^*.

To interpolate given function $h : \mathcal{I} \to \mathcal{O}$ using random program parts, two conditions have to be met:

1. Set of semantics S of random program parts must be linearly independent,
2. Target $\mathbf{t}$ must be expressible by linear combination of semantics from S.

The first condition is there, because the set of semantics that can be expressed by linearly combining the semantics from S is exactly the same like for S with linearly dependent semantics discarded. The second condition is the necessary one to solve the program induction problem Π. The above conditions can be met by making S a basis of semantic space $\mathcal{S}$.

Below, we present Linear Combination (LC) algorithm. Although the general idea behind LC is maintained across domains, a particular realization is domain-dependent. For each domain, we divide the presentation of LC into two parts:

---

[2] The distinction between linear combination and interpolation is made to emphasize that linear combination is any weighted sum of programs, while interpolation is a weighted sum that goes through the certain points.

a way of constructing a basis, and a way to interpolate given set of points. Note that the presented algorithm is naive, and cannot be considered optimal in any sense, except that it is guaranteed to solve program induction problem Π.

3.1   Symbolic Regression

Basis Construction

Theorem 2. *Let* $S = \mathbb{R}^n$, *and* $I \subseteq \mathcal{I}$ *be a set of program inputs on which semantics of a program is calculated,* $|I| = n$. *Then,* $\mathbf{in_i} \in I$ *is a vector of inputs. Basis of* S *is set* S *of semantics of programs* $p_1, p_2, ..., p_n$ *given by the formula:*

$$p_i = e^{-i(\mathbf{x}-\mathbf{in_i})\cdot(\mathbf{x}-\mathbf{in_i})}, \tag{1}$$

where $\mathbf{x}$ *is a vector of program arguments and* $\cdot$ *is dot product of vectors.*

Proof. Proof comes from linear independence of exponential functions $e^{-iy(x)}$, where $y(x)$ is a non-constant function.

The term $-i$ (negative integer) in the exponent in Eq. (1) makes sure that p_is are linearly independent, and the dot product causes the entire formula to equal 1 only for $\mathbf{x} = \mathbf{in}_i$.

Interpolation. Interpolation of h using abovementioned programs can be done by solving the following system of linear equations:

$$[s(p_1)\, s(p_2) ... s(p_n)]\, \mathbf{w} = \mathbf{t},$$

where $\mathbf{w}$ is a vector of weights to be calculated and $\mathbf{t}$ is the target. The solution can be found, e.g., using Gaussian elimination [7] in $O(n^3)$ time. Then, the optimal program p^* is given by:

$$p^* = \sum_{i=1}^{n} w_i p_i. \tag{2}$$

The programs p_i, for which $w_i = 0$ can be omitted from the sum to keep final program size at bay.

Example 1. Consider programs with two arguments $\mathbf{x} = [x_1, x_2]^T$. Let set of program inputs $I = \{[1,2]^T, [3,4]^T, [5,6]^T\}$ and target $t = [7,5,3]^T$. Then, the basis consists of p_is given by equations:

$$p_1 = e^{-1([x_1,x_2]-[1,2])\cdot([x_1,x_2]-[1,2])} = e^{-1((x_1-1)^2+(x_2-2)^2)}$$
$$p_2 = e^{-2([x_1,x_2]-[3,4])\cdot([x_1,x_2]-[3,4])} = e^{-2((x_1-3)^2+(x_2-4)^2)}$$
$$p_3 = e^{-3([x_1,x_2]-[5,6])\cdot([x_1,x_2]-[5,6])} = e^{-3((x_1-5)^2+(x_2-6)^2)}$$

Next, by solving the system of linear equations:

$$\begin{bmatrix} 1 & 1.13 \times 10^{-7} & 2.03 \times 10^{-42} \\ 3.35 \times 10^{-4} & 1 & 3.78 \times 10^{-11} \\ 1.27 \times 10^{-14} & 1.13 \times 10^{-7} & 1 \end{bmatrix} \mathbf{w} = \begin{bmatrix} 7 \\ 5 \\ 3 \end{bmatrix}$$

and we obtain weights: $\mathbf{w} = [6.99, 4.99, 2.99]^T$. A weighed sum of the basis programs is the final one: $p^* = 6.99 p_1 + 4.99 p_2 + 2.99 p_3$.

3.2   Boolean Function Synthesis

Basis Construction

Theorem 3. *Let $\mathcal{S} = \{0, 1\}^n$, and $I \subseteq \mathcal{I}$ be a set of program inputs on which semantics of a program is calculated, $|I| = n$. Then, $\mathbf{in_i} \in I$ is a vector of inputs. Basis of $\mathcal{S}$ is set S of semantics of programs $p_1, p_2, ..., p_n$ given by the formula:*

$$p_i = \bigwedge_{j=1}^{n} y_j \tag{3}$$

$$y_j = \begin{cases} x_j & in_{ij} \\ \overline{x_j} & \overline{in_{ij}} \end{cases},$$

where x_js are program arguments.

Proof. Each program p_i outputs 1 for only one setting of x_js. Thus, each semantics $s(p_i)$ is a vector of all, but one 0s, and the only non-zero locus is different for each $s(p_i)$.

In other words, each p_i is a minterm, where each argument x_j is negated if vector of inputs $\mathbf{in}_i$ is 0 on locus j. This causes p_i to return 1 only for $\mathbf{x} = \mathbf{in}_i$.

Interpolation. The program p^* that interpolates h using p_is is given by the formula:

$$p^* = \bigvee_{i=1}^{n} y_i \tag{4}$$

$$y_i = \begin{cases} p_i & t_i \\ 0 & \text{otherwise.} \end{cases}$$

In other words, for each locus i in $\mathbf{t}$, if t_i equals 1, p_i is a part of alternative in p^*, and p^* is a disjunctive normal form. Zeros can be dropped from the alternative for simplification.

Example 2. Consider programs with two arguments $\mathbf{x} = [x_1, x_2]^T$. Let set of program inputs $I = \{[1, 0]^T, [0, 1]^T, [1, 1]^T\}$ and target $t = [1, 0, 1]^T$. Then, the basis consists of p_is given by equations:

Algorithm 1. Construction of the optimal program p^* in classifier induction. PICK(P) picks an element from set P, p_is and $cond_i$s are defined in Eq. (5).

1: $P = \{p_1, p_2, ..., p_n\}$
2: **while** $|P| \neq 1$ **do**
3:     $p_i \leftarrow$ PICK(P)
4:     $p_j \leftarrow$ PICK($P\backslash\{p_i\}$)
5:     $p_k \leftarrow IF\ cond_i\ THEN\ p_i\ ELSE\ p_j$
6:     $cond_k \leftarrow cond_i \lor cond_j$
7:     $P \leftarrow P\backslash\{p_i, p_j\} \cup \{p_k\}$
8: **return** $p^* \in P$

$$p_1 = x_1\overline{x_2}$$
$$p_2 = \overline{x_1}x_2$$
$$p_3 = x_1x_2$$

By substituting p_is in Eq. (4) we obtain the final program: $p^* = x_1\overline{x_2} \lor x_1x_2$.

3.3   Classifier Induction

Basis Construction

Theorem 4. *Let $\mathcal{S} = \{c_1, c_2, ..., c_k\}^n$, and $I \subseteq \mathcal{I}$ be a set of program inputs on which semantics of a program is calculated, $|I| = n$. Then, $\mathbf{in_i} \in I$ is a vector of inputs. Basis of $\mathcal{S}$ is set S of semantics of programs $p_1, p_2, ..., p_n$ given by formula:*

$$p_i = IF\ cond_i\ THEN\ t_i\ ELSE\ \varnothing, \tag{5}$$

$$cond_i = \bigwedge_{j=1}^{n}(x_j = in_{ij})$$

where x_js are arguments, t_i is a class at locus i in t, and $\varnothing$ is a 'null' (no decision) symbol.

Proof. Each program p_i outputs a class $c \in \{c_1, c_2, ..., c_k\}$ for only one setting of x_js. Thus, each semantics $s(p_i)$ is a vector of all, but one $\varnothing$s, and the only non-zero locus is different for each $s(p_i)$.

In other words, $cond_i$ is a logical conjunction of comparisons of arguments x_j to the respective inputs in vector $\mathbf{in}_i$.

Interpolation. The programs $p_1, p_2, ..., p_n$ should be treated as a set of decision rules, where each one is executed against given example data. Thanks to the construction of basis shown in Eq. (5), at most one p_i actually outputs a class. Algorithm 1 shows how to merge $p_1, p_2, ..., p_n$ into a single program p^* using only IF instructions. The algorithm begins with a set of all programs P. Then, in a loop it picks two programs p_i and p_j from P. Next, the algorithm merges p_i and p_j into a one program p_k using IF instruction and creates condition $cond_k$ for p_k that may be used in merges in further loops. The program p_k replaces p_i and p_j in the set P. The loop terminates when P contains only one program, which is finally returned. Time complexity of the algorithm is $O(n)$.

Example 3. Consider programs with two arguments $\mathbf{x} = [x_1, x_2]^T$. Let set of program inputs $I = \{[a, b]^T, [c, a]^T, [b, c]^T\}$ and target $t = [c_1, c_2, c_1]^T$. Then, the basis P consists of p_is given by equations:

$$p_1 = IF\, x_1 = a \wedge x_2 = b\, THEN\, c_1\, ELSE\, \varnothing$$
$$p_2 = IF\, x_1 = c \wedge x_2 = a\, THEN\, c_2\, ELSE\, \varnothing$$
$$p_3 = IF\, x_1 = b \wedge x_2 = c\, THEN\, c_1\, ELSE\, \varnothing$$

Next, we run Algorithm 1. First, we pick programs p_1 and p_2 and combine them into $p_4 = IF\, x_1 = a \wedge x_2 = b\, THEN\, p_1\, ELSE\, p_2$, and define new condition $cond_4 = cond_1 \vee cond_2$. Then, we remove p_1 and p_2 from P and add p_4. In the second iteration of the algorithm's loop we pick p_3 and p_4 and combine them into $p_5 = IF\, x_1 = b \wedge x_2 = c\, THEN\, p_3\, ELSE\, p_4$. Finally, we remove p_3 and p_4 from P and add p_5. Since p_5 is the only remaining program in P, p_5 becomes the final program p^*.

4 Experiment

4.1 Setup

We compare LC to GSGP to verify which algorithm provides more desirable characteristics: lower training- and test-set error, smaller size of programs and less computational cost.

Table 1 presents benchmark problems that come from [13, 16, 27]. In univariate symbolic regression 20 Chebyshev nodes[3] [1] and 20 uniformly picked points in the given range are used for training- and test-sets, respectively. For bivariate these numbers amount to 10 for each variable and Cartesian product of them constitutes a data set. In Boolean domain training-set incorporates all inputs and there is no test-set. For classifier benchmarks statistics of data set are presented in Table 1.

Table 2 shows parameters of evolution. Values not presented there are set to ECJ defaults [12]. We do not involve any simplification procedure for the produced programs. LC is run once for each problem, since it is deterministic. Presented CPU times are obtained on Intel Core i7-950 CPU and 6GB DDR3 RAM running in x64 mode under control of Linux and Java 1.8. The times exclude calculation of statistics.

---

[3] Points given by $x_k = \frac{1}{2}(a+b) + \frac{1}{2}(b-a)\cos(\frac{2k-1}{2n}\pi)$, $k = 1..n$, where $[a, b]$ is the range of training set, and n is number of data points. Using Chebyshev nodes minimizes the likelihood of Runge's phenomenon [29].

Table 1. Benchmark problems.

Symbolic regression benchmarks

Problem	Definition (formula)	Variables	Range	Size
R1	$(x_1 + 1)^3/(x_1^2 - x_1 + 1)$	1	$\langle -1, 1\rangle$	20
R2	$(x_1^5 - 3x_1^3 + 1)/(x_1^2 + 1)$	1	$\langle -1, 1\rangle$	20
R3	$(x_1^6 + x_1^5)/(x_1^4 + x_1^3 + x_1^2 + x_1 + 1)$	1	$\langle -1, 1\rangle$	20
Kj1	$0.3x_1 \sin(2\pi x_1)$	1	$\langle -1, 1\rangle$	20
Kj4	$x_1^3 e^{-x_1} \cos(x_1) \sin(x_1)(\sin^2(x_1)\cos(x_1) - 1)$	1	$\langle 0, 10\rangle$	20
Ng9	$\sin(x_1) + \sin(x_2^2)$	2	$\langle 0, 1\rangle^2$	100
Ng12	$x_1^4 - x_1^3 + \frac{x_2^2}{2} - x_2$	2	$\langle 0, 1\rangle^2$	100
Pg1	$1/(1 + x_1^{-4}) + 1/(1 + x_2^{-4})$	2	$\langle -5, 5\rangle^2$	100
Vl1	$e^{-(x_1-1)^2}/(1.2 + (x_2 - 2.5)^2)$	2	$\langle 0, 6\rangle^2$	100

Boolean benchmarks

Problem	Instance	Variables	Size
Even parity	Par5	5	32
	Par6	6	64
	Par7	7	128
Multiplexer	Mux6	6	64
	Mux11	11	2048
Majority	Maj7	7	64
	Maj8	8	128
Comparator	Cmp6	6	64
	Cmp8	8	256

Classifier benchmark

Problem	Attributes	Classes	Training	Test
Cancer	9	2	457	226

Table 2. Parameters of evolution.

	Symbolic regression	Boolean domain	Classifier induction
Number of runs	30		
Population size	1000		
Fitness function	L_2 metric	L_1 metric	Classification error
Termination condition	At most 50 generations or find of a program with fitness 0		
Instructions	$x_1, x_2, +, -, \times, /,$ $\sin, \cos, \exp, \log^{\text{a}}$	$x_1, x_2, ..., x_{11},^{\text{b}}$ nand, nor	$x_1, x_2, ..., x_9, c_1, c_2, \text{if}$

[a] log is defined as $\log|x|$; / returns 0 if divisor is 0.
[b] The number of inputs depends on a particular problem instance.

Table 3. Average and 95 % confidence interval of fitness of the best of run program on training set.

(a) Training-set fitness

Problem	LC	GSGP
R1	**0.00** ±0.00	2.71 ±0.27
R2	**0.00** ±0.00	0.50 ±0.06
R3	**0.00** ±0.00	0.19 ±0.02
Kj1	**0.00** ±0.00	0.35 ±0.01
Kj4	**0.00** ±0.00	0.96 ±0.03
Ng9	**0.00** ±0.00	0.65 ±0.04
Ng12	**0.00** ±0.00	0.37 ±0.02
Pg1	**0.00** ±0.00	1.25 ±0.08
Vl1	**0.00** ±0.00	1.03 ±0.02
Par5	**0.00** ±0.00	0.00 ±0.00
Par6	**0.00** ±0.00	0.00 ±0.00
Par7	**0.00** ±0.00	0.13 ±0.12
Mux6	**0.00** ±0.00	1.13 ±0.37
Mux11	**0.00** ±0.00	132.33 ±11.25
Maj7	**0.00** ±0.00	0.00 ±0.00
Maj8	**0.00** ±0.00	0.00 ±0.00
Cmp6	**0.00** ±0.00	0.00 ±0.00
Cmp8	**0.00** ±0.00	0.00 ±0.00
Cancer	**0.00** ±0.00	0.05 ±0.01
Rank:	1.16	1.84

(b) Test-set fitness

Problem	LC	GSGP
R1	**0.00** ±0.00	17.25 ±0.83
R2	**0.00** ±0.00	3.20 ±0.05
R3	**0.00** ±0.00	0.67 ±0.01
Kj1	**0.00** ±0.00	0.51 ±0.01
Kj4	1.15 ±0.00	2.25 ±0.07
Ng9	0.26 ±0.00	5.85 ±0.05
Ng12	0.11 ±0.00	2.11 ±0.01
Pg1	15.60 ±0.00	**7.34** ±0.07
Vl1	3.63 ±0.00	**3.35** ±0.02
Cancer	0.51 ±0.00	**0.04** ±0.01
Rank:	1.30	1.70

4.2   Results

Table 3a presents average and 95 % confidence interval of the best of run fitness. Since LC is guaranteed to construct the optimal program, it achieves 0 fitness and 0 confidence interval in each problem. In turn, GSGP is able to find the optimum only in 6 out of 19 problems. Wilcoxon signed rank test results in p-value of 1.66×10^{-3}, thus LC is statistically better than GSGP.

Table 3b shows average and 95 % confidence interval of test-set fitness of the best of run program on training-set. In 4 out of 10 problems LC achieves 0 fitness and is better than GSGP in 7 out of 10 problems. Wilcoxon test results in p-value of 0.13, thus at significance level $\alpha = 0.05$, the difference of generalization abilities of LC and GSGP is insignificant.

Table 4a compares numbers of nodes in programs produced by LC and GSGP. In all 19 problems LC produces programs smaller by 1 up to 14 orders of magnitude than GSGP. Wilcoxon test results in p-value of 3.82×10^{-6}, which only confirms our observation that LC produces smaller programs. Note that the numbers of nodes are equal in LC programs for all univariate, and all bivariate except Vl1 symbolic regression problems, respectively. This comes from the way how LC constructs the final program: basis for all problems with the same number of variables is exactly the same, only weights of the sum in Eq. (2) differ. LC program in Vl1 is smaller, because some of the weights in Eq. (2) are zero and the respective program parts are dropped.

Table 4. Average and 95 % confidence interval of:

(a) number of nodes in the best of run
program; values $> 10^4$ are rounded to an
order of magnitude

(b) total CPU time (seconds)

Problem	LC	GSGP
R1	**259.00** ±0.00	10^{16} ±10^{15}
R2	**259.00** ±0.00	10^{15} ±10^{15}
R3	**259.00** ±0.00	10^{16} ±10^{15}
Kj1	**259.00** ±0.00	10^{15} ±10^{15}
Kj4	**259.00** ±0.00	10^{15} ±10^{15}
Ng9	**2099.00** ±0.00	10^{16} ±10^{15}
Ng12	**2099.00** ±0.00	10^{15} ±10^{14}
Pg1	**2099.00** ±0.00	10^{15} ±10^{14}
Vl1	**1448.00** ±0.00	10^{15} ±10^{15}
Par5	**199.00** ±0.00	10^{4} ±1327.87
Par6	**479.00** ±0.00	10^{6} ±10^{5}
Par7	**1119.00** ±0.00	10^{15} ±10^{15}
Mux6	**215.00** ±0.00	10^{16} ±10^{15}
Mux11	**6776.00** ±0.00	10^{16} ±10^{15}
Maj7	**1049.00** ±0.00	10^{5} ±10^{4}
Maj8	**1719.00** ±0.00	10^{5} ±6049.50
Cmp6	**419.00** ±0.00	2467.60 ±350.10
Cmp8	**2399.00** ±0.00	10^{6} ±10^{5}
Cancer	10^{6} ±0.00	10^{16} ±10^{15}
Rank:	1.00	2.00

Problem	LC	GSGP
R1	**0.03** ±0.00	1.13 ±0.02
R2	**0.03** ±0.00	1.13 ±0.03
R3	**0.03** ±0.00	1.17 ±0.03
Kj1	**0.02** ±0.00	1.18 ±0.04
Kj4	**0.03** ±0.00	1.20 ±0.03
Ng9	**0.08** ±0.00	1.37 ±0.04
Ng12	**0.08** ±0.00	1.38 ±0.04
Pg1	**0.09** ±0.00	1.34 ±0.03
Vl1	**0.08** ±0.00	1.36 ±0.05
Par5	**0.04** ±0.00	0.31 ±0.01
Par6	**0.06** ±0.00	0.46 ±0.01
Par7	**0.06** ±0.00	0.74 ±0.07
Mux6	**0.05** ±0.00	1.06 ±0.17
Mux11	**0.79** ±0.00	3.75 ±0.22
Maj7	**0.08** ±0.00	0.49 ±0.01
Maj8	**0.13** ±0.00	0.52 ±0.02
Cmp6	**0.05** ±0.00	0.32 ±0.01
Cmp8	**0.07** ±0.00	0.53 ±0.02
Cancer	**0.19** ±0.00	25.89 ±0.44
Rank:	1.00	2.00

Last, but not least, Table 4b shows average and 95 % confidence interval of
CPU time required to finish the run. In each problem, LC run takes less than
0.8 second and is faster than the GSGP run. In contrast, GSGP requires 1 or 2
orders of magnitude more time, depending on a problem. Wilcoxon test reports
p-value of 3.82×10^{-6}, hence LC is significantly faster than GSGP.

5   Discussion

GSGP builds overgrowth programs [18,27] that, due to its size, are difficult to
interpret by humans, require a lot of a storage memory and possibly execute
longer than smaller semantically equal programs. Since program simplification
is NP-hard in general [5], it is doubtful that any simplification procedure can
reduce programs produced by GSGP (cf. Table 4a) to a human-interpretable size.
Additionally, a program built by GSGP is a linear combination of the given program parts. This means that virtually any, random, unrelated conglomerates
of instructions can be synthesized, and combined together using constant coefficients. The progress of evolution in GSGP is limited to adaptation of these
coefficients, leaving the given program parts intact. This limits also the program's ability to properly model a relation of input and output hidden in the
training data, thus to properly operate on previously unseen data.

The presented LC algorithm also constructs a program using a linear combination of arbitrarily synthesized program parts. This means that LC and GSGP share some of the drawbacks of this way of constructing programs, e.g., overfitting to the training data. Nevertheless, LC has advantages over GSGP. LC is two-step, deterministic and exact algorithm, i.e., it guarantees construction of *the optimal program* w.r.t. the given program induction problem in a polynomial time. On the other hand, GSGP iteratively and stochastically combines given programs, to gradually converge to the optimum, without a guarantee of reaching it or terminating. This difference makes the final programs constructed by LC smaller than by GSGP in all our experiments.

LC is a naive approach that does not go beyond a toy example. In the field, there are many methods that construct either linear, or non-linear models that perform and generalize well in the considered class problems. For instance, in symbolic regression one can use Fast Function Extraction heuristic [14] or even assume a specific model and use one of classic regression or interpolation methods for it. In Boolean domain, the common approach is to use Karnaugh map minimization [8] that obviously produces smaller programs than LC. In classifier induction domain, one can use, e.g., decision trees or probabilistic classifiers [6].

This wide range of simple well performing methods, inclines us to claim that GSGP is overkill for problems of program induction stated as learning of function $h : \mathcal{I} \to \mathcal{O}$.

Where Is GSGP Useful? Though, it is important to say, that GSGP is not entirely doomed to fail. We can distinguish at least two features that makes GSGP useful in certain situations.

First, consider target to be unknown and fitness function to be black-box that fulfils requirements of metric. All algorithms discussed in the previous section, except GSGP, require access to the target. This is because GSGP operators conduct strictly syntactic manipulations that have well-defined impact on program semantics, however the semantics itself is not used by the geometric operators.

Second, GSGP can be split into broad theoretic framework with multiple achievements in the recent years [17, 19–21, 24–26] and algorithms of geometric operators. The main weakness of GSGP lies not in the former, but in the latter ones. GSGP theory does not define how to build offspring using parent programs, it only poses requirements to be met by the offspring (cf. Definitions 1 and 3). It is the algorithms are responsible for building linear combination of random program parts, code growth and poor generalization. To the date, we do not have other algorithms than the presented in Definitions 2 and 4 that fulfil requirements of Definitions 1 and 3. However, there were multiple attempts to create approximate algorithms that on average or in the limit can fulfil these requirements, e.g., [10, 11, 22–24, 28]. Nevertheless, future work is need in designing new exact algorithms for geometric operators that do not construct offspring by linearly combining parents.

6   Conclusions

Program induction problem in Geometric Semantic Genetic Programming is stated by means of learning function $h : \mathcal{I} \to \mathcal{O}$. We demonstrated that GSGP in an attempt to solve this problem constructs a linear combination of random parts, however GSGP is not guaranteed to solve this problem optimally in finite time. We showed that the linear combination of random program parts can be constructed by much simpler means than GSGP with guarantee of solving the problem optimally. The optimal program is smaller than these of GSGP, and time consumed by the proposed algorithm is shorter and predetermined (polynomial).

This, does not preclude practical use of GSGP, however a future work has to be done to put GSGP on the right track. We need new algorithms for GSGP operators that fulfil definitions of geometric operators (cf. Definitions 1 and 3), but do not operate by linearly combining parent programs in offspring.

Acknowledgements. This work is funded by National Science Centre Poland grant number DEC-2012/07/N/ST6/03066.

References

1. Burden, R., Faires, J.: Numerical Analysis. Cengage Learning (2010). http://books.google.pl/books?id=Dbw8AAAAQBAJ
2. Castelli, M., Castaldi, D., Giordani, I., Silva, S., Vanneschi, L., Archetti, F., Maccagnola, D.: An efficient implementation of geometric semantic genetic programming for anticoagulation level prediction in pharmacogenetics. In: Correia, L., Reis, L.P., Cascalho, J. (eds.) EPIA 2013. LNCS, vol. 8154, pp. 78–89. Springer, Heidelberg (2013)
3. Castelli, M., Henriques, R., Vanneschi, L.: A geometric semantic genetic programming system for the electoral redistricting problem. Neurocomputing **154**, 200–207 (2015). http://www.sciencedirect.com/science/article/pii/S0925231214016671
4. Castelli, M., Vanneschi, L., Silva, S.: Prediction of high performance concrete strength using genetic programming with geometric semantic genetic operators. Expert Syst. Appl. **40**(17), 6856–6862 (2013). http://www.sciencedirect.com/science/article/pii/S0957417413004326
5. Dershowitz, N., Jouannaud, J.P.: Rewrite systems. In: Handbook of Theoretical Computer Science. Formal Models and Sematics, vol. B, pp. 243–320 (1990)
6. Flach, P.: Machine Learning: The Art and Science of Algorithms that Make Sense of Data. Cambridge University Press, New York (2012)
7. Gentle, J.E.: Numerical Linear Algebra for Applications in Statistics. Statistics and Computing. Springer, New York (1998). http://opac.inria.fr/record=b1098288
8. Karnaugh, M.: The map method for synthesis of combinational logic circuits. Trans. Am. Inst. Electr. Eng. Part I: Commun. Electron. **72**(5), 593–599 (1953)
9. Koza, J.R.: Genetic Programming: On the Programming of Computers by Means of Natural Selection. MIT Press, Cambridge (1992). http://mitpress.mit.edu/books/genetic-programming
10. Krawiec, K., Lichocki, P.: Approximating geometric crossover in semantic space. In: Raidl, G., et al. (eds.) Proceedings of the 11th Annual Conference on Genetic and Evolutionary Computation. GECCO 2009, pp. 987–994. ACM, Montreal, 8–12 July 2009

11. Krawiec, K., Pawlak, T.: Locally geometric semantic crossover: a study on the roles of semantics and homology in recombination operators. Genet. Program. Evolvable Mach. **14**(1), 31–63 (2013)
12. Luke, S.: The ECJ Owner's Manual - A User Manual for the ECJ Evolutionary Computation Library, zeroth edition, online version 0.2 (edn.), October 2010. http://cs.gmu.edu/eclab/projects/ecj/docs/
13. Mangasarian, O.L., Street, W.N., Wolberg, W.H.: Breast cancer diagnosis and prognosis via linear programming. Oper. Res. **43**, 570–577 (1995)
14. McConaghy, T.: FFX: fast, scalable, deterministic symbolic regression technology. In: Riolo, R., Vladislavleva, E., Moore, J.H. (eds.) Genetic Programming Theory and Practice IX. Genetic and Evolutionary Computation, pp. 235–260. Springer, New York (2011). http://trent.st/content/2011-GPTP-FFX-paper.pdf
15. McDermott, J., Agapitos, A., Brabazon, A., O'Neill, M.: Geometric semantic genetic programmingfor financial data. In: Esparcia-Alcázar, A.I., Mora, A.M. (eds.) EvoApplications 2014. LNCS, vol. 8602, pp. 215–226. Springer, Heidelberg (2014)
16. McDermott, J., et al.: Genetic programming needs better benchmarks. In: Soule, T., et al. (eds.) Proceedings of the Fourteenth International Conference on Genetic and Evolutionary Computation Conference. GECCO 2012, pp. 791–798. ACM, Philadelphia, Pennsylvania, USA, 7–11 July 2012
17. Moraglio, A.: Abstract convex evolutionary search. In: Beyer, H.G., Langdon, W.B. (eds.) Foundations of Genetic Algorithms, pp. 151–162. ACM, Schwarzenberg, Austria, 5–9 January 2011
18. Moraglio, A., Krawiec, K., Johnson, C.G.: Geometric semantic genetic programming. In: Coello, C.A.C., Cutello, V., Deb, K., Forrest, S., Nicosia, G., Pavone, M. (eds.) PPSN 2012, Part I. LNCS, vol. 7491, pp. 21–31. Springer, Heidelberg (2012)
19. Moraglio, A., Mambrini, A.: Runtime analysis of mutation-based geometric semantic genetic programming for basis functions regression. In: Blum, C., et al. (eds.) Proceeding of the Fifteenth Annual Conference on Genetic and Evolutionary Computation Conference. GECCO 2013, pp. 989–996. ACM, Amsterdam, The Netherlands, 6–10 July 2013
20. Moraglio, A., Mambrini, A., Manzoni, L.: Runtime analysis of mutation-based geometric semantic genetic programming on boolean functions. In: Neumann, F., De Jong, K. (eds.) Foundations of Genetic Algorithms, pp. 119–132. ACM, Adelaide, Australia, 16–20 January 2013. http://www.cs.bham.ac.uk/axm322/pdf/gsgp_foga13.pdf
21. Moraglio, A., Sudholt, D.: Runtime analysis of convex evolutionary search. In: Soule, T., Moore, J.H. (eds.) GECCO, pp. 649–656. ACM (2012). http://dblp.uni-trier.de/db/conf/gecco/gecco2012.html#MoraglioS12
22. Nguyen, Q.U., Pham, T.A., Nguyen, X.H., McDermott, J.: Subtree semantic geometric crossover for genetic programming. Genet. Program. Evolvable Mach., 1–29. Online first
23. Pawlak, T.P.: Combining semantically-effective and geometric crossover operators for genetic programming. In: Bartz-Beielstein, T., Branke, J., Filipič, B., Smith, J. (eds.) PPSN 2014. LNCS, vol. 8672, pp. 454–464. Springer, Heidelberg (2014)
24. Pawlak, T.P.: Competent Algorithms for Geometric Semantic Genetic Programming. Ph.D. thesis, Poznan University of Technology, Poznan, Poland, 21 September 2015. http://www.cs.put.poznan.pl/tpawlak/link/?PhD
25. Pawlak, T.P., Krawiec, K.: Progress properties and fitness bounds for geometric semantic search operators. Genet. Program. Evolvable Mach., 1–19. Online first

26. Pawlak, T.P., Krawiec, K.: Guarantees of progress for geometric semantic genetic programming. In: Johnson, C., Krawiec, K., Moraglio, A., O'Neill, M. (eds.) Semantic Methods in Genetic Programming. Ljubljana, Slovenia 13 September 2014 (Workshop at Parallel Problem Solving from Nature 2014 Conference). http://www.cs.put.poznan.pl/kkrawiec/smgp2014/uploads/Site/Pawlak.pdf
27. Pawlak, T.P., Wieloch, B., Krawiec, K.: Review and comparative analysis of geometric semantic crossovers. Genet. Program. Evolvable Mach. **16**(3), 351–386 (2015)
28. Pawlak, T.P., Wieloch, B., Krawiec, K.: Semantic backpropagation for designing search operators in genetic programming. IEEE Trans. Evol. Comput. **19**(3), 326–340 (2015). http://dx.doi.org/10.1109/TEVC.2014.2321259
29. Runge, C.: Über empirische funktionen und die interpolation zwischen äquidistanten ordinaten. Z. Math. Phys. **46**, 224–243 (1901)
30. Vanneschi, L., Silva, S., Castelli, M., Manzoni, L.: Geometric semantic genetic programming for real life applications. In: Riolo, R., Moore, J.H., Kotanchek, M. (eds.) Genetic Programming Theory and Practice XI. Genetic and Evolutionary Computation, pp. 191–209. Springer, Heidelberg (2013)
31. Zhu, Z., Nandi, A.K., Aslam, M.W.: Adapted geometric semantic genetic programming for diabetes and breast cancer classification. In: IEEE International Workshop on Machine Learning for Signal Processing (MLSP 2013), September 2013

Semantic Geometric Initialization

Tomasz P. Pawlak$^{(\boxtimes)}$ and Krzysztof Krawiec

Institute of Computing Science, Poznan University of Technology, Poznan, Poland
{tpawlak,krawiec}@cs.put.poznan.pl

Abstract. A common approach in Geometric Semantic Genetic Programming (GSGP) is to seed initial populations using conventional, semantic-unaware methods like Ramped Half-and-Half. We formally demonstrate that this may limit GSGP's ability to find a program with the sought semantics. To overcome this issue, we determine the desired properties of geometric-aware semantic initialization and implement them in Semantic Geometric Initialization (SGI) algorithm, which we instantiate for symbolic regression and Boolean function synthesis problems. Properties of SGI and its impact on GSGP search are verified experimentally on nine symbolic regression and nine Boolean function synthesis benchmarks. When assessed experimentally, SGI leads to superior performance of GSGP search: better best-of-run fitness and higher probability of finding the optimal program.

Keywords: Geometric semantic genetic programming · Semantic initialization · Population

1 Introduction

Geometric Semantic Genetic Programming (GSGP) [16] exploits the spatial properties of program semantics in order to improve the effectiveness of program synthesis. The operators proposed within this branch of genetic programming (GP) have well-understood effects in terms of program behavior on tests, and some of them even guarantee producing programs with semantics that remain in certain geometric relationship with the parent(s). As a result, the dynamics of a GSGP search process is in general more predictable than for conventional GP, GSGP methods are often superior in terms of performance [3,4,16,20,23] and lend themselves conveniently to theoretical analysis [17–20,22].

The majority of research effort in GSGP focuses on search operators, which is not surprising given that successful program synthesis is directly contingent on them. However, like in case of other evolutionary computation algorithms, the performance of GSGP depends also on the starting point of a search process, i.e., on the contents of the initial population. It is often assumed that the exploratory capabilities of evolutionary search weaken that dependency. This claim can be however questioned in GSGP, because of the mentioned above more predictable, more 'directional' and targeted behavior of search operators.

M. Heywood et al. (Eds.): EuroGP 2016, LNCS 9594, pp. 261–277, 2016.
DOI: 10.1007/978-3-319-30668-1_17

Our case in point in this study is the convergent character of semantic geometric crossover (SGX). The exact variant of this search operator [16] is guaranteed to produce an offspring with semantics located in the segment connecting parents' semantics. This, as we showed in [20], implies that for a GP run equipped with SGX only, the set of all semantics that can be reached in a search process is determined by the initial population. A scenario in which the initial population precludes arriving at the program with desired semantics (the *target*) is plausible, and the odds for it grow with the dimensionality of semantics (the number of tests (fitness cases)). This risk was largely ignored in past studies on GSGP, in part due to widespread use of mutation along with crossover. Nevertheless, this effect deserves better understanding. Also, as we will argue further, it may be worth addressing this issue even if mutation accompanies crossover as a search operator.

The main contribution of this study is the observation that an alternative (to mutation) remedy to SGX's high susceptibility to initial conditions is to construct the starting population more carefully. We propose Semantic Geometric Initialization (SGI), a semantically aware initialization method that designs the initial population with the search target in mind. A population initialized with SGI is guaranteed to make the target semantics *achievable* with SGX. SGX, due to its stochastic nature and oblivion to target, is still not guaranteed to synthesize the correct program when initialized with SGI; however, such a success becomes much more likely, as we demonstrate in experimental part of this study.

The following Sect. 2 briefly introduces the necessary formalisms. Section 3 uses that formal framework to identify the problem signaled above, i.e., that initial population imposes strict constraints on the set of semantics that can be reached with SGX. Section 4 presents the SGI algorithm and justifies its design. Section 5 argues that SGI is fundamentally different from population initialization methods proposed in the past (including the semantic-aware ones), and Sect. 6 demonstrates SGI's usefulness empirically on a suite of well-known GP problems. Section 7 discusses the main results, and Sect. 8 summarizes this study and outlines the potential follow-up directions.

2   Background

We define a program $p \in \mathcal{P}$ in a programming language $\mathcal{P}$ as a function that maps a set of inputs $\mathcal{I}$ into a set of outputs $\mathcal{O}$, which we denote by $o = p(in)$, where $in \in \mathcal{I}$ and $o \in \mathcal{O}$. We consider only deterministic programs that feature no side effects, nor memory persistent across executions. Semantics $s \in \mathcal{S}$ is a vector $s = [o_1, o_2, \dots, o_n] \in \mathcal{O}^n$, where we refer to $\mathcal{O}^n$ as *semantic space* (a vector space), and o_i corresponds to ith element in a given n-tuple of program inputs from $\mathcal{I}^n$ that defines the considered program synthesis task. Semantics $s(p)$ of a program p is a vector of p's outputs when executed on a fixed set of inputs $I \subset \mathcal{I}$, i.e., $s(p) = [p(in_1), p(in_2), \dots, p(in_n)]$, $in_i \in I$.

The concept of semantics allows reasoning about program behavior in terms of n-dimensional spaces. Each program has a well-defined semantics, i.e., a point in semantic space $\mathcal{O}^n$. The desired outputs given by a specific synthesis problem uniquely determine a point t in that space called *target* (or *target semantics*). If a fitness function happens to be a metric (which is almost always the case in GP), the fitness landscape defined over $\mathcal{O}^n$ is a unimodal cone with the apex in t [22]. These properties open the door to defining spatial relationships between program semantics, and investigating whether particular search operators obey them or not and what is the impact of those properties on the efficiency of search. The concepts of particular importance here are geometric mutation and geometric crossover.

Definition 1. *Given a parent program p, an r-geometric mutation is an operator that produces an offspring p' with semantics $s(p')$ in a ball of radius r centered in $s(p)$, i.e., $\|s(p), s(p')\| \leq r$.*

Definition 2. *Given parent programs p_1, p_2, a geometric crossover is an operator that produces an offspring p' with semantics $s(p')$ in a segment between $s(p_1)$ and $s(p_2)$, i.e., $\|s(p_1), s(p_2)\| = \|s(p_1), s(p')\| + \|s(p'), s(p_2)\|$.*

A crossover operator with the above property has the ideal 'mixing' characteristics: the semantics of the offspring is located 'in between' of parents' semantics. This is in stark contrast to the highly unpredictable semantics of programs produced by conventional search operators (see, e.g., the arguments in [11]).

The quest for practical algorithms that implement geometric search operators lasted for several years. Among others, multiple approximately geometric crossovers have been proposed [10–12, 20, 23, 24]. The breakthrough came with publication of [16], where exact versions of geometric crossover and mutation have been proposed, defined as follows for particular domains.

Definition 3. *(Algorithms for geometric mutation, SGM) Symbolic regression: Given parent arithmetic program p, an offspring is a program $p' = p + r(m_1 - m_2)$, where m_1 and m_2 are random arithmetic programs that output values in range [0,1]. Boolean domain: Given Boolean parent program p, an offspring is a program $p' = m \vee p$ with probability 0.5, $p' = \overline{m} \wedge p$ otherwise, where m is a random minterm.*

Definition 4. *(Algorithms for geometric crossover, SGX) Symbolic regression: Given parent arithmetic programs p_1, p_2, an offspring is a program $p' = mp_1 + (1 - m)p_2$, where m is a random arithmetic program that returns values in range [0,1]. Boolean domain: Given Boolean parent programs p_1, p_2, an offspring is a program $p' = (p_1 \wedge m) \vee (\overline{m} \wedge p_2)$, where m is a random Boolean program.*

3   The Problem

The SGM and SGX operators are geometric by construction: the offspring is guaranteed to be geometric with respect to parents in the sense of Definitions 1 and 2. Expectedly, they deliver superior search performance in all domains, as shown experimentally in [16] and further studies. However, SGM is the key to that success: it is indispensable for good performance, while SGX used alone sometimes fails to converge to the sought program [16,23].

The reason for this state of affairs is the 'centripetal' character of SGX, which can produce only the offspring with semantics in the segment connecting parents' semantics. On one hand, this is highly desirable given that fitness landscape in semantic space is unimodal: an application of SGX to any pair of solutions in that space is guaranteed to produce an offspring with attractive properties – for instance for fitness and operator's metric being Euclidean distance, an offspring that is not worse that the worse of the parents [22]. On the other hand however, if SGX is the only search operator used for program synthesis, the set of semantics achievable from a given initial population $P \subset \mathcal{P}$ is limited to the convex hull spanning the semantics of programs in P, since that convex hull incorporates all segments between semantics of all pairs of programs in P. Formally (cf. [20]):

Theorem 1. *Consider a population P_1 of programs and a search process that starts from P_1 and uses SGX to generate subsequent generations of programs. A program having semantics t can be achieved in that search process iff the convex hull of P_1 includes t.*

Proof. Let P_g be population in generation $g \geq 1$, C_g be convex hull of semantics of programs in P_g. For all given pairs of parent programs $p_1, p_2 \in P_g$, a semantics $s(p')$ of an offspring p' is included in a segment of $s(p_1)$ and $s(p_2)$ that in turn is included in C_g. The set of all offspring $P_{g+1} \subseteq C_g$ constitutes a population of generation $g + 1$. By the non-decreasing property of convex hull, $C_{g+1} \subseteq C_g$. By induction, $C_{g+1} \subseteq C_g \subseteq ... \subseteq C_1$. Hence, semantics t can be achieved in generation g iff $t \in C_g \subseteq C_1$.

The choice of 't' as the symbol denoting the semantic in question is not incidental: the above theorem applies in particular to the target, with profound consequences. If t happens to be located outside the convex hull, applications of SGX to P, regardless how many, cannot lead to a program with semantics t. Unfortunately, this is relatively likely for populations initialized in conventional, semantic-unaware ways. For instance, for symbolic regression, the semantics of programs generated by means of the popular Ramped Half-and-Half (RHH) method [9] tend to initialize programs with relatively simple semantics, typically clustered around the origin of coordinate system of semantic space [13]. A target located far away from that origin is likely to be outside the convex hull and thus unreachable by the actions of SGX.

This observation, though rarely formalized in the above way in past literature, was one of the reasons for using mutation operators alongside with SGX (the other reason being SGX's propensity to produce large programs). The natural operator of choice is in this context SGM, as it is provably capable of reaching the target from arbitrary starting location in semantic space (even when used alone; see the semantic stochastic hill climber in [16]). However, it may require multiple iterations and by this token produce large programs. In the following, we propose an alternative way of making target achievable for SGX.

4   Semantic Geometric Initialization

In the light of Sect. 3 it becomes evident that the target becomes reachable for SGX once it belongs to the convex hull spanning the semantics of programs in initial population. In this section, we propose Semantic Geometric Initialization (SGI), a method that achieves that goal by means of semantic- and geometry-aware population initialization.

Algorithm 1. Calculation of a set of semantics such that their convex hull in $\mathcal{O}^n$ encloses the target t. $popsize$ is desired population size.

1: **function** WRAP($t, popsize$)
2:     $S = \emptyset$ ▷ Output set
3:     $n = |t|$ ▷ Dimensionality of semantic space
4:     $I = \{1, \ldots, n\}$ ▷ Indices of all dimensions
5:     $k = 1$ ▷ Number of dimensions to change in t
6:     **while** $k \leq n$ **do**
7:         **for** $I' \in \binom{I}{k}$ **do**   ▷ $I' = k$-element combination of dimensions
8:             **for** $b \in \{0, 1\}^k$ **do**   ▷ b = combination of directions on dimensions in I'
9:                 $s \leftarrow t$
10:                 **for** $i \in I'$ **do**
11:                     **if** $b_i = 1$ **then**
12:                         $s_i \leftarrow$ ADDONE(t_i)
13:                     **else**
14:                         $s_i \leftarrow$ SUBTRACTONE(t_i)
15:                 **if** $s \notin S$ **then**
16:                     $S = S \cup \{s\}$
17:                     **if** $|S| = popsize$ **then**
18:                         **return** S
19:         $k \leftarrow k + 1$
20:     **return** S ▷ $|S| < popsize$, population is not filled

---

The input to the method is the target t. The algorithm proceeds in two steps:

1. Use the function WRAP (Algorithm 1) to generate a set of semantics $S \subset \mathcal{O}^n$ such that the convex hull of S encloses target t, i.e., $t \in C(S)$,
2. For each semantics $s_i \in S$, synthesize a program p such that $s(p) = s_i$.

The realization of each step is domain-dependent. For the first step we provide an abstract Algorithm 1 that will be specialized in the following for symbolic regression and Boolean function synthesis domains. It calculates set of semantics that wrap t in their convex hull in $\mathcal{O}^n$ semantic space. The algorithm iteratively constructs new semantics by modifying k components (dimensions) of the target t. For each subset of k components of t, the algorithm attempts to construct 2^k semantics by applying to these components all combinations of two domain-dependent functions: ADDONE and SUBTRACTONE. We gather the resulting semantics in the output set S while discarding duplicates. We start with $k = 1$ and increment it each time all k-component sets have been considered.

For the **symbolic regression** domain, we define ADDONE and SUBTRACTONE functions using arithmetic operations:

$$\text{ADDONE}(t_{a_i}) \equiv t_{a_i} + 1 \qquad \text{SUBTRACTONE}(t_{a_i}) \equiv t_{a_i} - 1$$

The WRAP algorithm, when instantiated with these functions and invoked with $popsize \geq 2$, is guaranteed to construct a set of semantics S such that $t \in C(S)$. In other words, there exists a dimension i of t and semantics s^1 and s^2, such that $s_i^1 < t_i < s_i^2$ and $\forall_{j \neq i} s_j^1 = t_j = s_j^2$. Thus, under any metric, t is included in the segment between s^1 and s^2 – a degenerated case of a convex hull.

Concerning program synthesis, for each semantics $s \in S$ calculated by Algorithm 1, SGI constructs a program using multivariate polynomial interpolation as described in [26]. The set of points used in interpolation comes from the set of program inputs $in \in I$ on which program's semantics is to be calculated and corresponding components of s, i.e., (in_i, s_i), $k = 1, \ldots, n$.

For the **Boolean function synthesis**, the definitions of ADDONE and SUBTRACTONE are Boolean counterparts of the above arithmetic operations, limited to the corners of the unit hypercube $\{0, 1\}^n$. Since there are only two values in Boolean domain: 0 (false) and 1 (true), the Boolean addition of 1 results in 1 (i.e., $q \vee 1 \equiv 1$), the Boolean subtraction of 1 results in 0 (i.e., $q \wedge 0 \equiv 0$) for any term, and the functions reduce to constants:

$$\text{ADDONE}(t_{a_i}) \equiv 1 \qquad \text{SUBTRACTONE}(t_{a_i}) \equiv 0$$

For $popsize \geq 2$ WRAP is guaranteed to include the target t in $C(S)$, because there exists a dimension i of t and semantics s^1 and s^2, such that $s_i^1 = t_i \neq s_i^2$ or $s_i^1 \neq t_i = s_i^2$ and $\forall_{j \neq i} s_j^1 = t_j = s_j^2$. SGI synthesizes the Boolean programs p_i for semantics $s_i \in S$ using the following formula:

$$p_i = \bigvee_{j=1..n \,:\, t_j=1} y_j, \text{ where } y_j = \bigwedge_{k=1..n} \begin{cases} x_k & \text{if } in_{jk} \\ \overline{x_k} & \text{if } \overline{in_{jk}} \end{cases}, \tag{1}$$

where in_{jk} is a value of kth variable of jth input used to calculate semantics, y_j is a minterm that is 1 for jth input and x_k is kth program argument.

5   Related Work

To our knowledge, SGI is the first semantic *and geometric* population initialization method in GP, however not the first semantic method of this kind.

Looks in [13] proposed Semantic Sampling (SS) heuristic that produces a population of semantically unique Boolean programs with uniformly distributed program sizes. SS partitions a population into bins by program size and fills them up to assumed capacity by semantically unique programs. SGI differs from SS in its awareness of geometry of semantic space. Also, SGI can operate in any

Table 1. Parameters of evolutionary algorithms.

	Symbolic regression	Boolean domain
Number of runs	30	
Population size	1000	
Termination condition	50 generations or fitness = 0	
Selection method	Tournament selection of size 7	
Fitness function	L_2 metric	L_1 metric
Instructions	$x_1, x_2, +, -, \times, /, \sin, \cos, \exp, \log^a$	$x_1, x_2, \ldots,^b$ and, or, nand, nor

^{a} log is defined as $\log|x|$; $/$ returns 0 if divisor is 0.
^{b} The number of inputs depends on a particular problem instance

Table 2. Benchmark problems.

	Problem	Definition (formula)	Variables	Range	Size
Symbolic regression	R1	$(x_1+1)^3/(x_1^2-x_1+1)$	1	$\langle-1,1\rangle$	20
	R2	$(x_1^5-3x_1^3+1)/(x_1^2+1)$	1	$\langle-1,1\rangle$	20
	R3	$(x_1^6+x_1^5)/(x_1^4+x_1^3+x_1^2+x_1+1)$	1	$\langle-1,1\rangle$	20
	Kj1	$0.3x_1\sin(2\pi x_1)$	1	$\langle-1,1\rangle$	20
	Kj4	$x_1^3 e^{-x_1}\cos(x_1)\sin(x_1)(\sin^2(x_1)\cos(x_1)-1)$	1	$\langle0,10\rangle$	20
	Ng9	$\sin(x_1)+\sin(x_2^2)$	2	$\langle0,1\rangle^2$	100
	Ng12	$x_1^4-x_1^3+\frac{x_2^2}{2}-x_2$	2	$\langle0,1\rangle^2$	100
	Pg1	$1/(1+x_1^{-4})+1/(1+x_2^{-4})$	2	$\langle-5,5\rangle^2$	100
	Vl1	$e^{-(x_1-1)^2}/(1.2+(x_2-2.5)^2)$	2	$\langle0,6\rangle^2$	100

	Problem	Instance	Variables	Size
Boolean domain	Even parity	Par5	5	32
		Par6	6	64
		Par7	7	128
	Multiplexer	Mux6	6	64
		Mux11	11	2048
	Majority	Maj7	7	64
		Maj8	8	128
	Comparator	Cmp6	6	64
		Cmp8	8	256

domain for which the ADDONE and SUBTRACTONE functions can be defined (and efficiently computed).

Beadle and Johnson [1] proposed Semantically Driven Initialization (SDI) that fills population with semantically unique programs. SDI was designed for Boolean and artificial ant problems and uses a reduced ordered binary decision diagram or a sequence of ant moves, respectively, as the representation of program's semantics. Like Ss, SDI does not engage geometric considerations.

An approach called Behavioral Initialization (BI) was proposed by Jackson [7]. BI is a wrapper on RHH that accepts a program created by RHH if it is semantically unique in the population being initialized. Although domain-independent, BI is oblivious to geometry of the semantic space.

Pawlak proposed Competent Initialization (CI) [20] for symbolic regression and Boolean domains. CI repetitively invokes SDI and accepts (i.e., adds to the population) the created program if its semantics expands the convex hull of the semantics of programs already present in the population. CI is geometric in the limit of population size approaching infinity, but in contrast to SGI does not guarantee including the target in the convex hull.

6 Experimental Verification

We compare SGI to Ramped Half-and-Half (RHH) [9] – the arguably most common population initialization method in GSGP and in GP in general. We run two groups of GSGP setups, with SGI and with RHH as the initialization operator, to determine the advantages and disadvantages of using SGI. In both groups, we consider a setup with SGX [16] as the only search operator to verify whether inclusion of the target in the convex hull of the initial population increases SGX's ability of reaching it (cf. Sect. 3). The second setup in each group uses both SGX and SGM [16], in a configuration that is most commonly practiced in GSGP. In addition, we run a canonical control setup, with RHH and traditional non-semantic search operators. Overall, there are thus five setups:

> SGIX – SGI accompanied with SGX only,
> SGIXM – SGI with SGX and SGM in proportions 50 : 50,
> RHHX – RHH with SGX only,
> RHHXM – RHH with SGX and SGM in proportions 50 : 50,
> RHHTXTM – RHH with tree crossover and tree mutation [9] in proportions 90:10.

Table 1 sums up the parameter settings; other parameters are set to ECJ [14] defaults.

We compare the setups on nine uni- and bi-variate symbolic regression problems, and nine Boolean function synthesis problems. The problems come from [15, 23] and are summarized Table 2. In univariate symbolic regression problems, 20 Chebyshev nodes[1] [2] are used for training, and 20 uniformly sampled points for testing. For bivariate problems 10 values are picked in analogous way for each input variable and the Cartesian product of them constitutes a data set. Points are selected from the ranges shown in the table. For the Boolean benchmarks, training sets enumerate all combinations of inputs and there are no testing sets.

Training Set Performance. Figure 1 and Table 3 present average and .95-confidence interval of the best-of-generation and the best-of-run program, respectively. Both SGI* setups begin from a relatively low fitness of about 1 in all problems in both problem domains. This phenomenon originates in the

---

[1] Points given by $x_k = \frac{1}{2}(a+b) + \frac{1}{2}(b-a)\cos(\frac{2k-1}{2n}\pi), k = 1..n$, where $[a, b]$ is the range of training set, and n is number of data points. Using Chebyshev nodes minimizes the likelihood of Runge's phenomenon [25].

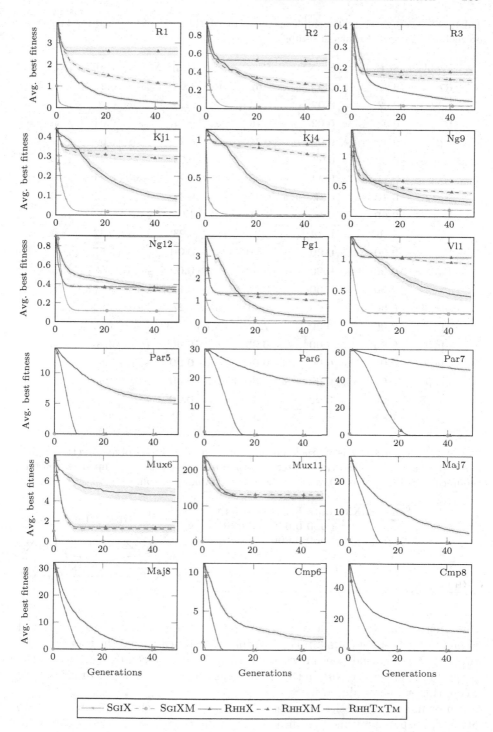

Fig. 1. Average and .95-confidence interval of the best-of-generation fitness.

Table 3. Average and .95-confidence interval of the best-of-run fitness. Last row presents the averaged ranks of setups.

Problem	SGIX	SGIXM	RHHX	RIIIIXM	RHHTXTM
R1	0.02 ±0.00	**0.02** ±0.00	2.62 ±0.20	1.05 ±0.08	0.24 ±0.07
R2	0.02 ±0.00	**0.02** ±0.00	0.53 ±0.06	0.26 ±0.02	0.20 ±0.03
R3	0.02 ±0.00	**0.02** ±0.00	0.18 ±0.02	0.14 ±0.01	0.04 ±0.01
Kj1	0.02 ±0.00	**0.02** ±0.00	0.34 ±0.02	0.29 ±0.02	0.08 ±0.02
Kj4	**0.02** ±0.00	0.02 ±0.00	0.95 ±0.03	0.79 ±0.03	0.26 ±0.04
Ng9	0.11 ±0.01	**0.11** ±0.01	0.59 ±0.05	0.39 ±0.03	0.24 ±0.06
Ng12	**0.12** ±0.01	0.12 ±0.01	0.37 ±0.02	0.33 ±0.02	0.35 ±0.04
Pg1	**0.09** ±0.00	0.10 ±0.00	1.31 ±0.08	1.00 ±0.06	0.28 ±0.08
Vl1	0.16 ±0.01	**0.14** ±0.01	1.03 ±0.03	0.93 ±0.02	0.42 ±0.08
Par5	**0.00** ±0.00	**0.00** ±0.00	**0.00** ±0.00	**0.00** ±0.00	5.57 ±0.56
Par6	**0.00** ±0.00	**0.00** ±0.00	**0.00** ±0.00	**0.00** ±0.00	17.93 ±0.90
Par7	**0.00** ±0.00	**0.00** ±0.00	0.17 ±0.19	0.10 ±0.11	47.77 ±1.22
Mux6	**0.00** ±0.00	**0.00** ±0.00	1.40 ±0.34	1.27 ±0.37	4.63 ±0.72
Mux11	**0.00** ±0.00	**0.00** ±0.00	125.50 ±9.63	132.33 ±11.25	122.73 ±6.43
Maj7	**0.00** ±0.00	**0.00** ±0.00	**0.00** ±0.00	**0.00** ±0.00	3.20 ±0.62
Maj8	**0.00** ±0.00	**0.00** ±0.00	**0.00** ±0.00	**0.00** ±0.00	0.67 ±0.32
Cmp6	**0.00** ±0.00	**0.00** ±0.00	**0.00** ±0.00	**0.00** ±0.00	1.40 ±0.41
Cmp8	**0.00** ±0.00	**0.00** ±0.00	0.03 ±0.06	**0.00** ±0.00	11.93 ±1.08
Rank:	1.89	1.72	4.08	3.36	3.94

Table 4. Post-hoc analysis of Friedman's test on Table 3: p-values of incorrectly judging a setup in a row as achieving better best-of-run fitness than a setup in a column. Significant values ($p < 0.05$) are visualized as outranking graph.

	SGIX	SGIXM	RHHX	RHHXM	RHHTXTM
SGIX			0.000	0.020	0.000
SGIXM	0.997		0.000	0.006	0.000
RHHX					
RHHXM			0.567		0.748
RHHTXTM			0.999		

construction of ADDONE and SUBTRACTONE formulas that cause the initial population to consists of multiple programs at distance 1 from the target. These superior starting conditions are especially evident in Boolean domain, where the programs produced by RHH in the initial generation are 1–2 orders of magnitude worse than those produced by SGI.

In symbolic regression problems GSGP clearly benefits from SGI. We observe steep improvement of fitness in the first ten generations that gradually slows down and finally stops after about 20 generations. The initial rate of

Table 5. Probability and .95-confidence interval of success over problems. Problems that were not solved at least once are not shown. Last row presents the averaged ranks of setups.

Problem	SGIX	SGIXM	RHHX	RHHXM	RHHTXTM
Ng9	0.00 ±0.00	0.00 ±0.00	0.00 ±0.00	0.00 ±0.00	**0.07** ±0.09
Par5	**1.00** ±0.00	**1.00** ±0.00	**1.00** ±0.00	**1.00** ±0.00	0.00 ±0.00
Par6	**1.00** ±0.00	**1.00** ±0.00	**1.00** ±0.00	**1.00** ±0.00	0.00 ±0.00
Par7	**1.00** ±0.00	**1.00** ±0.00	0.90 ±0.11	0.90 ±0.11	0.00 ±0.00
Mux6	**1.00** ±0.00	**1.00** ±0.00	0.17 ±0.13	0.23 ±0.15	0.00 ±0.00
Mux11	**1.00** ±0.00	**1.00** ±0.00	0.00 ±0.00	0.00 ±0.00	0.00 ±0.00
Maj7	**1.00** ±0.00	**1.00** ±0.00	**1.00** ±0.00	**1.00** ±0.00	0.03 ±0.06
Maj8	**1.00** ±0.00	**1.00** ±0.00	**1.00** ±0.00	**1.00** ±0.00	0.53 ±0.18
Cmp6	**1.00** ±0.00	**1.00** ±0.00	**1.00** ±0.00	**1.00** ±0.00	0.30 ±0.16
Cmp8	**1.00** ±0.00	**1.00** ±0.00	0.97 ±0.06	**1.00** ±0.00	0.00 ±0.00
Rank:	2.58	2.58	3.08	2.92	3.83

Table 6. Post-hoc analysis of Friedman's test on Table 5: p-values of incorrectly judging a setup in a row as having higher probability of success than a setup in a column. Significant values ($p < 0.05$) are visualized as outranking graph.

	SGIX	SGIXM	RHHX	RHHXM	RHHTXTM
SGIX	1.000	0.503	0.827		**0.001**
SGIXM		0.503	0.827		**0.001**
RHHX				0.119	
RHHXM			0.984		**0.029**
RHHTXTM					

improvement is slower for RHHX and RHHXM, which gradually improve in the first 5–8 generations and then saturate. In Boolean domain both SGI* setups find the optimum in 1–2 generations, while RHH* GSGP setups need 10–25 generations to solve all problems, except for multiplexers.

There is no noticeable difference in generational characteristic between both SGI* setups. In contrast, both RHH* GSGP setups differ noticeably: RHHXM fares better in most symbolic regression problems. In the Boolean domain, there are no significant differences. Best-of-generation fitness of RHHTXTM is in between those of the GSGP setups for greater part of runs for the symbolic regression problems, and worse than all GSGP setups for the Boolean problems.

Friedman's test [8] on the best-of-run fitness signals significant differences between setups ($p = 9.68 \times 10^{-6}$), so we conduct post-hoc analysis with symmetry test [6]. Table 4 presents the p-values for the hypothesis that a setup in a row is better than a setup in a column, and the graph of significant outrankings. The setups initialized with SGI are significantly better than all other setups.

Table 5 presents the empirical probability of solving problems by particular setups, which we define as achieving best-of-run fitness lower than 2^{-23}

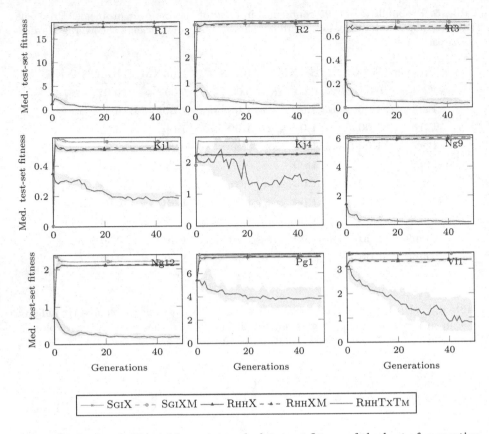

Fig. 2. Median and .95-confidence interval of test set fitness of the best-of-generation program on training set.

Table 7. Median and .95-confidence interval of test set fitness of the best-of-run program on training set. Last row presents the averaged ranks of setups.

Problem	SGIX		SGIXM		RHHX		RHHXM		RHHTXTM	
R1	18.40	±0.00	18.40	±0.00	17.53	±0.25	18.39	±0.12	**0.21**	±0.07
R2	3.38	±0.00	3.38	±0.00	3.29	±0.08	3.34	±0.02	**0.10**	±0.06
R3	0.71	±0.00	0.72	±0.00	0.66	±0.02	0.69	±0.01	**0.03**	±0.02
Kj1	0.56	±0.00	0.56	±0.00	0.51	±0.02	0.52	±0.02	**0.18**	±0.04
Kj4	2.66	±0.00	2.66	±0.00	2.24	±0.03	2.25	±0.04	**1.38**	±0.89
Ng9	6.08	±0.00	6.08	±0.00	5.91	±0.06	6.02	±0.03	**0.16**	±0.07
Ng12	2.21	±0.00	2.21	±0.00	2.10	±0.02	2.13	±0.03	**0.19**	±0.06
Pg1	7.55	±0.00	7.55	±0.01	7.37	±0.10	7.45	±0.19	**3.79**	±0.90
Vl1	3.62	±0.01	3.62	±0.01	3.38	±0.04	3.37	±0.06	**0.80**	±0.57
Rank:	4.44		4.56		2.11		2.89		1.00	

Table 8. Post-hoc analysis of Friedman's test on Table 7: p-values of incorrectly judging a setup in a row as achieving better test set fitness than a setup in a column. Significant values ($\alpha = 0.05$) visualized as outranking graph.

	SGIX	SGIXM	RHHX	RHHXM	RHHTXTM
SGIX		1.000			
SGIXM					
RHHX	**0.015**	**0.009**		0.835	
RHHXM	0.226	0.166			
RHHTXTM	**0.000**	**0.000**		0.569	0.083

Table 9. Average and .95-confidence interval of number of nodes in the best of run program. Values $\geq 10^4$ are rounded to an order of magnitude. Last row presents the averaged ranks of setups.

Problem	SGIX		SGIXM		RHHX		RHHXM		RHHTXTM	
R1	10^{17}	$\pm 10^{16}$	10^{17}	$\pm 10^{16}$	10^{14}	$\pm 10^{14}$	10^{14}	$\pm 10^{14}$	**103.73**	± 13.76
R2	10^{17}	$\pm 10^{16}$	10^{17}	$\pm 10^{16}$	10^{14}	$\pm 10^{14}$	10^{15}	$\pm 10^{14}$	**69.90**	± 12.79
R3	10^{17}	$\pm 10^{16}$	10^{17}	$\pm 10^{16}$	10^{14}	$\pm 10^{14}$	10^{15}	$\pm 10^{14}$	**123.37**	$+16.20$
Kj1	10^{17}	$\pm 10^{16}$	10^{17}	$\pm 10^{16}$	10^{14}	$\pm 10^{14}$	10^{15}	$\pm 10^{14}$	**115.70**	± 11.87
Kj4	10^{17}	$\pm 10^{16}$	10^{17}	$\pm 10^{16}$	10^{15}	$\pm 10^{14}$	10^{15}	$\pm 10^{14}$	**127.37**	± 14.44
Ng9	10^{17}	$+10^{16}$	10^{17}	$+10^{16}$	10^{15}	$\pm 10^{14}$	10^{15}	$\pm 10^{14}$	**65.07**	± 10.01
Ng12	10^{17}	$\pm 10^{16}$	10^{17}	$\pm 10^{16}$	10^{15}	$\pm 10^{14}$	10^{15}	$\pm 10^{14}$	**52.50**	± 12.88
Pg1	10^{17}	$\pm 10^{16}$	10^{17}	$\pm 10^{16}$	10^{15}	$\pm 10^{14}$	10^{15}	$\pm 10^{14}$	**77.07**	± 10.70
Vl1	10^{17}	$\pm 10^{16}$	10^{17}	$\pm 10^{16}$	10^{14}	$\pm 10^{14}$	10^{15}	$\pm 10^{14}$	**97.23**	± 13.71
Par5	**199.00**	± 0.00	**199.00**	± 0.00	10^4	± 1905.85	10^4	± 1458.59	321.80	± 26.53
Par6	694.90	± 25.58	700.10	± 20.84	10^6	$\pm 10^5$	10^6	$\pm 10^5$	**353.27**	± 34.98
Par7	1458.30	± 42.05	1461.20	± 42.64	10^{15}	$\pm 10^{15}$	10^{15}	$\pm 10^{15}$	**378.27**	± 39.02
Mux6	354.30	± 5.46	354.63	± 9.73	10^{15}	$\pm 10^{15}$	10^{16}	$\pm 10^{15}$	**186.13**	± 30.18
Mux11	9563.97	± 18.39	9416.97	± 219.98	10^{16}	$\pm 10^{14}$	10^{16}	$\pm 10^{15}$	**159.13**	± 19.26
Maj7	1400.20	± 23.91	1401.30	± 24.95	10^5	$\pm 10^4$	10^5	$\pm 10^4$	**406.13**	± 35.64
Maj8	1739.00	± 0.00	1739.00	± 0.00	10^5	± 5010.51	10^5	± 5096.79	**313.20**	± 34.53
Cmp6	579.77	± 13.04	578.87	± 9.58	1899.90	± 294.67	2350.37	± 354.01	**245.73**	± 25.83
Cmp8	2949.40	± 9.08	2892.53	± 38.50	10^{14}	$\pm 10^{14}$	10^5	$\pm 10^5$	**249.87**	± 38.20
Rank:	3.28		3.61		3.17		3.83		1.11	

Table 10. Post-hoc analysis of Friedman's test on Table 9: p-values of incorrectly judging a setup in a row as producing smaller programs than a setup in a column. Significant values ($\alpha = 0.05$) visualized as outranking graph.

	SGIX	SGIXM	RHHX	RHHXM	RHHTXTM
SGIX		0.970		0.828	
SGIXM				0.993	
RHHX	1.000	0.916		0.711	
RHHXM					
RHHTXTM	**0.000**	**0.000**		**0.001**	**0.000**

(the difference between 1.0 and the closest IEEE754 single precision number). Both SGI* setups solve all Boolean problems, while the setups using RHH for population initialization do not solve Mux11 in any run, and do not always solve the other 3 out of 9 Boolean problems. The probabilities are a bit higher for the setup that uses mutation. On the other hand, in symbolic regression, none of the GSGP setups solved any of problems and only Ng9 problem is solved by RHHTxTM. Friedman's test ($p = 6.73 \times 10^{-4}$) and post-hoc analysis in Table 6 show that both SGI* setups and RHHXM are better than RHHTxTM, however the evidence is too weak to conclude about the differences between SGI and RHH in GSGP.

Test-Set Performance. Figure 2 and Table 7 present the median and .95-confidence interval of test-set fitness of the best-of-generation and the best-of-run program on the training set, respectively. All GSGP setups significantly overfit to training data: the test set fitness quickly increases in early generations and remains high for the rest of runs. The values are slightly worse for the SGI* setups. This observation is consistent with previous studies, [20,23] and can be attributed to the well-known *bias-variance dilemma* [5] in machine learning. The use of SGI leads to better adaptation to training data, and thus reduces bias. That in turn increases the variance of performance on the unknown test data, and makes GP more prone to overfitting. The conventional RHHTxTM setup is the only one that generalizes well to the test set. Friedman's test ($p = 2.18 \times 10^{-5}$) and post-hoc analysis in Table 8 confirm: RHHTxTM and RHHX are better than both SGI* setups.

Program Size. Table 9 presents the average and .95-confidence interval of the number of nodes in the best-of-run programs. The exponential growth of SGX's offspring is clearly visible in the data. For the setups that involve that operator, we report the total number of nodes/instructions in 'unrolled' trees. The actual number of unique program nodes held in memory is many orders of magnitude lower, because a given program may refer to its ancestor programs multiple times, due to the 'aggregative' nature of exact semantic operators (see Definitions 3 and 4).

In contrast, RHHTxTM produces programs smaller than 500 nodes. For the setups initialized with SGI, it is worth noting that they produce large programs only for problems that they failed to solve in some of runs (cf. Table 5). For the remaining problems, the average number of nodes in a program is smaller, however still bigger than of RHHTxTM. An exception is Par5 problem, for which both SGI* setups produce the smallest programs. Fortunately, use of SGI increases the probability of success and thus rises the likelihood of producing small programs. Friedman's test ($p = 2.74 \times 10^{-6}$) and post-hoc analysis in Table 10 confirm that RHHTxTM produces smallest programs, but there is no sufficient evidence to reveal the differences between the remaining setups.

7   Discussion

The results presented above clearly corroborate the importance of population initialization for GSGP. In particular, the geometric and semantic-aware initialization offered by SGI brings to zero the differences of the setups that use and do not use SGM (i.e., SGIX and SGIXM), on virtually all performance indicators (fitness, test-set performance, program size). In other words, the convex hull of semantics built around the target by SGI makes the use of SGM optional. We hypothesize that this characteristics may be convenient in scenarios where SGM is slow at traversing search space, and where SGX may be better in that respect. Concerning low generalization capabilities, they are not due to SGI, but are common to all setups that involve SGM and SGX, and reveal the more fundamental problem of semantic geometric GP, i.e., its inherent lack of bias resulting in high variance [5] (Sec. 6).

A vigilant reader might have noticed the seemingly paradoxical feature of the proposed initialization technique. SGI employs *exact* techniques to synthesize the programs that support the convex hull around the target (multivariate polynomial interpolation for the symbolic regression domain and disjunctive normal forms for the Boolean domain). Then it relies on *heuristic* GP search to synthesize the program that solves the original task, i.e., has semantics in the target. It does not take long to realize that the above exact techniques could be directly used to synthesize the sought program, without using GP altogether.

Note however that the above paradox applies to the entire domain of GP, and not only to SGI or this particular study. Our goal was to verify the relevance of geometric and semantic-aware initialization for search conducted by means of GSGP, and the empirical evidence gathered here confirms the theoretical suppositions. We explore the possibility of one-step construction of a perfect program from a population in another study published in this very volume [21].

SGI offers certain advantages for program size in the Boolean domain. As it follows from Table 9, GSGP starting from traditionally initialized populations (with RHH) may grow monstrously large programs before reaching the target (even when using SGM, which is known to increase program size only by fixed factor in every application, compared to the exponential growth of SGX). When an initial population forms a convex hull around the target, a few moves of SGM and/or SGX may be sufficient to solve a synthesis task. We hypothesize thus that the positive impact of SGI is not only due to its convex hull property, but also due to the proximity of the initial population to the target.

8   Conclusions

We have brought theoretical evidence that the possibility of finding a program with the optimal semantics by GSGP running solely geometric crossover depends on whether the convex hull spanning the programs in the initial population includes the search target. Experimental verification has shown that the above is true also for GSGP equipped with crossover and mutation. The commonly used

RHH initialization does not provide this guarantee. To overcome this problem, we provided the SGI algorithm that seeds the initial population with programs that form an appropriate convex hull.

Further work is needed to develop SGI for other domains than those considered in this paper, e.g., for the categorical one. Another interesting research topic is the influence of the initial distribution of programs' semantics on the analogous distributions in subsequent populations and search performance. Last but not least, the convex hull property is only the *necessary* condition to reach the target. It remains an open question how to efficiently prevent GSGP operators from losing the target from the population's convex hull.

Acknowledgements. T. Pawlak acknowledges support from grant no. DEC-2012/07/N/ST6/03066, K. Krawiec from grant no 2014/15/B/ST6/05205, both funded by the National Science Centre, Poland.

References

1. Beadle, L., Johnson, C.G.: Semantic analysis of program initialisation in genetic programming. Genet. Program Evolvable Mach. **10**(3), 307–337 (2009)
2. Burden, R., Faires, J.: Numerical Analysis. Cengage Learning, Boston (2010)
3. Castelli, M., Henriques, R., Vanneschi, L.: A geometric semantic genetic programming system for the electoral redistricting problem. Neurocomputing **154**, 200–207 (2015)
4. Castelli, M., Vanneschi, L., Silva, S.: Prediction of the unified Parkinson's disease rating scale assessment using a genetic programming system with geometric semantic genetic operators. Expert Syst. Appl. **41**(10), 4608–4616 (2014)
5. Geman, S., Bienenstock, E., Doursat, R.: Neural networks and the bias/variance dilemma. Neural Comput. **4**(1), 1–58 (1992)
6. Hothorn, T., Hornik, K., van de Wiel, M.A., Zeileis, A.: Package 'coin': conditional inference procedures in a permutation test framework (2015). http://cran.r-project.org/web/packages/coin/coin.pdf
7. Jackson, D.: Phenotypic diversity in initial genetic programming populations. In: Esparcia-Alcázar, A.I., Ekárt, A., Silva, S., Dignum, S., Uyar, A.Ş. (eds.) EuroGP 2010. LNCS, vol. 6021, pp. 98–109. Springer, Heidelberg (2010)
8. Kanji, G.: 100 Statistical Tests. SAGE Publications, London (1999)
9. Koza, J.R.: Genetic Programming: On the Programming of Computers by Means of Natural Selection. MIT Press, Cambridge (1992)
10. Krawiec, K.: Medial crossovers for genetic programming. In: Moraglio, A., Silva, S., Krawiec, K., Machado, P., Cotta, C. (eds.) EuroGP 2012. LNCS, vol. 7244, pp. 61–72. Springer, Heidelberg (2012)
11. Krawiec, K., Lichocki, P.: Approximating geometric crossover in semantic space. In: GECCO 2009, 8–12 July 2009, pp. 987–994. ACM, Montreal (2009)
12. Krawiec, K., Pawlak, T.: Locally geometric semantic crossover: a study on the roles of semantics and homology in recombination operators. Genet. Program Evolvable Mach. **14**(1), 31–63 (2013)
13. Looks, M.: On the behavioral diversity of random programs. In: GECCO 2007, 7–11 July 2007, vol. 2, pp. 1636–1642. ACM Press, London (2007)

14. Luke, S.: The ECJ Owner's Manual - A User Manual for the ECJ Evolutionary Computation Library, zeroth edition, October 2010. http://cs.gmu.edu/~eclab/projects/ecj/docs/
15. McDermott, J., White, D.R., Luke, S., Manzoni, L., Castelli, M., Vanneschi, L., Jaskowski, W., Krawiec, K., Harper, R., De Jong, K., O'Reilly, U.M.: Genetic programming needs better benchmarks. In: GECCO 2012, 7–11 July 2012, pp. 791–798. ACM, Philadelphia, Pennsylvania, USA (2012)
16. Moraglio, A., Krawiec, K., Johnson, C.G.: Geometric semantic genetic programming. In: Coello, C.A.C., Cutello, V., Deb, K., Forrest, S., Nicosia, G., Pavone, M. (eds.) PPSN 2012, Part I. LNCS, vol. 7491, pp. 21–31. Springer, Heidelberg (2012)
17. Moraglio, A., Mambrini, A.: Runtime analysis of mutation-based geometric semantic genetic programming for basis functions regression. In: GECCO 2013, 6–10 July 2013, pp. 989–996. ACM, Amsterdam, The Netherlands (2013)
18. Moraglio, A., Mambrini, A., Manzoni, L.: Runtime analysis of mutation-based geometric semantic genetic programming on Boolean functions. In: FOGA, 16–20 Jan 2013, pp. 119–132. ACM, Adelaide, Australia (2013)
19. Moraglio, A., Sudholt, D.: Runtime analysis of convex evolutionary search. In: Soule, T., Moore, J.H. (eds.) GECCO, pp. 649–656. ACM (2012)
20. Pawlak, T.P.: Competent algorithms for geometric semantic genetic programming. Ph.D. thesis, Poznan University of Technology, Poznan, Poland, 21 September 2015. http://www.cs.put.poznan.pl/tpawlak/link/?PhD
21. Pawlak, T.P.: Geometric semantic genetic programming is overkill. In: Heywood, M., McDermott, J., Castelli, M., Costa, E., Sim, K. (eds.) EuroGP 2016. LNCS, vol. 9594, pp. 246–260. Springer, Switzerland (2016)
22. Pawlak, T.P., Krawiec, K.: Properties of progress and fitness bounds for geometric semantic genetic programming. Genet. Program Evolvable Mach., 1–19 (2015) (Online first)
23. Pawlak, T.P., Wieloch, B., Krawiec, K.: Review and comparative analysis of geometric semantic crossovers. Genet. Program Evolvable Mach. 16(3), 351–386 (2015)
24. Pawlak, T.P., Wieloch, B., Krawiec, K.: Semantic backpropagation for designing search operators in genetic programming. IEEE Trans. Evol. Comput. 19(3), 326–340 (2015)
25. Runge, C.: Über empirische funktionen und die interpolation zwischen äquidistanten ordinaten. Zeitschrift für Mathematik und Physik 46, 224–243 (1901)
26. Saniee, K.: A Simple Expression for Multivariate Lagrange Interpolation, pp. 1–9. Society for Industrial and Applied Mathematics (2007)

Patterns for Constructing Mutation Operators: Limiting the Search Space in a Software Engineering Application

Thomas Kühne[1], Heiko Hamann[2(✉)], Svetlana Arifulina[1], and Gregor Engels[1]

[1] Department of Computer Science, University of Paderborn, Paderborn, Germany
kuehne@mail.upb.de, {s.arifulina,engels}@upb.de
[2] Heinz Nixdorf Institute, Department of Computer Science,
University of Paderborn, Paderborn, Germany
heiko.hamann@upb.de

Abstract. We apply methods of genetic programming to a general problem from software engineering, namely example-based generation of specifications. In particular, we focus on model transformation by example. The definition and implementation of model transformations is a task frequently carried out by domain experts, hence, a (semi-)automatic approach is desirable. This application is challenging because the underlying search space has rich semantics, is high-dimensional, and unstructured. Hence, a computationally brute-force approach would be unscalable and potentially infeasible. To address that problem, we develop a sophisticated approach of designing complex mutation operators. We define 'patterns' for constructing mutation operators and report a successful case study. Furthermore, the code of the evolved model transformation is required to have high maintainability and extensibility, that is, the code should be easily readable by domain experts. We report an evaluation of this approach in a software engineering case study.

Keywords: Model transformations · Mutation operators · Software engineering

1 Introduction

A well-known challenge in genetic programming is to find appropriate combinations of a genetic representation and genetic operators [9,13]. The choice of a genetic representation usually depends on the particular problem. The genetic representation can have a considerable impact on the task difficulty and consequently on the performance of the evolutionary approach [5,8]. Besides simple direct encoding a number of sophisticated approaches have been reported, such as generative representations [5,8] that try to leverage regularities in the problem, developmental representations [1], gene regulatory networks [3,4], and grammatical evolution [14]. Once an appropriate genetic representation is found the genetic operators are expected to be effective. For example, the encoding

© Springer International Publishing Switzerland 2016
M. Heywood et al. (Eds.): EuroGP 2016, LNCS 9594, pp. 278–293, 2016.
DOI: 10.1007/978-3-319-30668-1_18

should be robust to mutation such that a mutated individual has roughly the same fitness (cf. causality, local search) [13]. Criteria for a good genetic representation include *completeness* (any solution can be encoded) and *closure* (any genotype represents a meaningful solution) [7].

In the following, we apply genetic programming to a problem from software engineering. We argue that for our domain finding an appropriate combination of a genetic representation and genetic operators is particularly difficult especially concerning closure and the effect of neutral mutations. We apply evolutionary algorithms to the software-engineering problem of 'model-to-model transformations'. Model here means formal representations of software (e.g., finite state machines, UML class diagrams) and the transformation is the process of converting an instance of one model to the corresponding instance of another model. The input is pairs of corresponding example instances of both models. The output should be an algorithm that transforms these examples but hopefully also generalizes to correctly transform instances not seen before (cf. supervised learning). These examples are assumed to be provided by a domain expert. This is the problem of (semi-)automatic, example-based generation of model-to-model transformations, which otherwise requires a human transformation designer.

Applying genetic programming to the example-based generation of model transformations imposes mainly two challenges that are also relevant for the field of genetic programming itself. First, it is difficult to find an appropriate genetic representation for the transformations, because the associated search space is loaded with rich semantics. The search space does not show regularities that could be represented easily, instead most model transformation steps create interdependencies between operations, and the search space is unlimited in its size. The options of how to relate elements of instances from different models are manifold and context-dependent. Hence, any trivial approach generates a vast number of semantically incorrect instances in a combinatorial explosion.

Second, the search space is discrete, high-dimensional, and there is always a vast number of options for what the evolutionary operators should change. Here, the design of the mutation operator requires special attention and is challenging because the semantics should be respected. Otherwise, there is a combinatorial explosion of possible offspring that could be generated by mutations. Even worse, most of these mutations are neutral in terms of fitness because the mutations are either without effect or generate semantically meaningless transformations. Based on our experience it seems that no simple solution based on different representations exists.

In addition to the motivation for an approach, that pushes towards closure, there are the requirements of readability and maintainability of the generated transformation code. We want to allow for an optional semi-automatic approach and to generate model transformations that are easily understood by a domain expert. The problem of automatically generating model transformations is very complex and hence we expect that final changes by a domain expert are useful or possibly even required.

In this paper we present our approach to the challenges of finding appropriate genetic representations and of designing appropriate mutation operators for the application of example-based model-to-model transformations. Our main idea is to design a number of different mutation operators that ensure meaningful changes to the genome. To guide the design of these operators we define, what we call, 'transformation patterns'. Next we explain the application and related work, followed by a description of our approach, and a discussion of the results.

2   Model Transformations by Example and Related Work

In software engineering many problems require tedious manual tasks which is also true for model-to-model transformations. Their design requires domain experts to create transformation rules, which allow to convert between different model types. This task is especially relevant, when designing software systems with the model driven development methodology.

Model Transformations by Example as Task. The starting point for a model-transformations-by-example problem is a domain specific language which is close to the problem domain, but semantically far from the technical solution. Therefore, a transformation from the domain-specific model to a more technical model is required. Creating these transformations manually is difficult. Hence, automating the creation of model-to-model transformations is desirable. One approach to define the transformation problem is to provide example instances for both models. This is called model transformation by example (MTBE) [10]. Depending on the definition not only example instances are provided, but also information about their relations. In our approach only the examples are required, because providing additional information is challenging in complex scenarios.

The search space for our evolutionary algorithm is the infinite set of all possible model-to-model transformations. A transformation relates elements of the one model to elements of the other model. and must conform to a common model language called 'Meta Object Facility'[1]. All transformations are based on the Meta Object Facility. Roughly, this language defines classes with properties and associations which relate classes. The search space is large because instance of a class and property can be transformed to instances of associations, instances of properties, etc. Instances can be of type string which means that transformations might be required to concatenate several strings correctly to implement the right naming convention etc. In addition, a transformation might be required to depend on information of a particular instance which means that decision-making is required at runtime.

The example presented in Fig. 1 shows a transformation on the model level. Both models describe the same behavior. On the left, the instance of a hierarchical-state machine is shown. It has states (e.g., *Machine Off*), transitions (e.g., *Switch on*), and composite states that can contain nested states. The state machine and each composite state require an initial state. *Machine off* is

---

[1] OMG – MOF http://www.omg.org/spec/MOF/2.4.1/, 2015/09/09.

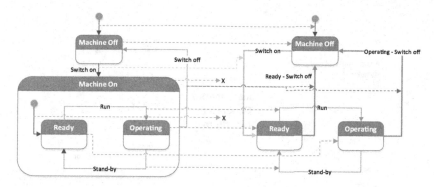

Fig. 1. Transformation scenario with example models and indicated relations, a hierarchical-state machine is transformed into a flat-state machine.

the general initial state and *Ready* is the initial state of the composite state *Machine On*. On the right, the flat state machine is shown, which models the same behavior without composite states. A correct transformation has to solve several tasks here. There are simple transformation steps, for example, all non-composite states (*Machine Off*, *Ready*, *Operating*) are transformed directly. That is, however, not possible for composite states and their initial states. Transitions pointing to composite states are transformed into transitions pointing to the composite state's previous start state. *Switch On* points to *Machine On* but should point to *Ready*. Also, transitions originating from previous composite states are transformed into a transition for each previous inner state. *Switch off* originates from *Machine On*. Thus, the transformation has to create a transition *Ready/Switch off* originating from *Ready* and *Operating/Switch off* from *Operating*. For this transformation also a string concatenation is required for the transition labels. The correct, context-dependent choice needs to be made out of multiple possible transformations for transitions.

In the technical terms of the software-engineering community the instance of a state machine M_a is called a model and the set of all possible state machines MM_a is called a meta-model (see Fig. 2). Hence, the overall task here is to generate a generic $MM_a \rightarrow MM_b$ transformation that transforms instances of the meta-model MM_a into instances of the meta-model MM_b. Following the MTBE methodology, a domain expert provides one or more example instances M_a of MM_a with expected output M_b^* conforming to MM_b. Hence, we have a challenging MTBE that provides little information as guidance. The search space grows fast with problem size. The main challenge is to handle these large search spaces and to enable the convergence on a suitable solution. Also no practicable generic approach to the problem of model-to-model transformation exists. Only transformation languages have been developed that focus on particular aspects of different domains. Hence, we have to define our own approach to structure the search space and allowed changes to genotypes.

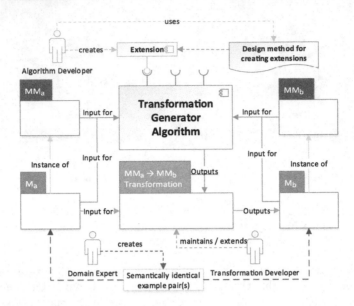

Fig. 2. Problem and requirement definition, generation of maintainable $MM_a \rightarrow MM_b$ transformations using semantically identical input/output pairs; definition of extensible algorithm and design method for creating extensions.

We require our evolutionary approach to create $MM_a \rightarrow MM_b$ transformations based on given meta-models MM_a and MM_b. It is guided by manually created (M_a, M_b) example pairs. Due to the complexity of model-to-model transformations, the evolutionary algorithm cannot be guaranteed to always converge on an appropriate solution. Hence, we require our approach to be extensible by providing generic extension parameters. A design method for creating extensions must be defined. It guides the algorithm developer to solve this challenging task. Generated $MM_a \rightarrow MM_b$ transformations might not be complete solutions and should therefore be maintainable by a transformation developer. A duration of evolutionary runs of a few hours is acceptable. We investigate two example scenarios derived from representative software engineering problems. The first scenario is to transform a simplified Unified Modeling Language (UML) class diagram[2] to a relational schema. In the second scenario hierarchical-state machines are transformed into flat-state machines.

Related Work. An overview over work that is related to our application, model transformation by example, is provided by Kappel et al. [10]. Here, we limit our discussion to two closely related approaches. We follow the approach by Faunes et al. [6] and Baki et al. [2] and apply genetic programming to derive a model transformation from examples. An initial population is created randomly using predefined patterns to generate model transformations. In order to evaluate the

---

[2] OMG – UML http://www.omg.org/spec/UML/2.4.1/, 2015/09/09.

resulting model transformations, they are executed on the source models given in the examples. As fitness function, the difference between the produced output (transformed source model) and the expected output (given target model) is calculated. In contrast to our work, Faunes et al. apply both a crossover operator and a mutation operator, and explicitly create a control flow in their model transformations. Our approach to the control flow is simpler as it is controlled by the execution engine. Their algorithms are based on a self-defined, simple transformation language. However, this language is used internally whereas the actual output conforms to a rule language. This output language does not reflect all concepts of the transformation language in a comprehensible way. Hence, the created transformations are not maintainable. The genetic operators are based on the internal language and have the capability to use all aspects of this language to create transformations. Due to the complexity of the search space and the combinatorial explosion of applicable changes to the genotype, their evaluation results indicate that the algorithms are likely to operate on neutral fitness landscape areas frequently. Complex transformations like the concatenation of property values are not supported. Due to the complexity of model-to-model transformations and the lack of a precise, generic problem definition, a fully generic approach is difficult. Going beyond these two works [2,6], we focus additionally on the extensibility and adaptability of our approach and the generated model-to-model transformations. Since convergence on a complete solution is not guaranteed, the evolved transformation should be readable and maintainable by a human developer.

Kessentini et al. [11,12] apply a variation of particle swarm optimization to find an optimal model transformation based on the given examples. This approach derives a target model directly from the source model without producing any model transformation rules. The approach tests different transformation possibilities and evaluates their fitness based on the given examples. In comparison, the advantage of our approach is the explicitly produced model transformation, which can be understood and refined by the modeler.

3 GP Approach to Evolve Model-to-Model Transformations

First, we explain the general concept followed by explanations of major parts of the solution. Figure 3 gives an overview of our general approach. The transformations are encoded in a special language called 'Epsilon Transformation Language' (ETL)[3], which is a regular programming language with standard control statements such as if-then-else and for-loops. This design decision is fundamental to provide maintainable solutions because it is a domain specific language targeted at humans. The search space is defined by the infinite set of all possible transformations representable by ETL. Due to the high complexity of our problem domain, we simplified the meta-meta-model language Meta Object Facility (MOF) to limit the number of possible transformations that are representable in

---

[3] Eclipse Foundation – ETL http://www.eclipse.org/epsilon, 2015/09/09.

our approach. Thereby, our genetic representation does not satisfy the completeness requirement [7] and also the number of possible meta-models is limited. The simplified MOF is the foundation for the systematic construction of the representative transformation scenarios. For example, solution D in Fig. 3 is a valid transformation but cannot be represented by our simplified MOF. Still, our simplified MOF can be incrementally extended towards the complete MOF.

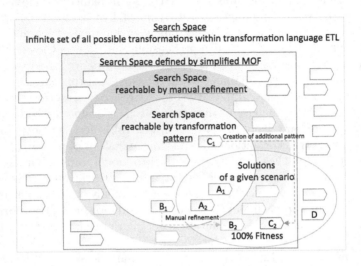

Fig. 3. Schematic representation of search space, our divide-and-conquer approach.

Following the standard GP approach, the genetic operators of the GP algorithm could be simple operations on trees because transformations conforming to the ETL could be represented as an abstract syntax tree (AST). However, such a common transformation language is similar to general purpose programming languages in terms of complexity. Here, many of these tree operations would generate changes that are neutral to fitness, that is, they could not be distinguished based on the fitness function. Instead, our approach is based on subsets of the ETL and the genetic operators are carefully defined to limit the possible changes. The design of the genetic operators is guided by what we call transformation patterns. Therefore, not all potential solutions are reachable for a given scenario. In Fig. 3, only solutions A_1 and A_2 can be generated by our approach. The idea of the transformation patterns is to create well-defined transformation concepts and to avoid the generation of incomprehensible solutions. For example, a transformation developer might easily extend transformation B_1 (an incomplete but partially correct solution candidate) into B_2 (a complete solution). In general, any transformation can be reached by manual modifications (e.g., C_2). However, the idea of our approach is that only small changes should be necessary to implement a reasonable semi-automatic system.

A major design goal of the transformation patterns is to help in limiting the search space. We try to decrease the size of neutral areas in the fitness landscape by forcing genetic operators to implement changes of transformations that also change the fitness. For an overview of the defined transformation patterns see Fig. 4(a). The pattern structure is derived from the simplified meta-meta-model language MOF. Thus, the general pattern categories are *Class Transformation*, *Property Transformation*, and *Association Transformation*. Some patterns are based on other patterns' results as seen in Fig. 4(a). For example, a property is part of a class, hence, a *Property Transformation* reuses classes transformed within the *Class Transformation*. This set of patterns serves as a starting point and will be refined and extended in future research.

In Fig. 4(b) we give an example of a transformation pattern, namely 'One-to-One Object.' It implements the mutation to achieve simple 1:1 transformations. In general, a pattern describes the selected elements on the left-hand side which are used from the input meta-model MM_a and the created elements in the output meta-model MM_b. In the presented example, the pattern transforms a single instance of *Source* into a single instance of *Target* and copies the value of the *Name* property. Every pattern by definition results in at least a pair of an add-mutation (i.e., adds a 1:1 transformation) and a remove-mutation (i.e., removes a 1:1 transformation).

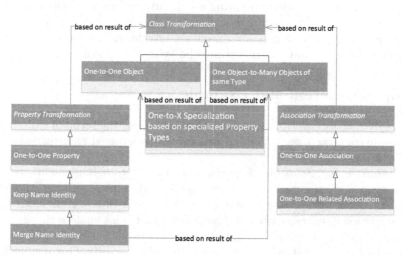

(a) Transformation Pattern Overview

(b) Example of the transformation pattern called 'One-to-One Object'.

Fig. 4. An overview over the transformation patterns and an example.

Based on this generic concept of transformation patterns multiple mutation operators are generated depending on features of the meta-models and models. In certain situations hundreds or even thousands of different mutation operators might be possible choices for the evolutionary algorithm. However, depending on the transformation and the mutation, the options of applicable mutation operators might not be evenly distributed over the different mutation types that were generated from different transformation patterns. This might introduce a bias if not handled differently. Therefore, we introduced mutation selection strategies to handle this challenge. We use two strategies. First, we use simple random selection. Second, we use a two-step random selection process that selects a mutation type followed by selecting a particular mutation instance.

A major goal of the pattern and mutation design is to ensure the maintainability of the generated transformations. This is achieved by adding constraints within the refinement step of the design process which reduce the probability of a combinatorial explosion. Redundant applications must be prohibited. In the above example, neither *Source* nor *Target* may be of a type which is already part of the $MM_a \rightarrow MM_b$ transformation. Defining several transformation processes that operate on the same elements several times within one transformation decreases readability and also creates unnecessary mutation instances. The second application of a mutation operating on already transformed elements might create contradictions. The results of such transformations depend on execution-flow specifics of the transformation execution engine, which are nontrivial and hence should be irrelevant to understand the evolved transformations.

4   Implementation and Example

Next we give a short description of our implementation, the setup of experiments for the evaluation, and we discuss an example of an evolved transformation.

Implementation and Setup for Experiments. In order to implement and analyze our evolutionary algorithm we developed a modular algorithm framework based on the Eclipse Modeling Framework (EMF)[4] and Java 8. It includes a relational evaluation database with a graphical front-end. This enables the algorithm developer to quickly evaluate the impact of changes within the transformation pattern, the fitness function and other configurable aspects. Our evaluation is based on two example scenarios: the previously introduced transformation of hierarchical-state machines into flat-state machines and the transformation of a simplified UML class diagram model into a simplified relational schema model. UML is a common language to describe the structure of software systems independently from the technical realization. Relational schema models are used to describe structures of database systems. Hence, the transformation enables a software designer to define an implementation-independent structure which is then transformed into a specific structure automatically. This is a realistic and common scenario in software engineering.

---

[4] Eclipse Foundation – EMF http://www.eclipse.org/emf/, 2015/09/09.

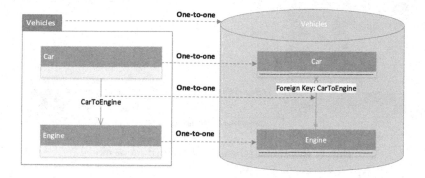

Fig. 5. Simple transformation scenario with a UML class diagram (left) and the equivalent relational schema (right).

Evolved Example. In Fig. 5 we give a simple example of a UML class diagram to a relational schema transformation scenario. The example is limited to simple 1:1 transformations only. The class diagram consists of a package 'Vehicles', the classes 'Car' and 'Engine', as well as an association between them. The relational schema has a database schema 'Vehicles', the tables 'Car' and 'Engine', and a foreign-key in table 'Car' which points to the table 'Engine'.

The example shown in Fig. 5 is one of the examples that our algorithm gets as input to evolve a general transformation. The example contains an instance of a UML class diagram and an instance of a relational schema along with the correct transformations. For both UML class diagrams and relational schemas we also have so-called meta-models (as discussed above) that describe the space of all possible instances. The reduced UML class diagram meta-model shown in Fig. 6 consists of a 'UmlPackage' that contains 'UmlClasses' and 'UmlAssociations'. Classes may be associated to other classes, hence, a 'UmlAssociation' has a 'Source' and 'Target' association.

The simplified relational schema meta-model is shown in Fig. 7 with a 'RelationalSchema' containing 'RelationalTables'. RelationalTables may refer to other RelationalTables via a 'RelationalForeignKey' with an association named 'ReferencedTable'. In difference to the UML class diagram, the 'ForeignKey' is only associated to the owning 'Table' and not to the 'RelationalSchema'.

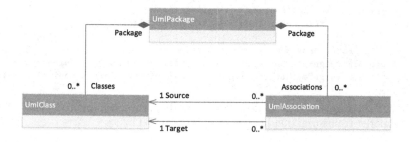

Fig. 6. UML class diagram – simplified meta-model

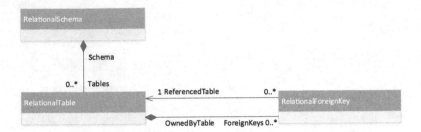

Fig. 7. Relational schema – simplified meta-model

A hand-coded transformation that converts any instance of the UML class diagram meta-model to the appropriate instance of the relational schema meta-model is presented in Listing 1.1. This is just one valid transformation but there are many other possible solutions. Listing 1.1 contains three rules, one for each MOF class (i.e., boxes in Fig. 6). All rules have a mapping for the 'Name' property. Additionally, 'UmlClass To RelationalTable' maps the associated 'UmlPackage' of the 'UmlClass' to the corresponding 'RelationalSchema' of the created 'RelationalTable' with the keyword 'equivalent'. The rule 'UmlAssociation To RelationalForeignKey' contains mappings for the 'Source' and 'Target' to the corresponding properties of the relational schema.

```
1   rule UmlPackageToRelationalSchema
2       transform umlPackage : Source!UmlPackage
3       to relationalSchema : Target!RelationalSchema {
4
5       relationalSchema.Name = umlPackage.Name;
6   }
7
8   rule UmlClassToRelationalTable
9       transform umlClass : Source!UmlClass
10      to relationalTable : Target!RelationalTable {
11
12      relationalTable.Name = umlClass.Name;
13      relationalTable.Schema = umlClass.Package.equivalent();
14  }
15
16  rule UmlAssociationToRelationalForeignKey
17      transform umlAssociation : Source!UmlAssociation
18      to relationalForeignKey : Target!RelationalForeignKey {
19
20      relationalForeignKey.Name = umlAssociation.Name;
21      relationalForeignKey.OwnedByTable = umlAssociation.Source.equivalent();
22      relationalForeignKey.ReferencedTable=umlAssociation.Target.equivalent();
23  }
```

Listing 1.1. Example of hand-coded UML class diagram to simple relational schema model-to-model transformation in concrete syntax of ETL.

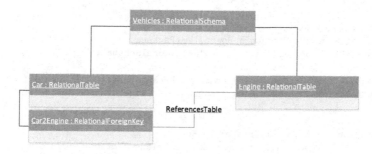

(a) UML class diagram example instance

(b) Relational schema example instance that is semantically identical to (a)

Fig. 8. Example instances of a UML class diagram and a relational schema.

The example shown in Fig. 5 is very simple. A slightly more complex instance of the UML class diagram meta-model along with its semantically identical instance that conforms to the relational schema meta-model is given in Fig. 8.

Before we discuss the results in the next section, we give an example of an evolved transformation for the UML-class–relational-schema transformation problem in Listing 1.2 that was obtained by using our approach with the above examples as input. The sequence of stepwise improvements, in which the code was created by applying six mutations, is color-coded (note that also the initial population is created from 'empty' individuals by mutations). A few of these mutations are initially neutral. For example, the third mutation (lines 6 and 7) introduces a transformation for 'referencedBy' which has no effect without the respective 'UmlAssociation2RelationalForeignKey' rule. That rule is then added by the next mutation which also increases the fitness. Such improvements, that have no immediate effect but prepare beneficial mutations, could be detected with a more sophisticated fitness function. However, we want to keep the fitness function conceptually simple and base fitness only on the transformation product. When comparing Listing 1.2 with the hand-coded transformation shown in Listing 1.1 we notice a similar approach but also a difference in how the reference to the table is defined in the transformation of associations to relational foreign keys. Still, the evolved transformation is correct.

```
1   rule UmlClass2RelationalTable
2     transform sourceUmlClass : Source!UmlClass
3     to targetRelationalTable : Target!RelationalTable {
4
5     targetRelationalTable.Name = sourceUmlClass.Name;
6     targetRelationalTable.ReferencedBy =
7                             sourceUmlClass.incomingRelations.equivalent();
8     targetRelationalTable.ForeignKeys =
9                             sourceUmlClass.outgoingRelations.equivalent();
10  }
11
12  rule UmlPackage2RelationalSchema
13    transform sourceUmlPackage : Source!UmlPackage
14    to targetRelationalSchema : Target!RelationalSchema {
15
16    targetRelationalSchema.Name = sourceUmlPackage.Name;
17    targetRelationalSchema.Tables = sourceUmlPackage.Classes.equivalent();
18  }
19
20  rule UmlAssociation2RelationalForeignKey
21    transform sourceUmlAssociation : Source!UmlAssociation
22    to targetRelationalForeignKey : Target!RelationalForeignKey {
23
24    targetRelationalForeignKey.Name = sourceUmlAssociation.Name;
25  }
```

Listing 1.2. Evolved transformation that correctly transforms simple UML class diagrams to simple relational schemas; the code was generated by genetic programming in the sequence: lines 1-5, 12-16, 6-7, 20-25, 17, and 8-9.

5   Results

We have tested a vast number of different parameter settings of the evolutionary algorithm. Configurable options are, for example, the selection strategy (e.g., roulette wheel, tournament), the replacement strategy, and the fitness function. Including both scenarios, about 10^3 different configurations were tested in total. For each configuration 50 independent evolutionary runs were done. Overall, this results in about 5×10^4 populations with a total of 3×10^8 individuals, a total runtime of about 32 days and 27 GB of data. Based on a comparison between the number of obtained correct solutions (model-to-model transformations that are correct with respect to the examples) and the required runtime, we identified the best configuration. 55 % of all configurations identified at least one correct solution. A configuration, that is optimized for runtime but affords intermediate unsuccessful evolutionary runs, returns a solution after about 42 min. A bigger number of evaluations allowed for 9 solutions after 3.7 h in a sequence of evolutionary runs. Figure 9 shows the fitness of 50 evolutionary runs over generations for the best system configuration. This configuration with a population size of 100 and 100 generations shows good quality in all 50 populations. The fitness saturates after an expected initial steep increase. Many of the evolved incomplete transformations are close to a correct solution and easily readable. Hence,

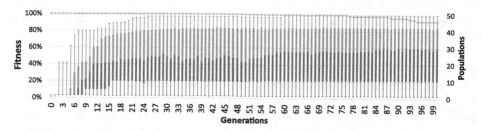

Fig. 9. Fitness over generations of 50 independent evolutionary runs. Execution terminates when a solution is found, which decreases the number of populations.

they could be extended manually to solve the scenario completely and correctly. The results could, of course, also be improved with more computational power.

In order to optimize the approach, we also tested different fitness functions. The fitness function has to evaluate the performance of an individual that is transforming a provided example instance M_a to the desired instance M_b. This is done by comparing so-called 'object graphs' (i.e., directed graphs representing such instances, cf. data modeling) of the individual's output M_b^* and the desired instance M_b. The best results were achieved with the following comparison approach. First, the graphs of M_b and M_b^* are compared to identify how many objects, properties, and associations are transformed correctly. Second, this information is used to calculate the ratio of correct matches compared to the expected matches for each of the three categories. The creation of wrong objects is avoided by imposing penalties. The score from the three categories are accumulated using configurable weights. The ability to partially match transformations of objects is important to avoid bootstrapping problems.

```
1   rule State2State
2     transform sourceState : Source!State
3     to targetState : Target!State {
4     targetState.Name = sourceState.Name;
5     targetState.FlatStateMachine =
6       sourceState.HierarchicalStateMachine.equivalent();
7   }
```

Listing 1.3. Generated Transformation Fragment – simple 1:1 transformation from the state of a hierarchical-state machine to the state of a flat-state machine

As previously stated, the automatically created transformations often need to be maintained manually, which is an important requirement. Listing 1.3 shows a fragment of a typical generated transformation. It is a simple 1:1 transformation from the state of a hierarchical-state machine to the state of a flat-state machine. In general, we focus on readability by assembling comprehensible identifiers and ensuring a properly structured code. For example, the rule name *State2State* reflects the general purpose, as well as the variables pointing to the *sourceState* of the M_a model and *targetState* of the M_b model. With our transformation patterns we define constraints that avoid the creation of solutions that lack comprehensibility. We also have implemented a cleanup mechanism that removes ineffective code from the transformations to increase readability.

As stated above, the two main challenges for this approach are the definition of an appropriate genetic representation and the design of the mutation operators that limit the effect of a combinatorial explosion because there is always a vast number of options for possible changes by the mutation operator. A main issue is the large number of changes that are neutral to the fitness which would generate large neutral plateaus in the fitness landscape and limit our evolutionary algorithm to a mere random search. As an indicator for neutrality we do a simple fitness landscape analysis and measure the effect of all possible mutations for a selection of 20 individuals from populations of a single evolutionary run for generations 1, 50, and 100, see Fig. 10. The results show that despite our efforts in decreasing the number of neutral mutations, the number of mutations without effect on the fitness are still very frequent. If we would use less complex mutation operators, the number of neutral mutations would be even higher and the proportion of improving mutations would probably be lower. Overall, this measurement confirms our hypothesis on the challenges in this application.

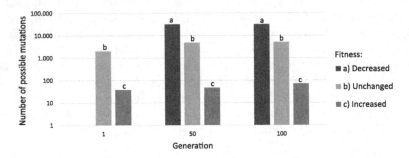

Fig. 10. Analysis of mutation effects, change in fitness for all applicable mutations for the top-20 individuals of generations 1, 50, and 100. Based on a single evolutionary run with 100 generations and a population size of 100; note logarithmic scale.

6   Conclusion

We have reported the successful application of genetic programming to the problem of model transformations by example. The challenges of developing an appropriate genetic representation and of designing useful mutation operators were addressed. Our focus is on the design of transformation patterns to define complex mutation operators. Our hypothesis is that a sophisticated approach is required due to special properties of the search spaces, such as the vast number of semantically incorrect instances and of possible changes to instances that are neutral to the fitness function. Also our experience with this system indicates that there is no simple genetic representation that could deal with this challenge. The idea of designing a set of sophisticated mutation operators is to limit the number and extent of neutral plateaus in the fitness landscape. Our hypothesis is also supported by measurements of mutation effects shown in Fig. 10. Our methodology has potential to generalize to other domains with search spaces of similar structure. Our future work will be to apply this approach to other problems of software engineering such as requirements specification by examples.

Acknowledgment. This work was partially supported by the German Research Foundation (DFG) within the Collaborative Research Centre 'On-The-Fly Computing' (SFB 901).

References

1. Astor, J.C., Adami, C.: A developmental model for the evolution of artificial neural networks. Artif. Life **6**(3), 189–218 (2000)
2. Baki, I., Sahraoui, H., Cobbaert, Q., Masson, P., Faunes, M.: Learning implicit and explicit control in model transformations by example. In: Dingel, J., Schulte, W., Ramos, I., Abrahão, S., Insfran, E. (eds.) MODELS 2014. LNCS, vol. 8767, pp. 636–652. Springer, Heidelberg (2014)
3. Banzhaf, W.: Artificial regulatory networks and genetic programming. In: Riolo, R., Worzel, B. (eds.) Genetic Programming Theory and Practice, pp. 43–62. Kluwer, Dordrecht (2003)
4. Bongard, J.: Evolving modular genetic regulatory networks. In: Proceedings of the World on Congress on Computational Intelligence, pp. 1872–1877. IEEE (2002)
5. Clune, J., Ofria, C., Pennock, R.T.: How a generative encoding fares as problem-regularity decreases. In: Rudolph, G., Jansen, T., Lucas, S., Poloni, C., Beume, N. (eds.) PPSN 2008. LNCS, vol. 5199, pp. 358–367. Springer, Heidelberg (2008)
6. Faunes, M., Sahraoui, H., Boukadoum, M.: Genetic-programming approach to learn model transformation rules from examples. In: Duddy, K., Kappel, G. (eds.) ICMB 2013. LNCS, vol. 7909, pp. 17–32. Springer, Heidelberg (2013)
7. Gruau, F.: Automatic definition of modular neural networks. Adapt. Behav. **3**(2), 151–183 (1994)
8. Hornby, G.S.: Generative representations for evolutionary design automation. Ph.D. thesis, Brandeis University (2003)
9. Hornby, G.S., Pollack, J.B.: Creating high-level components with a generative representation for body-brain evolution. Artif. Life **8**(2), 223–246 (2002)
10. Kappel, G., Langer, P., Retschitzegger, W., Schwinger, W., Wimmer, M.: Model transformation by-example: a survey of the first wave. In: Düsterhöft, A., Klettke, M., Schewe, K.-D. (eds.) Conceptual Modelling and Its Theoretical Foundations. LNCS, vol. 7260, pp. 197–215. Springer, Heidelberg (2012)
11. Kessentini, M., Sahraoui, H., Boukadoum, M., Omar, O.B.: Search-based model transformation by example. Softw. Syst. Model. **11**(2), 209–226 (2010)
12. Kessentini, M., Sahraoui, H.A., Boukadoum, M.: Model transformation as an optimization problem. In: Czarnecki, K., Ober, I., Bruel, J.-M., Uhl, A., Völter, M. (eds.) MODELS 2008. LNCS, vol. 5301, pp. 159–173. Springer, Heidelberg (2008)
13. Matarić, M.J., Cliff, D.: Challenges in evolving controllers for physical robots. Robot. Auton. Syst. **19**(1), 67–83 (1996)
14. O'Neill, M., Ryan, C.: Grammatical Evolution: Evolutionary Automatic Programming in an Arbitrary Language. Springer, New York (2003)

Iterative Cartesian Genetic Programming: Creating General Algorithms for Solving Travelling Salesman Problems

Patricia Ryser-Welch[✉], Julian F. Miller,
Jerry Swan, and Martin A. Trefzer

The University of York, Heslington, UK
{patricia.ryser-welch,julian.miller,jerry.swan,Martin.Trefzer}@york.ac.uk

Abstract. Evolutionary algorithms have been widely used to optimise or design search algorithms, however, very few have considered evolving iterative algorithms. In this paper, we introduce a novel extension to Cartesian Genetic Programming that allows it to encode iterative algorithms. We apply this technique to the Traveling Salesman Problem to produce human-readable solvers which can be then be independently implemented. Our experimental results demonstrate that the evolved solvers scale well to much larger TSP instances than those used for training.

Keywords: Iterative algorithms · Cartesian Genetic Programming · TSP

1 Introduction

Designing effective search algorithms for difficult problems has long been an intensive field of study in computer science [13]. Evolutionary algorithms have been used to optimise or design search algorithms and it is typical for such algorithms to operate on a human-designed template in which new operations are generated at fixed points in the template. Very few have evolved loop-based control flow and attempted to answer John Koza's question: *"Is it possible to automate the decision about [...] the particular sequence of iterative steps in a computer program?"* [16]. Such control flow can be implemented either via iteration or recursion. Recursive approaches to various problems of program induction have been presented in Yu and Clack [43] and Alexander [1] but we are not aware of any direct applications to search problems.

In this paper, we introduce a novel and extended form of a well-known graph-based form of Genetic Programming, Cartesian Genetic Programming (CGP), to encode iterative algorithms. The original motivation for using hyper-heuristics is that they do not require skilled practitioners. The use of CGP is significant in this regard, since it doesn't require knowledge of specialized bloat-handling techniques. We apply this technique to the well-known Traveling Salesman Problem

© Springer International Publishing Switzerland 2016
M. Heywood et al. (Eds.): EuroGP 2016, LNCS 9594, pp. 294–310, 2016.
DOI: 10.1007/978-3-319-30668-1_19

(TSP), a problem domain which has been extensively studied in mathematics and computer science and is often used to benchmark new techniques. Various search algorithms such as Iterative Local Search, Memetic Algorithms, Particle Swarm Optimization, Ant Colony algorithms and other novel metaheuristics have solved a wide variety of TSP benchmark instances that have often less than 2000 cities [10, 14, 29]. We use iterative CGP to generate new TSP algorithms: the quality of an algorithm during the evolution process is determined by a small training set of TSP instances ranging between 200 and 800 cities. Subsequently algorithms are validated on larger unseen instances varying in level of complexity. The easiest instances have 30 cities but the most challenging contain just under 25,000 cities. The contributions of the work presented here are three-fold:

1. Previously CGP has encoded as a *data-flow diagram*, in which information flows through the graph from inputs to outputs [25]. For this work, CGP has been adapted to provide a *flowchart* which represents an iterative algorithm using *"Decision"*, *"Process"* and *"Terminal"* elements.
2. Both instruction ordering and iterative control flow are evolved: groups of instructions can be repeated. The resulting algorithms find good solutions to unseen TSP instances.
3. CGP is used to generate hybrid search algorithms using combinations of local search and binary crossover. We evolve human-readable algorithms that can reach optimal TSP solutions and can be directly translated into other programming languages.

2   Optimisation of Algorithms

The goal is to improve some aspects of an algorithm in order to solve problems more efficiently or with fewer resources. It is useful to distinguish here between two distinct search spaces: we use the term *problem solutions* to refer to elements of the *underlying* problem space (e.g. permutations in the case of the TSP) problem and the *algorithm solution* to the generated algorithms. Early approaches in this area have been referred to as *"automatic programming systems"* [16]. The problem solutions are obtained using a solver generated by a technique such as Genetic Programming (GP). Research in this area [3, 16] has largely focused on results in terms of the performance on the underlying search problem, rather than human-readability of the algorithms themselves. More recently, a variety of search methods have been used to automatically configure algorithms via parameter optimization [12, 20]. The latest approaches in this area are increasingly general (e.g. [21, 22]), allowing entire component configurations to be treated as a parameter hierarchy. However, parameter tuning is still currently a rather limited way to optimise an algorithm. The evolution of iterative control flow with CGP offers a more general approach and additionally provides human-readable output without the need for explicit parsimony pressure to combat expression bloat.

A graph-based form of Genetic Programming (GP) has automated machine code with basic loops. This technique was promising, but it has not yet been applied to higher level programming languages [41].

The generation of search algorithms can be considered within the context of *hyper-heuristics*, which are defined as "a search method or learning mechanism for selecting or generating heuristics to solve computational search problems" [5]. Hyper-heuristics can be *selective* or *generative*. The popular conception of selective hyper-heuristics is exemplified by the HYFLEX framework, in which the selection is performed (via an opaque *domain barrier* [5]) from a collection of pre-existing operators (heuristics). There are number of alternative hyper-heuristic frameworks to HYFLEX (e.g. [36,37]), including selective frameworks with a less-restrictive notion of the domain barrier [4,38]. In contrast to the selective approach, generative hyper-heuristics create new operators [32] and tend to use nature-inspired mechanisms (such as Learning Classifier Systems or GP) to discover better quality algorithms. This can result in algorithms capable of addressing an entire class of problems [30]. What both approaches have in common is to combine human-designed search components in new ways with the goal of outperforming any individual components. A detailed review of the state-of-art in hyper-heuristics can be found [5,28]. Ryser-Welch et al. [33] and Ross [31] complement these reviews by focusing specifically on hyper-heuristic frameworks.

The automated design of sizeable algorithms without any external help is beyond the state-of-the art. Suitably expressive algorithms may never terminate or have over-long computations. It is therefore useful to consider an algorithm search-space as consisting or both feasible and infeasible algorithms [16]. When algorithm design is automated, these unwanted occurrences are usually prevented via some forms of constraint. For example, [17,18,21,35,42] restrict the structure of an algorithm to prevent unfeasible sequences being discovered. Syntactic rules control the pattern of the primitives that are combined to form the algorithm-solutions. The body of a loop, the initialisation step, the update step, and sometimes the termination criteria are influenced by evolution. For instance, [39] evolves the body of the loop of ant algorithms using Grammatical Evolution; these algorithms are human readable and strictly restricted to the syntactic rules. Although some good results have been obtained from many of these techniques, the resulting algorithms can be very challenging to understand. In some cases, the chosen algorithm representation (e.g. GP tree) can cause bloat during the evolution, resulting in very large complex algorithms. Other algorithm generation schemes do not express all three elements of a looping construct or restrict the algorithm-solutions to a limited collection of primitives. In the next section, we describe how an extension of CGP can take advantages of properties of this graph-based GP, to relax strict syntactic rules to produce compact iterative algorithms.

3   Iterative Cartesian Genetic Programming

We describe an extension of CGP to the generation of iterative algorithms. In contrast to 'traditional' GP, which operates on expression trees, CGP uses a directed acyclic graph. An integer-based encoding scheme is used to define a two-dimensional grid, representing the adjacency matrix of a set of user-defined nodes. A characteristic of CGP is that it encodes both active and inactive nodes. Inactive nodes are nodes that are not on any paths connecting inputs to outputs; in Fig. 1 the output connects to node 4, but node 5 is inactive (shaded in gray). Nodes may be activated or deactivated during evolution. Each node has a *function gene* indexing a primitive operation in a user-specified look-up table. Nodes are connected in a feed-forward manner from either a previous node or a program input, using at least one *node input*. The *output genes* can connect to any previous nodes or program inputs. The identification of all the active nodes starts from the nodes pointed to by the output genes and continues until an input is reached. All the active nodes are then processed from left to right. In Fig. 1, the decoding step identifies the active nodes 1,2,3 and 4; these are executed in ascending order (i.e. 1,2,3 and 4). The CGP-graph has a fixed length, but the number of active nodes can be anything from zero to the number of nodes (see Fig. 1). Unlike other Genetic Programming techniques, CGP has been shown not to bloat [24, 40].

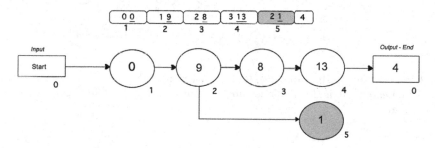

Fig. 1. This CGP graph encodes an algorithm made of 4 primitives, with 1 input and 1 output. All these active nodes (in white) constitute the "process" elements of the flowchart.

A directed graph can fully encode an iterative program: we call this an iterative CGP graph. This allows a cycle to be formed so that loops are possible. The stopping criterion, the iterative update step and the body of the loop are all alterable by evolution. To accomplish this it is necessary for every node to have at least four different types of genes:

Feed-forward. connections are standard feed-forward CGP connection genes. They connect the input to the current node with either a previous node or a program input. We refer to these nodes as *process nodes* as they represent a process element of the flowchart.

Branching. connections can point to a previous node, a program input, itself, or a suitable subsequent node. They are connection genes which determine the boundaries of the body of a loop and split a CGP graph into smaller subsequences. The first operation in the sub-sequence is the operation determined by the function gene of the current node. The last operation in the sub-sequence is the operation defined by the node pointed to by the branching gene of the current node. In these cases we refer to the current node as a *decision node* by analogy with a node in a flowchart which represents a "decision" element.

Function. genes are as in standard CGP and encode a primitive operation. Their values correspond with a function look-up table.

Condition. genes represent the stopping criteria of loops. A condition look-up table provides a set of Boolean primitives, these indicate whether a loop exits (and control subsequently moves to the next node following the last loop node) or continues to execute the next node inside the loop.

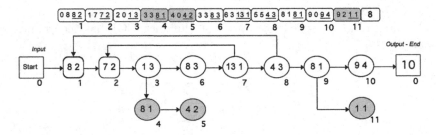

Fig. 2. An iterative CGP graph encodes an algorithm made of 8 primitives starting at node 1 and ending at node 10. Nodes no 4, 5 and 11 are non-coding genes, these are shaded in gray.

The distinction between decision and process nodes plays an important role in the decoding process of an iterative CGP graph. First all the active nodes are identified (by working backwards from the outputs), then the decision nodes are placed so that branching can happen during the decoding process; the index of the decision node is inserted after the last active node of the body of the loop. For example in Fig. 2, all the active nodes are executed in the following order: **1,2**,3,6,7,**2**,8, **1**,9,10. Also it is assumed that upon the second call of node 1, condition no 2 is also met, causing program execution to move to the next node (9) after the loop terminates at node (8).

1. When an iterative CGP graph does not encode any loops the value of any branching gene is free to point to any nodes and program inputs.
2. For any nodes inside an existing loop, their branching genes can only connect to node with a higher index that is inside the current loop or any previous nodes and program inputs. In Fig. 2, the branching gene of nodes 3, 4, 5, and 6 can be valid if its value is lower than the index of the node. It can also point to the right to a node with an index lower than 7.
3. For any nodes outside an existing loop, their branching genes can connect to a node that is outside any existing loops. A valid value for the branching gene of node 1 can only point to the input or nodes 9, 10 or 11.

Algorithm 1. The $(\mu + \lambda)$ evolutionary strategy used by both versions of CGP. Often there is one parent (μ) and four offspring (λ)

1: Randomly generate individual i
2: Select the fittest individual, which is promoted as the parent (algorithm)
3: **while** solution is not found **or** the generation limit is not reached **do**
4:    Mutate the parent to generate offspring
5:    Generate the fittest algorithm using the following rules:
6:    **if** offspring has a better or *equal* fitness than the parent **then**
7:        offspring is chosen as fittest
8:    **else**
9:        The parent remains the fittest
10:    **end if**
11: **end while**

CGP generally uses a $(1+\lambda)$ evolutionary strategy (this is shown in Algorithm 1). Either point or probabilistic mutation is traditionally used, crossover is not. If an offspring has an equal or better fitness than the parent it is promoted to the next generation [25]. Two basic grammatical rules ensure that either only nested loops are created or new loops do not overlap. This is ensured during the initialization of the iterative CGP population and the mutation of parents to produce new iterative CGP offspring.

4   Discovery of Iterative TSP Solvers

The goal of these experiments is to gain insight into how hybrid metaheuristics can be discovered with iterative CGP. We ran our generated algorithms on the TSP instances from well-known benchmarks[1]. The settings of Iterative CGP for the all the tests are given in Table 1. Our proposed method evolves merely a sequence of heuristics, but repeated sub-sequences (or loops). At the end of this process, a generated algorithm can then be extracted, and if desired, re-coded in some conventional programming language. Subsequently, the generated algorithms are evaluated in an independent process using an unseen test. The upper-level process is problem-domain independent and the specialised TSP heuristics used in the lower-layer are described in the next sub-section.

4.1   The Travelling Salesman Problem

This combinatorial problem seeks the shortest possible route that visits each of a list of cities exactly once and returns to the first city, i.e. a Hamiltonian

---

[1] d1291, u2152, usa23505 and d18512 are benchmarks from the well-known TSPLIB. The remaining instances are benchmarks from real-life geographical data; these are wi29, dj38, qa194, zi929, ca4663, ym7663, ja9874, gr9882, sw24978. All these instances can be found at http://www.math.uwaterloo.ca/tsp/world/countries.html and http://comopt.ifi.uni-heidelberg.de/software/TSPLIB95/STSP.html.

Table 1. Experimental parameters for iterative CGP

Parameter	Value
Length (no of nodes)	300
Levels-back (no of nodes)	100
Levels-forward (no of nodes)	100
Program inputs	1
Program outputs	1
$\mu + \lambda$	$1 + 1$
Mutation rate	0.10
Generations	1500

cycle on n cities. A route is are referred to as *a tour* and is typically represented as a permutation on n elements. The problem is naturally represented with a complete weighted graph $G = (V, E)$. Each edge of E defines the link connecting the cities and their weight indicates the distance between two cities u and v; these are retrieved by the distance function $d(u, v)$. The length of a tour is given by the sum of its edge weights. For training purposes, we wish to combine the results across multiple problem instances of different sizes, so we therefore use the popular normalized measure of *relative error* as the fitness value of a TSP solution. This uses the best tour length that is known *a priori*, using the formula $(\text{tourlength} - \text{knownoptimum})/\text{knownoptimum}$.

A wide variety of TSP-specific operators have been examined in the literature. *Lin-Kernighan heuristics* are the most studied methods for solving this problem effectively. In these, k edges are deleted and subsequently re-assembled to construct the sub-paths of a new tour with a lower minimum weight [11,19]. Traditionally those are often referred as *k-opt heuristics*. For example, when $k = 2$, edges between two pairs of cities are reconnected in a different way to obtain a new shorter tour. Solutions obtained from genetic operators (e.g. PMX, as below) can be further improved by local search [10]. Operators given below include those taken from recent work which merged a local search operator with genetic search [14], along with well-known crossover and mutation operators. Table 2 shows all the primitives we will be using for our experiments.

- **Order Based Crossover (OX)** chooses a subtour in one parent and imposes the relative order of the cities of the other parent [6].
- **Partially-Mapped Crossover (PMX)** copies an arbitrary chosen subtour from the first parent into the second parent, before applying minimal changes to construct a valid tour [8,9].
- **Voting Recombination Crossover (VR)** uses a randomized Boolean voting mechanism to decide from which parents each city is copied from [23].
- **Subtour-Exchange Crossover (SEC)** preserves randomly selected subtours from both parents to construct one new offspring [15].

- **Edge-Assembly Crossover (EAX)** assembles sub-tours together by building intermediary permutations. Each of these permutation are repeatedly minimised [26]
- **Stem-and-cycle ejection Crossover (SCX)** unusually asexually reproduces before improving the child solution using a "Stem-And-Cycle" Local Search approach [14].
- **Insertion Mutation (IM)** moves a randomly chosen city in a tour to randomly selected place [7].
- **Exchange Mutation (EM)** swaps two randomly selected cities [2].
- **Scramble Mutation (SM)** rearranges a random subtour of cities [6]. HYFLEX applies this mutation operator on a subtour and on the whole tour.
- **Simple Inversion Mutation (SIM)** implements a 2-opt Lin-Kernighan heuristic.
- **3-point Inverstion Mutation (3IM)** implements a 3-opt heuristic [14].

4.2   Automatic Design of Hybrid Metaheuristics

In our experiments, operators provided by the Hyflex cross-domain hyperheuristic framework[2,3] were chosen (see Table 2) as described in previous research [34]. The parameters for our generated metaheuristic were set to 2 offspring, 2 parents, and a maximum of 500 evaluations. The search depth for local search controls the number of iterations used in a local search operators; following preliminary experiments, it was set to 0.89. The intensity of mutation (which defines the numbers of cities shuffled in a permutation) was similarly set to 0.8. These standard parameters tune the performance in general of local searches and mutation operators for any problem domain provided in Hyflex. These parameters are standard choices for the Hyflex system.

Tables 2 and 3 provide the heuristics and termination criterion for the generated TSP solvers. For Conditions 2 and 3, the evolution is split into two stages: each phase uses half the available evaluations Condition 4 stops the search when all the evaluations have been used or no shorter tour has been found in the last 50 generations.

A predefined template (Algorithm 2) guarantees that the generated algorithm initializes and evaluates a population of permutations, before selecting parents (lines 1 to 3 of Algorithm 2). The code in lines 4 to 21 execute the iterative algorithm defined by the active nodes of an iterative CGP graph. The last line enforces that shorter tours are promoted in the population before the algorithm ends its run (see line 22 of Algorithm 2). The remaining lines apply the heuristics of the active process and branching nodes (see lines 4 and 23). The 'goto' statements can either jump to the start of a loop, the next heuristics (if there is one) or the first when the stopping criterion is met, or to the first heuristic of the metaheuristic.

---

[2] http://www.asap.cs.nott.ac.uk/external/chesc2011/.

[3] http://www.hyflex.org/chesc2014/.

Table 2. Function set: list of TSP heuristics used as primitives.

Index	TSP heuristics
0	InsertionMutation()
1	ExchangeMutation()
2	ScrambleWholeTourMutation()
3	ScambleSubtourMutation()
4	SimpleInversionMutation()
6	2-OptLocalSearch()
7	Best2-OptLocalSearch()
8	3-OptLocalSearch()
9	OrderBasedCrossover()
10	PartiallyMapCrossOver()
11	VotingRecombinationCrossOver()
12	SubtourExchangeCrossover()
13	ReplaceLeastFit()
	SelectParents()
15	RestartPopulation()

Table 3. Condition set: Boolean primitives chosen for the stopping criterion.

Index	TSP heuristics
1	Number of evaluations > 0
2	The evaluations fall in the first half of the evolution
3	The evaluations fall in the second half of the evolution
4	Number of evaluations > 0 or no improved neighbouring solutions are available

The testing phase was performed on TSP lib instances pr299, pr439 and rat783[4] having 299, 439 and 783 cities respectively. It is well-known that hyper-heuristic evaluation is computationally expensive, so this small subset of instances was chosen for their diverse clustering of cities. The fitness measure used for generated solvers during the training phase is obtained by averaging the relative error values obtained for these instances; each run had a budget of 500 evaluations. The fitness measure for testing is the relative error of the instance under test.

5   Experimental Results

Algorithms 3 and 4 show the best iterative algorithms evolved by iterative CGP. As discussed above, the first three lines and the last instruction of these algorithms are part of the template described in Algorithm 2. The remaining instructions of the algorithms were generated during the decoding phase of the iterative CGP graphs.

Algorithm 3 resembles a memetic algorithm; evolution has re-discovered a similar algorithm to the most effective sequential algorithm evolved in our previous research [34]. In fact, lines 6 to 8 cancel out the effect of restarting the population p, if no shorter tour has been found in 50 generations, then the population is initialized again. However the newly-created TSP solutions are replaced immediately by the offspring (t); if and only if the length of their tour is shorter than the new generated individuals in population p.

Algorithm 4 applies two loops that are carried out during the first half of the evolution and fewer evaluations are required. The first loop can be perceived as

---

[4] http://www.iwr.uni-heidelberg.de/groups/comopt/software/TSPLIB95/.

Algorithm 2. Template for a hybrid meta-heuristic, with main structure (line 4 to 21) being evolved by an Hyper-Heuristic algorithm.

```
 1: p₀ ← GenerateInitialSolution();
 2: p₀ ← EvaluatePopulation();
 3: t ← SelectParents();
 4: {Start of code generated by Iterative CGP}
 5: goto the first active node
 6: while Not the end of of evolved sequence of heuristics do
 7:     if The current node is a process node then
 8:         Apply the heuristic on t or p
 9:         goto the next active node
10:     else
11:         if the current node is a decision node and the last node a loop then
12:             goto the first node of the loop
13:         end if
14:         StoppingCriterion ← apply condition of the currentNode
15:         if StoppingCriterion is false then
16:             Apply the heuristic on the t or p
17:             Go to the next active node
18:         else
19:             Go to the first node after the loop
20:         end if
21:     end if
22: end while
23: {End of code generated by Iterative CGP}
24: p ← replaceLeastFit(t, p)
```

redundant, but its purpose is to execute only once two Lin-Kernighan operators; one before and one after searching more thoroughly. This occurs in a nested loop, constructed with the Best2-OptLocalSearch() to reduce the length of the tour, before applying ExchangeMutation heuristic. This heuristic should prevent the 3-opt-LocalSearch() finding no available neighbouring solutions and then finding the same local optima again. These new offspring then replace the least fit individuals in the population p.

Algorithms 3 and 4 were translated from their iterative CGP graph form and coded as TSP solvers in the programming language Java^{TM}. The same primitives were retained, but a different set of benchmarks was used for testing. As observed above, hyper-heuristics are notoriously computationally expensive and a representative subset (having different distributions of cities) was chosen so to allow the experiments to be performed within a reasonable time. After some initial experiments, we set our number of evaluations to 6000, so that the search can be performed in a reasonable amount of time. We are aware the search is likely to be short, however, it would be just a matter of increasing the evaluations to solve more instances. For direct comparison, the best performing sequential TSP solver obtained from previous research [34] and the memetic algorithm due to Özcan [27] were also coded in Java. Both algorithms apply the same set of

Algorithm 3. This algorithm is the outcome of applying Algorithm 2 on the iterative graph in Fig. 2

1: $p_0 \leftarrow GenerateInitialSolution()$;
2: $p_0 \leftarrow EvaluatePopulation()$;
3: $t \leftarrow SelectParents()$;
4: {Start of code generated by Iterative CGP}
5: **while** Number of evaluation left > 0 **do**
6:     $t \leftarrow$ 3-OptLocalSearch(t)
7:     $p \leftarrow$ restart population
8:     $p \leftarrow$ replaceLeastFit(t,p)
9:     $t \leftarrow$ SelectParents()
10:     $t \leftarrow$ ExchangeMutation(t)
11: **end while**
12: {End of code generated by Iterative CGP}
13: $p \leftarrow replaceLeastFit(t, p)$

Algorithm 4. This algorithm is the outcome of applying Algorithm 2 on the iterative graph in Fig. 2

1: $p_0 \leftarrow GenerateInitialSolution()$;
2: $p_0 \leftarrow EvaluatePopulation()$;
3: $t \leftarrow SelectParents()$;
4: {Start of code generated by Iterative CGP}
5: **while** The evaluations fall in the first half of the evolution (node 1) **do**
6:     $t \leftarrow$ 3-OptLocalSearch(t) (node 1)
7:     **while** The evaluations fall in the first half of the evolution (node 2) **do**
8:         $t \leftarrow$ Best2-OptionLocalSearch(t) (node 2)
9:         $t \leftarrow$ ExchangeMutation(t) (node 3)
10:         $t \leftarrow$ 3-OptionLocalSearch(t) (node 4)
11:         $p \leftarrow$ replaceLeastFit(t, p) (node 5)
12:         $t \leftarrow$ SelectParents() (node 5);
13:     **end while**
14:     $t \leftarrow$ SimpleInversionMutation(t) (node 6)
15: **end while**
16: $t \leftarrow$ 3-OptionLocalSearch(t) (node 7)
17: $t \leftarrow$ OrderBaseCrossover(t) (node 8)
18: {End of code generated by Iterative CGP}
19: $p \leftarrow replaceLeastFit(t, p)$

operators, with statistical comparison provided in Table 4, which gives the mean of the best obtained tour lengths over 30 runs and the mean relative error (and its standard deviation) from the best-known tour length.

We can see in Fig. 3 the evolution has constructed algorithms that enhance the strength and ameliorate the weaknesses of the heuristics and conditions listed in Tables 2 and 3. Both algorithms start their search with 3-Opt-LocalSearch, to reduce dramatically the length of the tours generated during the initialization process. In Fig. 3, the search descends sharply from around a relative error

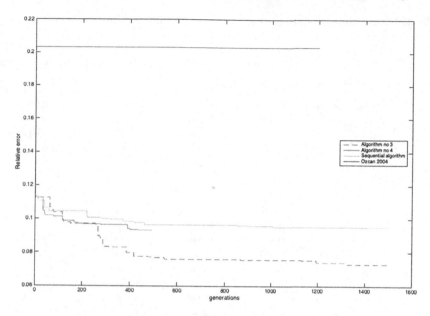

Fig. 3. A comparison of the four algorithms during the search for an optimum tour for the benchmark D1219

approximately around 0.20 from the known minimum to a relative error around 0.11 from generation 0 to 1. Hence it appears that evolution has rediscovered some elements of the template applied in Ryser-Welch et al. [34].

The iterative and the sequential algorithms achieved the best average fitness overall with a small standard deviation. For most benchmarks, the iterative algorithm scales well, finding good solutions to some benchmarks larger than the instances used during the training phase. Algorithm 3 has found the best solutions for the instances d1291, ym7663, usa13509 and sw24978; these instances are particularly hard to solve. Algorithm 4 uses many fewer evaluations; the termination criterion stop the loop when half of the evaluations have been used. The algorithm has found better tours than the sequential algorithms for the TSP instances d1291, zi929, ja9874 and usa13509.

We applied the Mann-Whitney U nonparametric test (for $p = 0.05$) to all pairs of algorithms, the results of which are in Table 5. The symbol = indicates that there is no significant difference between (the results of) Alg A and Alg B, > denotes that Alg A is significantly better than Alg B and < that Alg A is significantly worse than Alg B. In general, Algorithms 3 and 4 have found better or similar tours than related previous work [34]. Our approach has generated some iterative metaheuristics that have higher scalability than the best performing sequential TSP solver obtained from this previous research.

Table 4. Mean values of TSP solutions on 30 independent runs. The optimum value was either found by using Concorde or Lin-Kernighan

TSP instance	Known optimum[a]	Iterative Alg no 3	Iterative Alg no 4	Ryser-Welch 2015	Ozcan 2004
wi29	27,603	27,603	27,603	27,603	30,704
	relative error	0.000	0.000	0.000	0.001
	standard dev.	0.000	0.000	0.000	0.068
dj38	6,656	6,656	6,656	6,656	7,044
	relative error	0.000	0.000	0.000	0.002
	standard dev.	0.000	0.000	0.000	0.112
qa194	9,352	9,369	9,560	9,378	9,361
	relative error	0.002	0.022	0.004	0.001
	standard dev.	0.001	0.008	0.001	0.021
zi929	95,345	96,472	99,996	97,283	118071
	rel. error	0.011	0.048	0.019	0.240
	standard dev.	0.004	0.009	0.004	0.019
d1291	50,801	56,264	58,562	58,562	58,750
	relative error	0.081	0.112	0.121	0.200
	standard dev.	0.008	0.009	0.029	0.011
u2152	64,253	67,064	69,827	68,732	78,692
	relative error	0.043	0.086	0.069	0.223
	standard dev.	0.006	0.014	0.017	0.015
ca4663	1,209,319	1,277,495	1,331,639	1,304,901	1,547,992
	relative error	0.056	0.101	0.079	0.284
	standard dev.	0.004	0.024	0.015	0.022
ym7663	238,314	260,199	267,905	266,738	266,738
	relative error	0.091	0.124	0.119	0.281
	standard dev.	0.021	0.023	0.033	0.022
ja9874	491,924	533,304	555,201	564,581	625,035
	relative error	0.084	0.128	0.147	0.276
	standard dev.	0.018	0.033	0.046	0.011
gr9882	300,899	327,118	334,135	334,642	383087
	relative error	0.087	0.110	0.112	0.273
	standard dev.	0.018	0.022	0.021	0.023
usa13509	19,982,859	21,083,162	21,465,644	21,320,901	25,109,189
	relative error	0.055	0.074	0.066	0.251
	standard dev.	0.007	0.010	0.011	0.012
d18512	645,238	671,752	676,486	674,104	790,769
	relative error	0.041	0.048	0.044	0.225
	standard dev.	0.003	0.002	0.002	0.013
sw24978	855,597	912,915	927,663	928,355	1,075,056
	relative error	0.066	0.084	0.085	0.256
standard dev.	0.008	0.012	0.012	0.011	

[a]These tours are considered to be the best known and can be found at http://www.math.uwaterloo.ca/tsp/world/countries.html and http://comopt.ifi.uni-heidelberg.de/software/TSPLIB95/STSP.html

Table 5. Comparison of TSP solvers via Mann-Whitney U, $p = 0.05$.

Instance	Alg3 vs Alg4	Alg3 vs Ryser-Welch [34]	Alg4 vs Ryser-Welch [34]
wi29	=	=	=
dj38	=	=	=
qa194	>	=	>
zi929	>	>	<
d1291	>	>	=
u2152	>	>	>
ca4663	>	>	<
ym7663	>	>	<
ja9874	>	>	>
gr9882	>	>	=
usa13509	>	>	<
d18512	=	=	<
sw24978	>	>	=

6   Conclusion

We have presented a novel approach to evolving metaheuristics, which generates new metaheuristic variants containing evolved looping constructs. We evolved two novel TSP solvers and applied them to benchmark instances of the Travelling Salesman Problem. We show that not only that the method can produce human-readable algorithms (our sequence of operations was readily re-coded in Java), but it can also rediscover effective algorithms and generate new ones. The results of our experiments are promising: from a small training set, solutions equal or close to the actual known optima have been found for the benchmark instances under test. Our next step will be to apply this type of evolutionary hyper-heuristic to other problem domains as well to generate new hybrid metaheuristics and to demonstrate the generality and scalability of the proposed method. For example, personnel scheduling, vehicle routing and numerical optimisation will be considered with a larger range of instance sizes, allowing the potential of this technique to be fully evaluated.

Acknowledgements. The N8 HPC computer cluster used to host our evolutionary cross-domain hyper-heuristics and test their performance was provided and funded by the N8 consortium and EPSRC (Grant No.EP/K000225/1). The Centre is co-ordinated by the Universities of Leeds and Manchester.

References

1. Alexander, B., Zacher, B.: Boosting search for recursive functions using partial call-trees. In: Bartz-Beielstein, T., Branke, J., Filipič, B., Smith, J. (eds.) PPSN 2014. LNCS, vol. 8672, pp. 384–393. Springer, Heidelberg (2014)
2. Banzhaf, W.: The "molecular" traveling salesman. Biol. Cybern. **64**(1), 7–14 (1990)
3. Brave, S.: Evolving Recusive Programs for Tree Search. MIT Press, Cambridge (1996)
4. Brownlee, A.E., Swan, J., Özcan, E., Parkes, A.J.: Hyperion^2: A toolkit for Meta-, Hyper- heuristic research. In: Proceedings of the 2014 Conference Companion on Genetic and Evolutionary Computation Companion, GECCO Comp 2014, NY, USA, pp. 1133–1140. ACM, New York (2014)
5. Burke, E.K., Gendreau, M., Hyde, M., Kendall, G., Ochoa, G., Özcan, E., Qu, R.: Hyper-heuristics: a survey of the state of the art. J. Oper. Res. Soc. **64**(12), 1695–1724 (2013)
6. Davis, L., et al.: Handbook of Genetic Algorithms, vol. 115. Van Nostrand Reinhold, New York (1991)
7. Fogel, D.B.: An evolutionary approach to the traveling salesman problem. Biol. Cybern. **60**(2), 139–144 (1988)
8. Goldberg, D.E., Lingle, R.: Alleles, loci, and the traveling salesman problem. In: Proceedings of an International Conference on Genetic Algorithms and Their Applications, vol. 154, Lawrence Erlbaum, Hillsdale, NJ (1985)
9. Grefenstette, J.J.: Incorporating problem specific knowledge into genetic algorithms. Genet. Algorithms Simulated Annealing **4**, 42–60 (1987)
10. Gutin, G., Karapetyan, D.: A memetic algorithm for the generalized traveling salesman problem. Nat. Comput. **9**(1), 47–60 (2010)
11. Helsgaun, K.: An effective implementation of the lin-kernighan traveling salesman heuristic. Eur. J. Oper. Res. **126**(1), 106–130 (2000)
12. Hoos, H.H.: Programming by optimization. Commun. ACM **55**(2), 70–80 (2012)
13. Kant, E.: Understanding and automating algorithm design. IEEE Trans. Softw. Eng. **SE−11**(11), 1361–1374 (1985)
14. Kasturi, E., Narayanan, S.L.: A novel approach to hybrid genetic algorithms to solve symmetric TSP. Int. J. **2**(2) (2014)
15. Katayama, K., Sakamoto, H., Narihisa, H.: The efficiency of hybrid mutation genetic algorithm for the travelling salesman problem. Math. Comput. Model. **31**(10), 197–203 (2000)
16. Koza, J.R., Andre, D.: Evolution of iteration in genetic programming. In: Evolutionary Programming, pp. 469–478 (1996)
17. Langdon, W.B.: Genetic programming and data structures. Ph.D. thesis, University College London (1996)
18. Larres, J., Zhang, M., Browne, W.N.: Using unrestricted loops in genetic programming for image classification. In: 2010 IEEE Congress on Evolutionary Computation (CEC), pp. 1–8. IEEE (2010)
19. Lin, S., Kernighan, B.: An effective heuristic algorithm for the traveling-salesman problem. Oper. Res. **21**(2), 498–516 (1973)
20. López-Ibánez, M., Dubois-Lacoste, J., Stützle, T., Birattari, M.: The irace package, iterated race for automatic algorithm configuration. Technical report, Citeseer (2011)
21. López-Ibánez, M., Stützle, T.: The automatic design of multiobjective ant colony optimization algorithms. IEEE Trans. Evol. Comput. **16**(6), 861–875 (2012)

22. Mascia, F., López-Ibáñez, M., Dubois-Lacoste, J., Stützle, T.: Grammar-based generation of stochastic local search heuristics through automatic algorithm configuration tools. Comput. Oper. Res. **51**, 190–199 (2014). http://dx.doi.org/10.1016/j.cor.2014.05.020
23. Miihlenbein, H., Kindermann, J.: The dynamics of evolution and learning-towards genetic neural networks. Connectionism Perspect. pp. 173–197 (1989)
24. Miller, J.: What bloat? cartesian genetic programming on boolean problems. In: 2001 Genetic and Evolutionary Computation Conference Late Breaking Papers, pp. 295–302 (2001)
25. Miller, J.F. (ed.): Cartesian Genetic Programming. Springer, Heidelberg (2011)
26. Nagata, Y., Soler, D.: A new genetic algorithm for the asymmetric traveling salesman problem. Expert Syst. Appl. **39**(10), 8947–8953 (2012)
27. Ozcan, E., Erenturk, M.: A brief review of memetic algorithms for solving Euclidean 2D traveling salesrep problem. In: Proceedings of the 13th Turkish Symposium on Artificial Intelligence and Neural Networks, pp. 99–108 (2004)
28. Pillay, N.: A review of hyper-heuristics for educational timetabling. Ann. Oper. Res. pp. 1–36 (2014)
29. Rokbani, N., Abraham, A., Alimil, A.M.: Fuzzy ant supervised by PSO and simplified ant supervised PSO applied to TSP. In: 2013 13th International Conference on Hybrid Intelligent Systems (HIS), pp. 251–255. IEEE (2013)
30. Ross, P.: Hyper-heuristics. In: Burke, E.K., Kendall, G. (eds.) Search Methodologies, pp. 529–556. Springer, US (2005)
31. Ross, P.: Hyper-heuristics. In: Burke, E.K., Kendall, G. (eds.) Search Methodologies, pp. 611–638. Springer, US (2005)
32. Ross, P., Schulenburg, S., Marín-Blázquez, J.G., Hart, E.: Hyper-heuristics: learning to combine simple heuristics in bin-packing problems. In: GECCO 2002, Proceedings of the Genetic and Evolutionary Computation Conference, pp. 942–948. Morgan Kaufmann Publishers Inc., San Francisco, CA, USA (2002)
33. Ryser-Welch, P., Miller, J.F.: A review of hyper-heuristic frameworks. In: Proceedings of the 50th Anniversary Convention of the AISB, London, 1–4 April 2014
34. Ryser-Welch, P., Miller, J.F., Asta, S.: Generating human-readable algorithms for the travelling salesman problem using hyper-heuristics. In: GECCO Companion 2015, Proceedings of the Companion Publication of the 2015 on Genetic and Evolutionary Computation Conference, pp. 1067–1074. ACM, New York, NY, USA (2015). http://doi.acm.org/10.1145/2739482.2768459
35. Shirakawa, S., Nagao, T.: Graph structured program evolution with automatically defined nodes. In: Proceedings of the 11th Annual Conference on Genetic and Evolutionary Computation, pp. 1107–1114. ACM (2009)
36. Swan, J., Woodward, J.R., Özcan, E., Kendall, G., Burke, E.K.: Searching the hyper-heuristic design space. Cogn. Comput. **6**(1), 66–73 (2014)
37. Swan, J., Burles, N.: Templar - a framework for template-method hyper-heuristics. In: Machado, P., et al. (eds.) Genetic Programming. Lecture Notes in Computer Science, vol. 9025, pp. 205–216. Springer, Switzerland (2015)
38. Swan, J., Özcan, E., Kendall, G.: HYPERION – a recursive hyper-heuristic framework. In: Coello, C.A.C. (ed.) LION 2011. LNCS, vol. 6683, pp. 616–630. Springer, Heidelberg (2011)
39. Tavares, J., Pereira, F.B.: Automatic design of ant algorithms with grammatical evolution. In: Moraglio, A., Silva, S., Krawiec, K., Machado, P., Cotta, C. (eds.) EuroGP 2012. LNCS, vol. 7244, pp. 206–217. Springer, Heidelberg (2012)
40. Turner, A.J., Miller, J.F.: Neutral genetic drift: an investigation using cartesian genetic programming. Genet. Program. Evolvable Mach. **16**(4), 531–558 (2015)

41. Walker, J.A., Liu, Y., Tempesti, G., Timmis, J., Tyrrell, A.M.: Automatic machine code generation for a transport triggered architecture using cartesian genetic programming. Int. J. Adapt. Resilient Auton. Syst. (IJARAS) **3**(4), 32–50 (2012)
42. Wijesinghe, G., Ciesielski, V.: Evolving programs with parameters and loops. In: 2010 IEEE Congress on Evolutionary Computation (CEC), pp. 1–8. IEEE (2010)
43. Yu, T., Clack, C.: Recursion, lambda-abstractions and genetic programming. In: Poli, R., Langdon, W.B., Schoenauer, M., Fogarty, T., Banzhaf, W. (eds.) Late Breaking Papers at EuroGP 1998: The First European Workshop on Genetic Programming, CSRP-98-10, pp. 26–30. The University of Birmingham, UK, Paris, France, 14–15 April 1998

Author Index

Printed in the United States
By Bookmasters